Nanoscale Semiconductor Devices

Fundamentals and Applications

Nanoscale Semiconductor Devices

Fundamentals and Applications

Editors
Dr. Balwinder Raj
Dr. Monirujjaman Khan
Dr. Shailendra Singh
Dr. Shashi Bala

CWP
Central West Publishing

This edition has been published by Central West Publishing, Australia

For more information about the books published by Central West Publishing, please visit https://centralwestpublishing.com

Disclaimer

Every effort has been made by the publisher, editors and authors while preparing this book, however, no warranties are made regarding the accuracy and completeness of the content. The publisher, editors and authors disclaim without any limitation all warranties as well as any implied warranties about sales, along with fitness of the content for a particular purpose. Citation of any website and other information sources does not mean any endorsement from the publisher, editors and authors. For ascertaining the suitability of the contents contained herein for a particular lab or commercial use, consultation with the subject expert is needed. In addition, while using the information and methods contained herein, the practitioners and researchers need to be mindful for their own safety, along with the safety of others, including the professional parties and premises for whom they have professional responsibility. To the fullest extent of law, the publisher, editors and authors are not liable in all circumstances (special, incidental, and consequential) for any injury and/or damage to persons and property, along with any potential loss of profit and other commercial damages due to the use of any methods, products, guidelines, procedures contained in the material herein.

NATIONAL LIBRARY OF AUSTRALIA

A catalogue record for this book is available from the National Library of Australia

ISBN (print): 978-1-922617-58-3

Preface

For today's ICs, switching time is on the order of nanoseconds, minimum feature size is on the order of nanometers, transistor count is on the order of billions and cost is on the order of a few dollars. Scaling of conventional MOSFETs towards smaller sizes is near to end because of limited mobility, large Short Channel Effects (SCEs), high leakage currents and power dissipation. To mitigate these challenges of conventional MOSFET devices and to scale the semiconductor device further into nanoscale domain, alternative materials and device structures are required.

Nanoscale Semiconductor Devices are important sector whose applications in VLSI design have huge impact for bringing revolutionary advancements in nano-electronic circuits and systems due to improved electronic properties of the nano devices. Nano semiconductor devices are the potential devices for VLSI circuits and system applications and are highly preferred in novel nanoscale devices due to their high mobility and high driving capacity. Although currently the knowledge and resources for the fabrication of nanowires are limited but it is predicted by International Technology Road Map for Semiconductors (ITRS) that, with advancement of technology the conventional fabrication flow can be used for the nanoscale devices specifically nanowires.

Chapter 1 presents gradually CMOS technology advances into the nanoscale (nanometer regime); the conventional complementary-MOS devices started facing many problems like difficulty in increase of Ion current, lower reliability and yield, large parameter variations, more manufacturing cost, increase in leakage currents and so forth because of recent developments in nanotechnology, demand for smaller and more efficient nanodevices is increasing day by day. Hence numerous innovations in complementary-MOS materials as well as devices-structures has been proposed. Among this, recent development in NW- growth have resulted in invention of various nanowire-materials (e.g. Silicon, Germanium, III-V composite etc.) with accurate control of electrical properties, composition and morphology. Till date many novel nano composite-materials has been presented, like carbon nanotube, porous nanowires, core-shell nanowires etc. This paper gives brief review of homogeneous, heterostructures, III-V Compound nanowires, their advantages, limitations and applications. The

Graphene nano-ribbon, Junction-less nanowire, TFET based nanowires, their merits and applications are briefly summarized in this paper.

Chapter 2: This study supported the use of TFETs for better ON-current and steeper sub-threshold (SS) swings by utilising charge plasma (CP) and negative capacitance (NC). Recent research has shown that every metal oxide semiconductor is experiencing an increase in power dissipation (CMOS). Therefore, it is imperative to create new technology that can solve these problems. Then negative capacitance enters the picture, which can reduce the supply voltage and subthreshold slope while also minimizing the threshold voltage and subthreshold slope to some extent. In this book chapter, the performance of the device (CP-Ferro-VTEFT), is examined using a ferroelectric capacitor, which causes a negative capacitance effect as the device begins to operate. Device performance is assessed both with and without the use of negative capacitance. A comparison with different dioxide material (SiO2 and Si3N4) and has been used to engineering the presence the effect of negative capacitance. The use of charge plasma and doping less silicon nanowire is improving electronics engineering. As a result, the device's CP and NC both produce better electrical properties, a reduced thermal budget, and process variation immunity. Investigations are also conducted into how differences in ferroelectric density affect device conducts such polarised charge and threshold voltage.

Chapter 3: Carbon Nanotube Field-Effect Transistors (CNFETs) are brilliant possibility for building exceptionally energy-productive computerized frameworks. Notwithstanding, flaws intrinsic in carbon nanotubes (CARBON NANO TUBEs) present critical obstacles to acknowledging down to earth CNFET circuits. To accomplish CNFET VLSI frameworks within the sight of these innate flaws, cautious organization of plan and handling is expected: from gadget handling and circuit reconciliation, the entire way to enormous scope framework plan and streamlining. In this chapter, we sum up the key thoughts that empowered the primary exploratory exhibition of CNFET math and capacity components. We likewise present an outline of a probabilistic structure to examine the effect of different CNFET circuit plan procedures and CARBON NANO TUBE handling choices on framework level energy. We show the way that this structure can be utilized to further develop the energy-delay-item (EDP) of CNFET-based advanced frameworks.

Chapter 4: The electronic industry has accomplished remarkable growth since last few decades. This is essentially because of prompt expeditious developments in Integrated Circuit (IC) technologies and also in design of large-scale systems. There is no doubt that our routine lives are remarkably dependent on microelectronics and related technologies. This is verifiable on the domestic scene, in our professional exercise, in the workplace and in leisure actions. Today, electronics is characterized by low power dissipation, low cost, high reliability, incorporating with the ability to survive comfortably with high degree of refinement and complexity. Integrated circuit shave made feasible the design of robust and workable processors which present extremely intelligent and conformable devices for the users. Silicon CMOS technology has been considered superior from the past several years, for cost effective and high-performance systems. The requirement of image processing, quality improvement in multimedia content and future generation multimedia, together with the exposure of new and complex optical and photonics technologies being driven by microelectronics, put heavy demands on current CMOS technology integrated systems. Despite the fact that the technology is progressively developing to compose smaller systems with reduced power dissipation, the IC industry is facing crucial challenges due to limitations on power density and high dynamic and static power dissipation. The clue to vanquish these challenges reside in the refinement of design, material and manufacturing processes. The main issue faced by the VLSI designer is to find the effective technique for balancing the requirement of performance along with power dissipation.

Chapter 5: A Tunnel Field-Effect Transistor (TFET) is a type of field-effect transistor that uses the quantum mechanical phenomenon of tunneling to control the flow of current through the device. Unlike traditional metal-oxide-semiconductor field-effect transistors (MOSFETs), which use thermionic emission to inject electrons from the source to the channel, TFETs rely on tunneling, which allows for a more efficient and faster transfer of charge carriers. The unique properties of TFETs, such as their high transconductance, low subthreshold swing, and high current on/off ratio, make them attractive for a variety of applications. These include low-power and high-frequency digital logic, low-power analog circuits, low-power memory devices, and low-power sensor interfaces. Additionally, TFETs are also being investigated as a potential replacement for MOSFETs (see Fig.1) in complementary metal-oxide-semiconductor (CMOS) technology, which is the backbone of modern integrated circuits.

Chapter 6: As CMOS technology is facing many severe problems as a sensor application like- power scaling, short channel effects, sensitivity limitation and many fabrication issues etc. So, junction less transistor replaces CMOS technology because of its immunity to short channel effects and outstanding characteristics. But sensitivity is limited for junction less transistor devices either for charged biomolecule or for the neutral biomolecule. So, there is a need to enhance the sensitivity of the device for the detection of biomolecules that is why in this chapter work the main focus is to study the sensitivity of the TFET devices for the detection of the biomolecules. In the proposed device, a cavity is introduced on side of the gate metal to attain high sensitivity for biomedical applications. The immobilization of biomolecules within the cavity induces variation in surface potential.

Chapter 7: In this chapter the various types of non-volatile memory using Artificial intelligence has been discussed. In this chapter various non-volatile memory like FeFET, PCRAM, RRAM, MRAM has been discussed. Conventional memories like DRAM, SRAM, and NAND Flash have been progressively being overwhelmed by the enormous demand for information storage and transmissions, and the process miniaturization of conventional memories is getting more challenging. This is forcing the semiconducting company to create new memories with increased storage efficiency, cheaper prices, and system miniaturization. The MRAM, RRAM or PCRAM memory types stood out.

Chapter 8: Perovskite SC (solar cell) & Coming-generation renewable sources of energy have transformed the photovoltaic industry & new research. After initial research perovskite solar cell have a remarkable Power Conversion Efficiency (PCE) of 24.2%, which is higher than the solar cell based on thin films of multi-crystalline silicon, thin films of crystalline silicon, cu, In, Ga, Se and cadmium telluride, based on plastic substrates, all of which are suitable for transitioning into flexible solar cell. Organic plus inorganic -hybrid perovskite material revolutionized solar energy conversion systems. PSCs are popular because of their high efficiency, balanced charge transport, robust optical absorption, and longer lengths for carrier diffusion. However, perovskite solar cells' poisonous lead prevents commercialization. Tin based perovskite solar cell, which is lead free, might solve this problem; however, several technical challenges must be resolved before they can equal the power-conversion efficiency (PCEs) & stability of lead-based ones. This article explores the evolution of PCEs of conception into current developments. Discussed were device architectures and working, potential and horizons of lead-free low toxicity perovskite solar cell and fabrication techniques.

Chapter 9: This study developed an SRAM cell based on carbon nanotube field effect transistor (CNTFET) technology and evaluated its performance using different technologies. A comparison has been made between the proposed SRAM cell and established CMOS-based SRAM cell technologies. HSPICE simulation software is used for SRAM cell design and performance analysis. According to the simulation results, the proposed CNTFET-based SRAM cell outperforms the conventional CMOS-based SRAM cell technology regarding speed, noise immunity, and power consumption. The study also discusses several techniques that can be used to successfully implement the proposed SRAM cell. Finally, Analysis is performed on CNTFET-based SRAM cells and was compared with 6T SRAM cells in terms of SRAM cell performance metrics viz. static noise margin (SNM), read SNM, write SNM, read delay, write delay and static power consumption.

Acknowledgements

I wish to express my gratitude to my loving family members for their support and blessings throughout my life. I would like to extend my sincerest thanks to Director, NIT Jalandhar for providing me this opportunity and helping environment for initial start of this book project. I also that to Dr. Ramesh Kumar Sunkaria, head, Department of Electronics and Communication Engineering, National Institute of Technology, Jalandhar for guidance and support. I have been fortunate in receiving support, assistance and encouragement from our colleagues and scholars in the Department of Electronics and Communication Engineering, NIT Jalandhar and National Institute of Technical Teachers Training and Research, Chandigarh, India. I acknowledge the contribution of authors of book chapters, which have been written very well with modifications number of times for improvement of quality of chapters and their timely response.

A special thanks to the reviewers for their valuable points, comments and suggestions, which have been greatly, helped us in giving final shape to the book. We gratefully acknowledge the contribution of each of them, in particular, Dr. S Dasgupta, IIT Roorkee, Dr. Arun Kumar, PEC Chandigarh, Dr. Rajeevan Chandel, NIT Hamirpur, Dr. Ashwani Rana, NIT Hamirpur, Dr. Sunny Anand, Amity University, Dr. Jeetendra Singh, NIT Sikkim, Dr. Balwinder Singh, CDAC Mohali, Dr. Gurmohan Singh, CDAC Mohali, Dr. Karmjit Singh, Thapar University Patiala, Dr. Balwant Raj, SSG Panjab University RC Hoshiarpur, Dr. Sanjeev Kumar Sharma, Dr. Intekhab Amin, Jamia Millia Islamia, University Delhi, Mr. Inderjit Singh, DAV University Jalandhar, Mr. Girish Wadhwa, Ms. Tulika Chawla, and research scholars, Dept. of ECE NIT Jalandhar, India.

Thanks are also due to publisher, CWP Australia, its editorial and production teams for providing us with necessary support for overseeing various aspects of production of this book.

Balwinder Raj

Table of Contents

1

Nanowires Structures for Low Power Design

Dinesh B Borude[1], Balwant Raj[2], M Arun Kumar[3] and Balwinder Raj[4,]*
[1]Department of ECE, Polytechnic Collage, A.P
[2]Department of ECE, UIET PU Hoshiarpur Campus
[3]Department of ECE, GITAM University Bangalore
[4]Department of ECE, NIT Jalandhar
**rajb@nitj.ac.in*

Abstract: **As CMOS technology advances into the nanoscale (nanometer regime); the conventional complementary-MOS devices have started facing challenges like difficulty in increasing ion current, lower reliability and yield, large parameter variations, high manufacturing cost and increase in leakage currents. Hence numerous innovations in complementary-MOS materials as well as devices-structures have been proposed. Among this, recent development in NW- growth have resulted in invention of various nanowire-materials (e.g., Silicon, Germanium, III-V composite etc.) with accurate control of electrical properties, composition and morphology. Till date many novel nano composite-materials has been presented, like carbon nanotube, porous nanowires, core-shell nanowires etc. This paper gives brief review of homogeneous, heterostructures, III-V Compound nanowires, their advantages, limitations and applications. The Graphene nano-ribbon, Junction-less nanowire, TFET based nanowires, their merits and applications are briefly summarized.**

1.1 Introduction

High-performance, ultra-low power logical devices have been the key driving force for the continued CMOS scaling, which further enables the electronic devices such as smart phones/mobiles to be portable and more energy efficient. The continuous scaling of CMOS/MOSFET device dimensions has resulted in rapid growth in device performance as well as density [1-5].

However, the conventional CMOS structure has reached its scaling limits now. The various SCEs (short-channel-effects), static power-consumption, power-dissipation, quantum mechanical effects such as carrier confinement and tunneling, the issues related to variability and deteriorating effects of parasitic are serious-problems concerned with bulk-MOSFETs beyond the 22-nanometres technology-node.

Therefore, to extend Moore's Law further as well as for reduction in increased short-channel-effects, several strategies have been adopted for Complementary-MOS devices till date. Few of them are; use of multi-gate structures for higher drive current at the constant over-drive voltage, improving carrier mobility by means of adopting higher mobility channel-materials, strain engineering and reducing parasitic, Improvement in electrostatic control over the channel by continuous EOT (equivalent oxide thickness) scaling with high-k/metal gate stack [6-10].

Because of continuous growth in nanotechnology, the demand for smaller and more efficient Nano-technology devices is increasing day by day. Therefore, for searching an alternative device to the sustain Moore's law, many research studies have been carried out till date with respect to device-domain and various types of nanotechnology-devices (beyond MOSFETs/Complementary-MOS) has been presented and introduced. Nowadays CMOS/MOSFET-devices can be replaced by various Nanoelectronic-devices. The promising Nanoelectronic-devices may include single electron transistors, spin transistors, Carbon nano-tube transistors (CNTs) or Nanowires, TFET based nanowires and graphene-FETs [11-16]. These types of nano-devices offer feature sizes in the range of few nanometers and are self-assembled. The silicon nanowires (Si-NWs) have become challenging material for future developments in Nanoscale optoelectronics information-applications because of vital role of silicon in conventional photovoltaic as well as microelectronic fields. The Silicon/Germanium core-shell composite NWs have dragged attention because of their excellent performance. The combination of Ge-QDs (germanium quantum dots) and Si-NWs integrates the features of both one dimensional and zero dimensional nono-material instead of Germanium-Nano-film layer. This type of integrated structures have been used in various promising applications like Nanoscale-photoelectric detectors, LEDs, thermoelectric energy conversion and field-effect transistors due to stronger quantum confinement effect of Ge-QDs. The multiple-gates NW- FETs architecture permits the supreme short-channel control and drive Moore's law down to sub-5nanometer gate length [17-22].

This paper includes five sections; section 1.1 introduces the topic, whereas section 1.2 illustrates the necessity of nanodevices beyond MOSFET. Section 1.3 gives brief ideas about various emerging nanowire devices along with their merits, demerits and applications. Section 1.4 gives an idea of the comparative study of various nanowire devices. Finally review of nanowire devices is concluded in section 1.5.

1.2 Necessity of beyond MOSFET

Due to conventional bulk MOSFET downscaling limitations in the range of nanometer regime; there is need to find remedial nanodevices which will overcome problem associated with bulk MOSFET and will be compatible with classical MOSET with below capabilities

i) Technological innovations in device materials/structures of MOSFET which leads to
Achieve high performance through very low power consumption.

ii) Decreased dynamic-power of devices that further results in circuit-power reduction.

iii) Reduction in switched capacitance which leads to Increase in device switching speed.

iv) Achievable device downscaling in nanometer regime with decreased cost.

1.3 Recent Nanodevices Beyond CMOS

1.3.1 Nanowires

The structure having a diameter or within range of nanometer scale and high aspect ratio are called as Nanowires. It means, for nanowires the diameter may range from tenths of nanometers to several hundred nanometers and the aspect ratio can be on the order of 10 or higher. Normally NWs have symmetry in the two smaller-dimensions and can have regular polyhedral or round cross section. The NWS have good potential to behave as interconnects among functional Nanoscale-components that could be fabricated sequentially into the same wire [23-30]. Also, the nanowires are becoming potentially interesting for sensing applications and controlled solid-surface reactions, because of their higher surface-to-volume ratio as compared to similar volume-structures which have three-dimensions of same scale. Thus, the Semiconducting-NWs have become new class of materials and got shifted from traditional 2-D bulk thin-films to three dimensional devices.

Many key factors have considerable contribution in the recent nanowire research boom as follows:

i. Semiconductor-NWs could be produced in mass along with reproducible electronic- properties as needed for large scale integrated-systems.
ii. The bottom-up Synthesized NW-material offers fine controlled size in at least one device dimension, the channel-width which is at or beyond the limit of lithography as compared with top-down nanofabricated device-structures [31-34].
iii. The capability to produce radial and axial nanowire heterostructures, smooth surfaces as well as crystalline structure can further reduce scattering which ultimately results into higher carrier mobility as compared to Nano-fabricated samples with similar size.
iv. The electrical integrity of nanowire-based electronics devices could be maintained even though gate length is abruptly scaled because of the body thickness /diameter of nanowires could be controlled down to below10 nanometers 35-38].
v. Because of nanowires perfect surrounding gate to gate capacitive coupling, nanowire devices exhibit a superior scalability and therefore is a promising candidate at sub-45 nm node.

The nanowires can be formed by using various materials like Germanium, Silicon, compound materials like ZnSe, CdSe, III-V compound semiconductor materials like GaN, InP, GaAs [3]. Ability to suppress short channel effect, Low off Leakage current (Ioff), High on current (Ion), a smaller number of fabrication processes are some of the features of Nanowires [39-42].

Nanowire FETs

Recently Nanowire FETs, GAA (Gate All around)-FETs along with very thinner nanowire- channel has attracted attention and which are being preferred for continuous CMOS scaling. Because of the Nanowire's non-planar geometry, they provide excellent electrostatic-control of the channel as compared to traditional planar-structure [43-47]. As compared to traditional bulk silicon-MOSFETs, in the NW-FETs the channel has been replaced with semiconductor material-nanowire to form a new-structure. Nanowire diameter ranges from 5 nm to 0.5 nm. Because of the Nanowire's smaller diameters, the inversion charge changes from surface inversion to bulk inversion due to quantum confinement. The nanowire-FET threshold voltage may get varied due to the variations in nanowire diameters. Hence reduction in variability has been a major-challenge in making NW-FETs a feasible-technology. Further the complexity in the modelling of nanowire transistors has been increased due to quantum confinement effects [48-54].

Figure 1.1 shows a construction diagram of Nanowire-FETs. Figure shows that Si-substrate of traditional bulk MOSFETs acts as base material over which silicon-dioxide layer is produced for purpose of isolation by means of oxidation process. Once the source and drain areas are diffused, then NWs are fabricated between drain and source terminals. Semiconducting NW-FETs can be fabricated by using two methods; the first good-established method for producing nanowires nano-cluster catalyzed VLS- growth methods, wherein nanocluster defines nucleation and subsequent NW-growth. The litho and dry/wet etch (top-down lithography process) is a second method [55-61]. This method includes lithography as well as etching processes for formation of semiconductor-NWs followed by stamping or printing process. The field effect transistors-fabricated by VLS-grown NWs would offer much better-size uniformity as compared to etching method at the range of smaller diameters i.e. (less than 5 nanometers), because of controlled chemical-synthesis [62-66]. The device speed and uniformity improvement are still key challenges in Nanowire transistors.

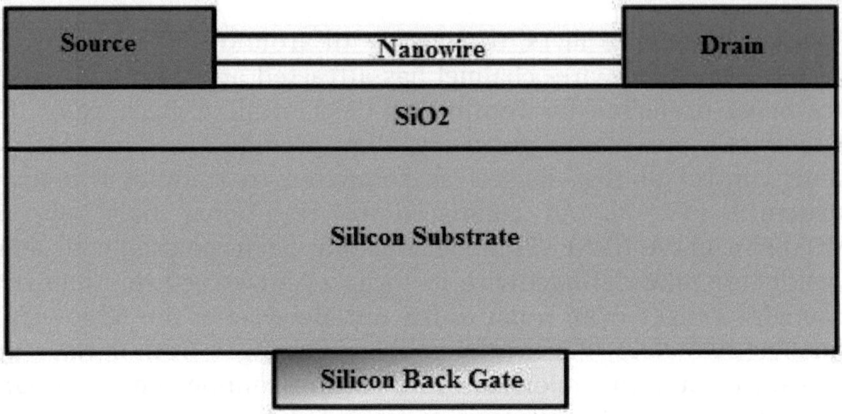

Figure 1.1 *Nanowire-FETs construction diagram.*

The Improvement in surface treatment and passivation techniques is still needed to minimize surface-defects and surface-roughness that further leads to device performance degradation [67-71].

1.3.2 Homogeneous Nanowire-Based Devices

Silicon Nanowires

A *Si-NW* is a very thin-wire of silicon materials having diameter ranging from few nanometers to few tenths of nanometers. Silicon nanowires have become popular in electronics industry, due to the drive for smaller electronic devices like diodes, resistors, capacitors and transistors [72-76].

The silicon whiskers had been synthesized successfully using VLS method in1964. Further it becomes basis for the synthesis of Silicon nanowires. Nowadays Si-NWs could be fabricated/manufactured using top-down or bottom-up approaches. The various dry and wet etching techniques are used in the top-down fabrication method while the bottom-up fabrication method mainly consists of chemical vapor deposition (CVD) and laser ablation [77-83].

The development on preparation and structure of Si-NWs during the recent research is elaborated as follows:

- The Si-NWs arrangement could be controlled from random-distribution to order array. In case of ordered array, the diameter of Si-NWs can be controlled up to 10 nanometers.
- Si-NWs having larger aspect ratios and smaller diameters can be prepared now days.
- Si-NWs with variety of structures have been proposed and introduced, such as ultra-thin structures, tapered nanowires, porous nanowires, p-n nanowires super lattices, kinked structures and zigzag structures etc. [84-88].

Due to unique physical and chemical properties the silicon nanowires have become promising candidates for wide range of applications such as FETs, Nanoelectronic storage devices, flash memory, chemical or biological sensors, nanowire batteries etc.
The control over the electrical properties and doping-level of Si-NWs have been improved due to recent development in CVD-based nanowire growth. While developing a Si-NWFETs, Si-NW is deposited over the degenerately-doped silicon substrate or on a Si-O_2, wherein the Silicon-substrate acts as back-gate. Later the metal-electrodes usually fabricated by means of e-beam/photolithography that further acts as source and drain-electrodes which completes FETs-structure. The uniform device performance and desired doping level can be achieved by improving the control on doping during growth [89-93].

To form a P-channel Si-NW FETs structure, acceptor impurities are introduced into Si-NWs by adding diborane during synthesis. The N-Channel Si-NW FETs structure can be formed by introducing n-type dopant (Phosphorous) into Si-NWs by adding phosphine as a reactant during NW-synthesis [94-97].

Germanium Nanowires

Because of Germanium's intrinsically high hole and electron mobilities in comparison with silicon, the Germanium-NWs have become attractive material for NW-based electronic applications. Ge-NWs synthesized with fluid-liquid-solid method can be assembled into single NW-transistor device structure. Once the deposition of NWs from solution on a Oxidized Si-substrates is done, then it can be electrically connected to drain and source- electrodes by using one of the further three-approaches: FIB(Focused-Ion-Beam)-assisted

CVD,FEB(Focused-Electron-Beam) assisted-CVD or EBL(Electron-Beam-Lithography).It is observed that Focused-Electron-Beam - deposited platinum-drain/source electrode on NWs having organic-monolayer-surface-treatment exhibits least contact-resistance as well as lesser nanowire-damages as compared to EBL and FIB [98-102].Germanium-NW devices are desired to have least contact-effects due to small Germanium-band gap that produces a low Schottky-barrier at metal/NW interfaces.

Metal oxide nanowires

The metal oxides like In2O3, ZnO and SnO_2 is a new class of nanowire that has been widely studied recently. For metal-oxide wires growth several methods such as hydrothermal metal–organic CVD, molecular-beam epitaxy thermal evaporation/vapor transport and pulsed laser deposition have been proposed and used. Among these methods, thermal evaporation/vapor transport deposition is mostly preferred due to lower equipment and operation cost. Undoped snO2 nanowires behave like poor conductor of electricity, because it forms Schottky contacts with metal electrodes. The Undoped/lightly nanowires used for detection of UV light and hence finds its application in ultraviolet photo detectors. The moderately doped SnO2 nanowires can be used for development of transparent transistor-devices (having mobility up to 550 $cm^2/V.s$). On other side highly dopedIn2O3 and SnO2 nanowires possesses metallic properties having resistivity below 10^{-4} Ω.cm. These types of nanowires can be used in various applications like optoelectronics devices; ultra-sensitive nanoscale gas sensors field emitters, ultraviolet photo detectors, field-effect transistors etc. [103-105].

1.3.3 Heterostructure Nanowire Devices

The semiconductor nanowires are being used for various applications as stated above. Most of application needs heterostructures. The heterostructures means integration of two or more materials into same NW-structure. The nanowire-substrate, axial and radial heterostructures are some of the promising applications of epitaxial NW-heterostructures. The heterostructure semiconductor nanowire device consists of two portions of dissimilar compound semiconductor materials. The first and second compound semiconductor materials are chosen such that at least one portion would be substantially metallic as a result of a strain between both compound semiconductor materials [106-108].

Depending on the materials employed, the contact formed by the first and second compound semiconductor materials may be Schottky-contacts or ohmic contacts. The core-shell nanowires heterostructure consists of two portions; the first portion is core portion, and second portion consists of a shell portion epitaxially grown on and substantially surrounds the core portion. The Core-shell structure provides semiconductor nanowires with reliable, reproducible electrical contacts whenever the shell portion would be metallic, and the core portion would be nonmetallic.

In the most of homogenous nanowire devices, the metal Source/Drain contacts yields Schottky-barrier types of contacts. Annealing further could improve contacts, which leads to Apparent-ohmic behavior. Still this method is only useful for heavily doped materials. On other side, in heterostructure NW-devices transparent ohmic-contacts having lower contact resistance to lightly doped/intrinsic channel can be achieved by means of band-structure engineering [109-110].

Even though different fabrication techniques of nanowires have been researched and introduced, still the shell nanowires face various challenges like lack of thickness uniformity, poor quality and doped shell layer as well as low yield.

Germanium/Silicon-Core/Shell heterostructure Nanowires

Nanowire-heterostructures and Silicon-nanowires are distinctive form of nano-silicon, which are grown by bottom-up approach (direct synthesis) instead of top-down (conventional lithography patterning) approach. Because of the higher hole-mobility and strong silicon-compatibility, the Si-Ge is used as the material for ultra-thin shell-channels [111-113]. In the composite structure of Germanium/Silicon core-shell nanowire, generally 3 to 5 nanometers silicon-shell is epitaxially grown on the top of ⍰10 nanometer diameter Germanium-NW core. As the huge valence band offset (⍰0.5 eV) is present between Germanium and Silicon, fermi level pinned inside the silicon-bandgap is under the Germanium Valence-band. This band results in a negative Schottky barrier i.e., ohmic contact to the Germanium channel and favors electrical-injection of holes to the Germanium core from the contacts [114-115].

The elimination of dopant scattering and formation of transparent i.e., negative Schottky contacts to the conduction channel are benefits of the Germanium/Silicon Core-Shell heterostructures as compared to homogeneous Ge/Si -NWs. The demonstration of ultra-high performance transistor devices based upon the Germanium/Silicon systems had been possible successfully due to these two advantages. Apart from this, many vital features are observed in AlGaAs/GaAs system such as formation of 0.7 structures at zero-bias and half-plateaus at higher-biases at comparatively higher temperature. The Ge/Si-NW system's heavy effective mass of holes, results in large interaction-parameters. This all together having a uniform and strong confinement-potential along the channel as well as potential to generate transparent-contacts have made the Germanium/Silicon-NW systems unique platform for the study of rich-physics in stronger interacting-systems [2].

The Germanium-core/Silicon-shell nanowires FET have been proposed for sake of achieving high current drive and hole-mobility making use of quasi- 1D transport in the Germanium-hole gas systems. For the p-type nanowire FET, the Germanium is used as the Shell-channel material due to its strong Si-compatibility and high hole-mobility. On a Si-nanowire core; an ultrathin Silicon-Germanium channel is constructed, where the channel region is made up of a co-axial Si/Ge-Shell for achieving high gate controllability above the greatly restricted channel region and higher current drivability. The Core/Shell heterostructure Nanowires shows superior performance because of various reasons like, small Sub-threshold slope (nearly 60 mV/dec), high Ion/Ioff ratio above 107 and very small DIBL in the range of only a few millivolts. The Si-Ge core/shell channel nanowire FETs have better potential for the applications to implement more ideal logic and a vital radio frequency component which is fully functional into Tetra hertz frequency-regime.

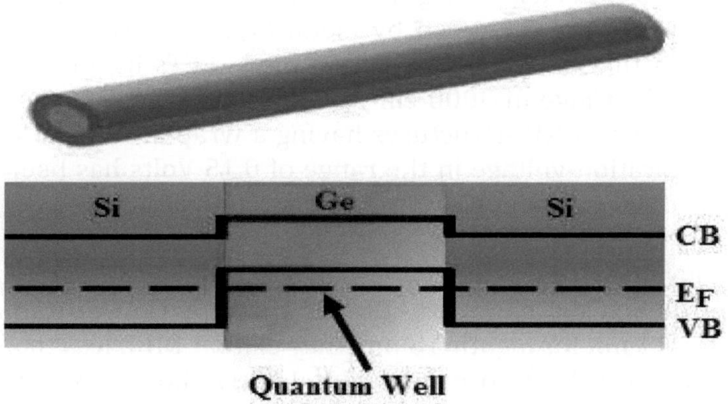

Figure 1.2 *The Ge/Si core-shell heterostructure cross section and the band diagram.*

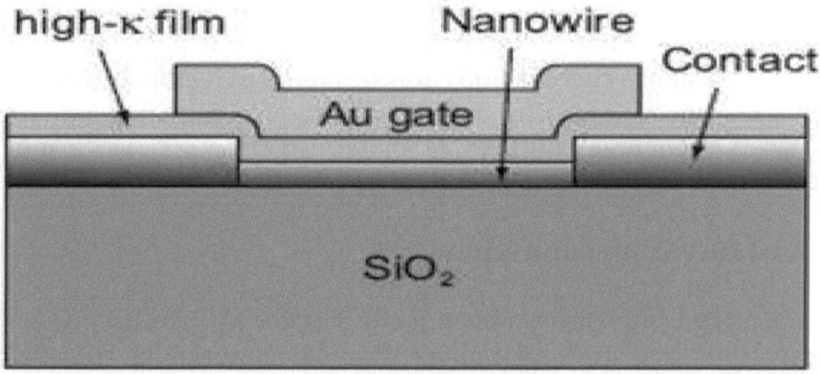

Figure 1.3 *Ge/Si Core/Shell heterostructure NW-FETs.*

InAs Nanowires

InAs-NWs have been studied widely as a basic building block for the n-type field effect transistors. Because of various reasons the InAs have become popular material for nanowire fabrication. InAs shows high electron mobility in bulk material due to its lesser effective electron-mass. At the surface of planar InAs the electron gas layer gets formed and this happens because of fermi-level pinning into conduction-band at the surface. Thus, the Electron gas formation integrated with smaller bandgap (up to 0.35 eV) will produces transparent-contacts to InAs NW-devices. The Nano cluster-catalyzed process is used for growing InAs-nanowires; wherein the reactant species are delivered by metal organic CVD or CBE. The Study shows that, n-channel depletion mode FETs having electron mobility in the range of 3000 cm^2/V can be produced. Demonstration of the vertical NW-structures having a wrap-around-gate with a lower saturation-voltage in the range of 0.15 Volts has been possible, due to the epitaxial-growths of InAs-Nanowires [2].

Merits of Nanowire devices

- Although the nanowire devices are scaled down to diameter-14 nm for legnth-18 nm, they exhibit DIBL less than 100 millivolts. Therefore, nanowire devices show very less DIBL effect among FINFET and Tri –state devices.
- Nanowire devices show small sensitivity to diameter changes, which implies good manufacturability and immunity of device dimension variation at the diameter range of 10 nm.
- Ability to reduce short channel effects, better sub threshold slope, high yield reproducible electronic properties, low off leakage current (Ioff) are the merits of nanowire devices.
- Due to Nanowire device's accurate surrounding gate to gate capacitive coupling, it shows superior scalability and has become promising candidate at sub-45 nanometer node [3].

1.3.4 III-V Compound Nanowires

The III-V composite semiconducting materials like InP, GaAs, GaP and InAs are used to replace channel material of conventional MOSFET. Some of NWs (e.g. InAs) have impressively higher electron-mobility (>20,000 $cm^2V^{-1}S^{-1}$); this makes NWs suitable for high frequency electronic applications which perhaps not easily achieved in silicon as well as other nanostructures [12].

These III-V Compound NW materials exhibit following major advantages than their bulk counterparts [5].

- The semiconductor materials downscaled into NW crystals will exhibit extra mechanical flexibility. Further it makes them supreme active materials for future flexible electronic applications.
- Large surface-to-volume ratio makes the Nanowires suitable for high performance sensors-applications.

- The application potential of the nanowires will be improved due to unique 1-D structures miniaturized co-axial-core multishell-heterojunctions.

Now a days, for the growth of many III–V nanowires the CVD (chemical vapour deposition) growth method, which utilizes the traditional tube furnace as well as solid powder source (i.e., solid-source CVD) has been used. As compared to rest sophisticated growth mechanisms like metal-organic CVD and MBE, the CVD (chemical vapour deposition) growth method has simple growth procedures, good controllability, relatively lower setup as well as operating cost and most importantly no-involvement of the toxic gas precursors. Arsenic (As) and Phosphorus (P) based III–V compound nanowire materials exhibit moderate band gaps hence it shows good photon response to infrared/visible lights. Also, this type of nanowire materials shows an excellent on/off ratio while configured into FETs therefore holds a great potential for applications in future optoelectronics. Among the all III–V compound nanowires Sb-based NW-materials have superior carrier mobility and smaller direct-bandgap, with GaSb and InSb exhibits highest hole-(p-type) and electron-(n-type) mobilities respectively. This makes them unique material for Quantum-devices, thermoelectric and longer wavelength Detectors and field-effect transistors [5].

III–VNW properties are affected by various factors such as diameter, surface roughness, the catalyst choice and so forth. The catalyst material needs to be selected carefully because it influences the morphology and composition as well as it affects the electronic-transport properties through in-situ doping and vacancy formation or defect. InAs Nanowires are usually synthesized by means of SSCVD methods which use various metal nano-clusters e.g. Au or Ni as catalysts. The research study shows that the Nanowires with Ni-catalyst are comparatively uniform and straight on substrates, in which approximately 98% of the Nanowires show unvarying negative-type i.e., n-type behavior with strong gate dependence. It further gives high Ion/*I*off ratio (> 10^{4}) and lower OFF-current approximately between 10^{-8} to 10^{-10} A. On other side, the Au-catalyzed Nanowires exhibits tapered morphology as well as non-uniform plus unstable electrical properties. About 80% of Au-catalyzed Nanowires shows weak gate dependence features [5].

Another Nanowire characterization feature is diameter of NWs. The electronic properties of nanowires would be significantly influenced with variation in diameter which essentially results in surface-to-volume ratio variations. III-V compound Nanowires shows intrinsically n-channel conduction behavior for larger diameters i.e., $d>$ 70nanometres due to minimal space charge depletion effect. The Nanowires with small diameter ranges i.e., $d<$ 40 nanometers exhibit high surface to volume ratio and inverted p-channel behavior while III-V nanowires shows ambipolar conduction in low current magnitude for its diameter between 40 70 nanometers [5]. Nanowire diameters can help to reduce off state current and to enhance the gate electrostatic control. But on other side diameter-scaling of nanowires will naturally introduces higher surface-carrier recombination and more surface-scattering, which further determines the carrier-mobility of Nanowire transistors. Thus Nanowire diameter scaling might leads to degradation in carrier mobility.

The theoretical study shows that the surface roughness can introduce significant changes in; NW-electronic band-structures and their transport properties like mobility, carrier mean free paths, etc. The increase in Nanowire surface roughness leads to degradation of electronic mobility [5].

Applications of III–V Nanowires

The III–V compound Nanowires have become popular materials for upcoming Electronic- devices because of their super high carrier-mobility. The recent developments in nanowire contact printing technology have given a useful platform for nanowires assembly as well as integration. We may obtain enhancement mode (E-mode) and depletion mode (D-mode) FETs with shunt nanowire arrays at well-defined locations by integrating the contact printing method with "Metal-Cluster-Decoration" methods. On this basis, III–V NW-NMOS (n-channel metal oxide semiconductor) inverters are constructed, in which original Depletion-mode InAs-nanowire field effect transistor act as a load and Au decorated Enhancement-mode InAs-nanowire field effect transistor act as drivers. The dissipation of power in these III–V compound nanowire-NMOS inverters is low due to lower operating voltage and higher electron-mobility of InAs-nanowires. The GaAs-nanowire fabricated CMOS-inverters also can be constructed for the complex circuit design [5].

III-V nanowires can be explored in various device applications like nanooptics, Nanoelectronics, LEDs, laser diodes and FETs.

1.3.5 Graphene Nanoribbons (GNRs)

Graphene means single-atomic layer of carbon-atoms arranged in two dimensional-hexagonal lattices or else can be thought as unrolled carbon nano-tubes. It exhibits better potential for Nanoelectronic devices. It shows many key advantages; like higher carrier-mobility for distinct fabrication methods and temperatures, higher carrier-velocity leads to superb thermal conductivity and fast switching etc.

It exhibits higher integration-potential with traditional Complementary-MOS fabrication processes due to its ability to create wafer scale graphene- films having full-planar processing for the devices [1].

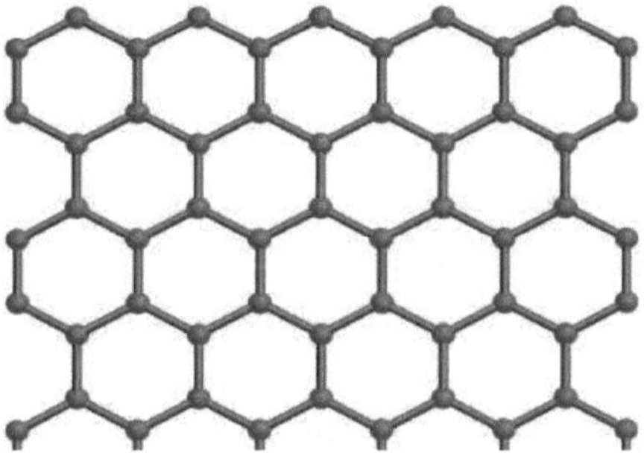

Figure 1.4 *Structure of graphene materials.*

By assuming the edges of the nanoribbon are passivated by hydrogen-atoms, the graphene nano-ribbons possesses two major types; therefore ZGNR (zigzag-edge) and armchair edge (AGNRs) [1]. The Zigzag GNRS are always metallic where armchair GNRs can be metallic or semiconducting AGNRs can be used in digital circuit applications like channel materials. Armchair-edge and zig-zag carbon nano-tubes both have nearly similar electronic structures.

The 2D-graphene exhibits zero bandage Semi-Metal characteristics. The bandgap could be induced like carbon nano-tube cases; by patterning-graphene in few nanometers (nm) wider GNRs. In the Graphene Nano ribbon energy bandgaps are inversely proportional to their widths. Therefore width- confinement under the sub-10 nanometer-scale is needed, to open the bandgap which is adequate for transistor operation at room temperature. As the Graphene materials shows zero bandgap between conduction and valence band, it leads to higher ON current (I_{ON}) and which further makes them ideal for higher speed radio frequency or analog applications. But when we deal with digital logic applications zero band gap becomes problem due to low off current. Therefore bi-layer graphene-sheet could be utilized as channel-materials for band gap improvement and an electric-field perpendicular to AB-stacked layers is applied. Other solution is to utilize GNRs as channel-materials which further leads to sufficient bandgap i.e., about 5eV with different ribbon widths [1].

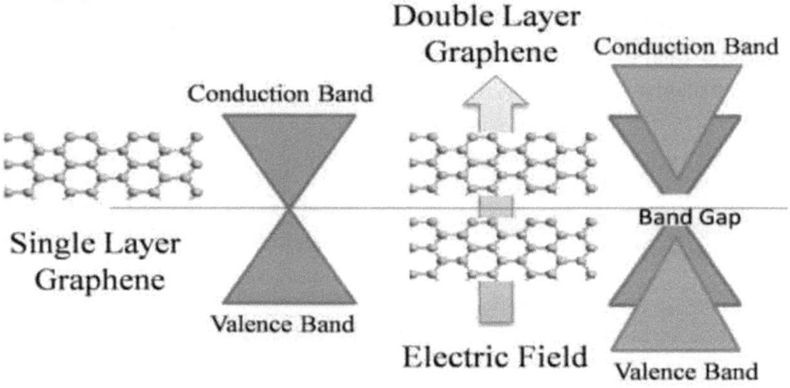

Figure 1.5 *Band gap in graphene materials.*

Even after the Graphene nanoribbon-materials promises fast, low energy and ultra-small FETs, both the key-effects of variability and defects i.e., leakage and lower noise-margins are considerable. E.g. the channel-width variation by a couple of angstrom leads to change in a leakage current by order of magnitude, while a charge impurity in the order of single Coulomb might lower the field effect transistor on-current by aprrox.30% [1].

Nano Ribbon FETs

In the Nano-Ribbon FETs Graphene sheet/graphene-materials acts as channel. The simple construction diagram of graphene-FET is shown in Figure 1.6. The silicon substrate-material is utilized as a basement over which two-regions are diffused for drain and source formation. In the Graphene FETS the channel is formed by fabricating two-sheets of graphene among source and drain below the gate-terminal. The oxide layer separates gate terminal. Terminal contacts are created by using metals.

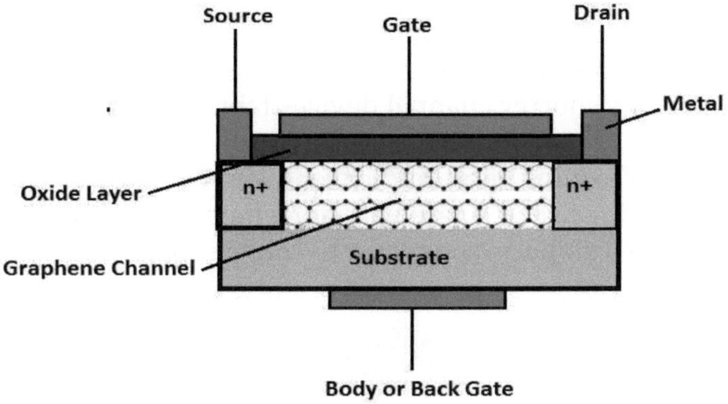

Figure 1.6 *Construction of nano-ribbon FET.*

Recently the Graphene FETs used new substrate-materials like, hBN (hexagonal boron-nitride) with two dimensional structures. Therefore, the mobility improvement can be seen at room-temperature in these types of devices. Further, by sandwiching Graphene among two hBN-flakes and edge-contacted with two Metal-electrodes, the mobility of such devices might be enhanced to great extent. Thus, the Graphene-FETs are promising candidates for continuing the Moore's-Law or for achieving higher circuit complexity for more than Moore-applications.

1.3.6 Junction less Nanowires

The Junction less Nanowires makes use of a simple silicon-nanowire surrounded by an electric-ring which forms a gate and control the electrons flow through the wires. The conventional transistors like BJT, MOSFET, JFET, and MESFET uses the semiconductor- junctions created by pouring dopant-atoms in the semiconducting materials. The higher doping concentration-gradients become necessary as distance between junctions drops under 10 nm in recent Nanoelectronic-devices.

For the semiconductor industries fabrication of such narrower junctions has become difficult challenge. Hence newer types of transistors i.e., JNTs, having no junctions as well as no doping-concentration gradient have been proposed.

The recently proposed junction-less nanowire transistors (gated resistor) have no junction and no doing concentration gradient. Gated resistors do not require costly ultra-fast annealing techniques but still permits shorter channel devices fabrications. These devices are made up of silicon-NWs and have full CMOS-functionalities. These devices exhibit very low leakage currents, lesser mobility - degradation with rise in a gate voltage/temperature and nearly ideal sub-threshold slope as compared to conventional-transistors [6].

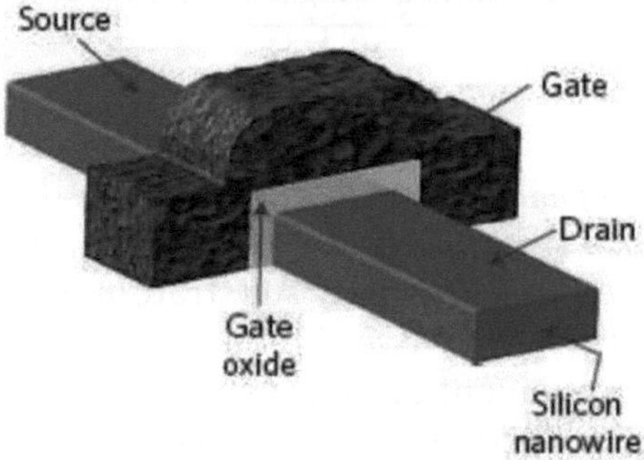

Figure 1.7 *Junction-less nanowire transistors (gated resistor).*

Key advantages of the junction-less Nanowire transistors (Gated register) over Classical MOSFETs

- Drain-current flows through the whole section of the Nano-ribbons rather than being confined into surface channel, in the highly-doped JNTs. In the JNTs, the conduction-path is located closer to center of the nanowires instead at silicon–SiO2 interface. This permits the electrons to travel through Silicon with bulk-mobility, which is least affected by scattering as compared to the Surface-Mobility experienced by conventional transistors [7].
- The gated resistor exhibits slower Transconductance degradation with increase in gate voltage which results in more current as well as high speed performance.

- The mobility degradation with temperature variations is lesser in gated-resistors as compared to trigate-field effect transistors, because the mobility is limited by Impurity-scattering instead of Phonon-scattering in the heavy doped gated-resistors or junction-less Nanowire transistors.

1.3.7 TFET based Nanowires

Recently TFETs (Tunnel field effect transistors) are replacing the MOS technology as band to band tunneling mechanism is leading to current conduction, it has low off state sub threshold swing 60mV/decade and have low off state current.

The tunnel-TFETs structure is identical to MOSFET only difference is instead of modulating thermionic-emission above the barrier as in conventional MOSFETs, tunnel-field effect transistors switch by modulating quantum-tunneling through a barrier [9]. Recent CMOS technology-generations are expecting lesser power consumption per transistors by scaling the supply voltage. TFETs have come up as better alternative for MOSFETs because of their minimized leakage current.

The Nanodevices having sub-threshold swing under 60 millivolts/decade limit of the MOSFETs is required to keep adequately larger Ion/Ioff ratio. Hence to fulfill these requirements TFETs have been Proposed [7].

TFETs are basically reverse-biased p-i-n diode which could be fabricated in standard complementary-MOS technology. Most of Tunnel field effect transistors are Si-based as they permit to full reuse of existing-expertise in fabricating higher quality Si-compatible gate-dielectric. The huge indirect-bandgap of Si results in; smaller band-to-band tunneling efficiency which further leads to lower Ion current of all Si-TFETs. Hence recent materials like III-V compounds or germanium having small effective mass and low band-gaps are used to improve tunneling efficiency. The various hetero-structures have been proposed and introduced, to improve Ion current by keeping a lower off state currents and to minimize effective tunnelling-barriers [7]. Although Ion current of Tunnel field effect transistors is not sensitive to gate-lengths, to improve the electric-field near the junction, the transistor-geometries yet require a gate having strong electrostatics control. Hence configurations/structures having feature-ultrathin body along with multiple gates and aggressively scaled EOT using higher gate-dielectric like GAA-NWs are needed. The lower ON current is a major limitation of TFETs. The I_{ON} current can be improved by using a double gate structure, multiple gate structures instead of a single gate structure. A GAA-structure have excellent gate -coupling and hence it has become promising alternative for ON current improvement *as* well as to improve $_{ION}/I_{OFF}$ current ratio [8].

GAA nanowire TFETs

A good solution for ON current improvement is to attain excellent electrostatic coupling by improving the gate channel-geometry by means of using multi-gate devices like finFETs, planar double-gate, triple gate or GAA-NW structures.

Figure 1.8 *GAA Nanowire TFET.*

Good gate-controllability and the increase in on chip device density both can be achieved by means of process integration of TFETs with gate–all-around structures. A GAA-TFETS can be designed to achieve high I_{ON} /$I_{OFF\ ratio}$ and steeper SS. by using the GAA-structure, the ION /I_{OFF} ratio in the range of 10^9 and sub-threshold swing below the 28.57 mV/decade can be obtained. Gate-All-Around structures have lesser leakage effects and superior gate-control as compared to single and double gate-structures [8].

1.4 Comparative table of different nano-material/devices, their merits and demerits

No.	Material/Device	Merits/features	Demerits
1.	Si/Ge-Nanowires	• Low off Leakage current (Ioff). • Ability to reduce short channel effects. • Nanowire devices exhibit lesser DIBL effect among FINFET and Tri –state devices. • Better subthreshold slope. • High-yield reproducible electronic properties. • Higher scalability because of accurate surrounding gate to gate capacitive coupling.	• Issues with uniformity in alignment of nanowires. • Issue of surface roughness and surface defects. • Off chip interconnect- capacitances limits operating speed of NWs. • Device threshold voltage may get varied in accordance with variations in NW-diameters.
`2.	III-V Compound Nanowire	• Holds higher carrier-mobility for high-speed device applications. • Ultra-high speed switching capability at low supply voltage. • Tunable direct bandgap. • Large surface to volume ratio. • Simple fabrication.	• Mobility variations with different materials combinations. • III-V Compound NW's properties may be influenced by various factors like the Nanowire diameters, surface roughness and catalyst choice.
3.	Graphene	• The higher carrier-mobility for sake of ballistic transport. • Fast switching due to higher carrier-velocity. • Thermal conductivity is superb. • Promising candidate for lower than 22nanometer design. • Ideal material for high frequency radio applica-	• During fabrication process variability may be arise due to difficulty in control of the GNRs width, oxide thickness and edge roughness. • Low Noise Margins and Leakage

		tions. • Can be used in logical applications by introducing band gap.	these two effects of Variability and defects are considerable.
4.	JNTs	• Shows superb short channel-behavior and lower leakage current for narrow gate length. • Simple structure makes fabrication process easier. • With rise in temperature and gate voltage JNTs exhibits lesser degradation of mobility as compared to classical- transistors.	• Faces the problem of mobility- degradation because of high doped channel, it further affects gain or the Transconductance (gm) and results in poor ON current.
5.	Nanowire TFETs	• Large Ion/Ioff current ratio. • Exhibits SS (Subthreshold swing) under 60milliVolts/decade. • Reduced Leakage-effects and Excellent gate control. • NW-TFETs exhibits low OFF-current (Ioff) as well as reduced short channel effects.	• considerably low ON current (I_{ON}) (can be improved by using multi-gate devices e.g., DG, GAA)

1.5 Conclusion

In this report/paper recent research progress on Nanoelectronic devices such as Si-NWs, nanowire homogeneous/heterostructures, Nanowire TFETs, GNRs, JNTs have been presented. These nanowire devices collectively have many key advantages over conventional bulk MOSFET like reduced short channel effects, High-yield reproducible electronic properties, large Ion/Ioff current ratio, low leakage currents, Large surface to volume ratio, Higher scalability due to accurate surrounding gate to gate capacitive-coupling, Small Subthreshold swing (SS) i.e. below the 60mV/decade. Therefore, these Nano-devices have become promising candidate for continuing the Moore's law for More Moore-applications as well as More than Moore-applications. Nanowire devices have identical fabrication-process as that of Classical MOSFETs. The comprehensive comparisons of different nanowire material/devices in terms of their features/merits and demerits have been presented towards the end of paper.

References

[1] Yong-Bin Kim, "Review Paper: Challenges for Nano scale MOSFETs and Emerging Nanoelectronics" IEEE Trans. Electr. Electron. Mater. 10(1) 21 (2009): G.-D. Hong *et al.,*2009

[2] Wei Lu, Member, IEEE, Ping Xie, and Charles M. LieberDd, "Nanowire Transistor Performance Limits and Applications",IEEE Transactions on Electron Devices, Vol. 55, No. 11, November 2008

[3] Chin-Yu Chen, Yi-Bo Liao, and Meng-Hsueh Chiang, "Scaling Study of Nanowire and Multi-Gate MOSFETs"IEEE,978-1-4244-2186-2/08 ,2008

[4] YasirHashim, and Othman Sidek "Investigation of Effects of Nanowires Numbers in N-channel and P-channel Nanowire Transistors on Nanowire-CMOSCharacteristics"RSM2011 Proc., 2011, Kota Kinabalu, Malaysia, IEEE,978-1-61284 846 4/11,2011

[5] Ming Fang, Ning Han, Fengyun Wang, Zai-xing Yang, Sen Po Yip, Guofa Dong, Jared J. Hou, YulunChueh, and Johnny C. Ho "III–V Nanowires: Synthesis, Property Manipulations, and Device Applications" Journal of Nanomaterials Volume 2014, http://dx.doi.org/10.1155/2014/702859,pp.1-15,2014

[6] Jean-Pierre Colinge, Chi-Woo Lee and others, "Nanowire transistors without junctions" Article *in* Nature Nanotechnology · DOI: 10.1038/nnano.2010.15 · Source: PubMed, February 2010

[7] Anne Vandooren, Alireza Alian, Anne Verhulst, Jacopo Franco, Rita Rooyackers, Quentin Smets, Devin Verreck, Niamh Waldron, Dan Mocuta, and Nadine Collaert, "Tunnel FETs for low power electronics" IEEE ,978-1-5090-4391-0/16 ,2016

[8] Ajith Ravindran, Abraham George, Praveen C S, NishaKuruvilla, "Gate All Around Nanowire TFET with High ON/OFF Current Ratio"Published by Elsevier Ltd.Proceedings 4 (2017), pp.10637–10642,2017

[9] M. T. Björk, J. Knoch, H. Schmid, H. Riel, and W. Riess "Silicon nanowire tunneling field-effect transistors" Appl. Phys. Lett.

92, 193504 (2008); doi: 10.1063/1.2928227, pp.193504-(1-4),2008

[10] Eunseon Yu1 and others"Si-core/Si-Ge-shell channel nanowire FET for sub-10-nm logic technology in the THz regime"ETRI Journal.2019; 41(6), pp.829–837,2019

[11] Anjana Bhardwaj, Pradeep Kumar, B. Raj, Sunny Anand, "Design and Performance Optimization of doping-less Vertical Nanowire TFET using Gate Stack Technique", Journal of Electronic Materials (JEMS) Springer, Vol. 41, Issuse 7, PP. 4005-4013, 2022.

[12] Jeetendra Singh, BRaj, "Tunnel Current Model of Asymmetric MIM Structure Levying Various Image Forces to Analyze the Characteristics of Filamentary Memristor", Applied Physics A, Springer, Vol. 125, No.3, pp. 203.1 to 203.11 Feb 2019.

[13] Candy Goyal, Jagpal Singh Ubhi and B Raj, Low Leakage Zero Ground Noise Nanoscale Full Adder using Source Biasing Technique, "Journal of Nanoelectronics and Optoelectronics", American Scientific Publishers, Vol. 14, PP. 360–370, March 2019.

[14] Gurmohan Singh, R. K. Sarin and B Raj, "A Novel Robust Exclusive-OR Function Implementation in QCA Nanotechnology with Energy Dissipation Analysis", Journal of Computational Electronics, Springer, Volume 15, Issue 2, PP. 455-465, June 2016.

[15] Tanu Wadhera, Deepti Kakkar, Girish wadhwa, B Raj, " Recent Advances and Progress in Development of the Field Effect Transistor Biosensor: A Review" Journal of ELECTRONIC MATERIALS, Springer, Volume 48, Issue 12, pp 7635–7646, December 2019.

[16] Girish Wadhwa, Priyanka Kamboj, Balwinder Raj" Design optimisation of junctionless TFET biosensor for high sensitivity" Advances in Natural Sciences: Nanoscience and Nanotechnology, Vol. 10, PP. 045001 (7), (2019).

[17] Priya Bansal, BRaj, "MEMRISTOR MODELING AND ANALYSIS FOR LINEAR DOPANT DRIFT KINETICS," Journal of Nanoengineering and Nanomanufacturing, American Scientific Publishers, Vol. 6, pp. 1–7, 2016.

[18] Amandeep Singh, Mamta Khosla, BRaj, "Circuit Compatible Model for Electrostatic Doped Schottky Barrier CNTFET,"Journal of Electronic Materials, Springer, Vol. 45, Issue 12, pp 4825-4835, 2016.

[19] Ashima, D Vaithiyanathan, B Raj," Performance Analysis of Charge Plasma induced Graded Channel Si nanotube", Journal of Engineering Research (JER), EMSME Special Issue pp. 146-154 Aug 2021.

[20] Abhishek Singh Tomar, Vijay Kumar Magraiya, and B Raj, Scaling of Access and Data Transistor for High Performance DRAM Cell Design, Quantum Matter, Vol. 2, PP. 412-416 Oct. 2013.

[21] Neeraj Jain, BRaj, "Parasitic Capacitance and Resistance Model Development and Optimization of Raised Source/Drain SOI FinFET Structure for Analog Circuit Applications, Journal of Nanoelectronics and Optoelectronins, ASP, USA, Vol. 13, PP. 531–539, Apl 2018.

[22] Shradhya Singh, S. K. Vishvakarma and B Raj, "Analytical Modeling of Split-Gate Junction-Less Transistor for a Biosensor Application", Sensing and Bio-sensing, Elsevier, Vol. 18, Pages 31-36, April 2018.

[23] Maisagalla Gopal, and Balwinder Raj, Low Power 8T SRAM Cell Design for High Stability Video Applications, ITSI Transaction on Electrical and Electronics Engineering, Vol. 1(5), pp. 91-97, 2013.

[24] Balwinder Raj, Jatin Mitra, Deepak Kumar Bihani, Rangharajan V, A.K. Saxena and S. Dasgupta, "Analysis of Noise Margin, Power and Process Variation for 32 nm FinFET Based 6T SRAM Cell" Journal of Computer (JCP), Academy Publisher, FINLAND, vol. 5, issue 6, PP.1-8, 2010.

[25] Divya Sharma, Rajesh Mehra and B Raj, "Comparative Analysis of Photovoltaic Technologies for High Efficiency Solar Cell Design", Superlattices and Microstructures, Elsevier, Volume 153, PP. 106861, May 2021.

[26] Pawandeep Kaur, Avtar Singh Buttar, B Raj, "A comprehensive Analysis of Nanoscale Transistor based Biosensor: A Review, Indian Journal of Pure and Applied Physics, Vol. 59, pp. 304-318, April 2021.

[27] Divya Yadav, Balwant Raj and B Raj, "Design and Simulation of Low Power Microcontroller for IoT Applications", Journal of Sensor Letters, ASP, Vol. 18, PP. 401–409, May 2020.

[28] Shailendra Singh, B Raj,"A 2-D Analytical Surface Potential and Drain current Modeling of Double-Gate Vertical t-shaped Tunnel FET", Journal of Computational Electronics, Springer, Vol. 19, PP.1154–1163, Apl 2020.

[29] Jeetendra Singh, B Raj, "An Accurate and Generic Window function for Non-linear Memristor Model" Journal of Computational Electronics, Springer, Volume 18 Issue 2, PP. 640-647 June 2019.

[30] Manjit Kaur; Neena Gupta; Sanjeev Kumar; BRaj; Arun Kumar Singh, "Comparative RF and Crosstalk Analysis of Carbon Based Nano Interconnects" IET Circuits, Devices & Systems, Vol.15(6), PP. 493-503, Feb 2021.

[31] Nehru Kandasamy, Firdous Ahmad, D. Ajitha, B Raj, and Nagarjuna Telagam "Quantum Dot Cellular Automata based Scan Flip Flop and Boundary Scan Register, IETE Journal of Research, Volume 66, PP. 535-548, 2020.

[32] Sharma, S. K.., Raj, B., Khosla, M., "Enhanced Photosensivity of Highly Spectrum Selective Cylindrical Gate In1-xGaxAs Nanowire MOSFET Photodetector, Modern Physics letter-B, Vol. 33, No. 12, PP. 1950144 (2019).

[33] Jeetendra Singh, BRaj, "Design and Investigation of 7T2M NVSARM with Enhanced Stability and Temperature Impact on Store/Restore Energy", IEEE Transactions on Very Large Scale Integration Systems, Vol. 27, Issure 6, PP. 1322 - 1328 June 2019.

[34] Anil Kumar Bhardwaj, Sumeet Gupta, BRaj, Amandeep Singh, "Impact of Double Gate Geometry on the Performance of Carbon Nanotube Field Effect Transistor Structures for Low Power Digital Design", Computational and Theoretical Nanoscience, ASP, Vol. 16, PP. 1813–1820, 2019.

[35] Neeraj Jain, BRaj, "Thermal Stability Analysis and Performance Exploration of Asymmetrical Dual-k underlap Spacer (ADKUS) SOI FinFET for Security and Privacy Applications", Indian Journal of Pure & Applied Physics (IJPAP), Vol. 57, pp. 352-360, May 2019.

[36] Amandeep Singh, Mamta Khosla, BRaj, "Design and Analysis of Dynamically Configurable Electrostatic Doped Carbon Nanotube Tunnel FET"," Microelectronics Journal, Elesvier, Volume 85, Pages 17-24, March 2019.

[37] Neeraj Jain and Balwinder Raj, "Dual-k Spacer Region Variation at the Drain Side of Asymmetric SOI FinFET Structure: Performance Analysis towards the Analog/RF Design Applications", communicated in "Journal of Nanoelectronics and Optoelectronics", American Scientific Publishers, Vol. 14, PP. 349–359, March 2019.

[38] Jeetendra Singh, Sanjeev Sharma, B Raj, Mamta Khosla Analysis of barrier layer thickness on performance of In1-xGaxAs based Gate Stack Cylindrical Gate Nanowire MOSFET," JNO, ASP, Vol. 13, PP 1473-1477, Oct 2018.

[39] Neeraj Jain, B Raj, "Analysis and Performance Exploration of High-k SOI FinFETs Over the Conventional Low-k SOI FinFET toward Analog/RF Design", Journal of Semiconductors (JoS), IOP Science, Vol.39, Issue 12, PP- 124002-1-7, Dec 2018.

[40] Candy Goyal, Jagpal Singh Ubhi and BRaj, A reliable leakage reduction technique for approximate full adder with reduced ground bounce noise, Journal of Mathematical Problems in Engineering, Hindawi, Volume 2018, Article ID 3501041, 16 pages, 15 Oct 2018.

[41] Anuradha, Jeetendra Singh, B Raj and Mamta Khosla, "Design and Performance Analysis of Nano-scale Memristor-based Nonvolatile SRAM" Journal of Sensor Letter", American Scientific Publishers, Vol. 16, PP 798–805, Oct 2018.

[42] Girish Wadhwa, BRaj, "Parametric Variation Analysis of Charge-Plasma-based Dielectric Modulated JLTFET for Biosensor Application" IEEE Sensor Journal, VOL. 18, NO. 15, AUGUST 1, PP. 6070 – 6077, 2018 (SCI)

[43] Jeetendra Singh, BRaj, "Comparative Analysis of Memristor Models for Memories Design", JoS, IoP, Vol. 39, No. 7, PP. 074006-1-12, July 2018.

[44] Divya Yadav, Shailesh Singh Chouhan, Santosh Kumar Vishvakarma and BRaj, "Application Specific Microcontroller Design for IoT based WSN", Sensor Letter, ASP, Vol. 16, PP. 374–385, May 2018..

[45] Gurmohan Singh, R. K. Sarin and BRaj, "Fault-Tolerant Design and Analysis of Quantum-Dot Cellular Automata Based Circuits", IEEE/IET Circuits, Devices & Systems, Vol. 12, PP. 638 – 64, 2018.

[46] Jeetendra Singh, BRaj, "Modeling of Mean Barrier Height Levying Various Image Forces of Metal Insulator Metal Structure to Enhance the Performance of Conductive Filament Based Memristor Model", IEEE Nanotechnology, Vol. 17, No. 2, PP. 268-267, March 2018.

[47] Aakash Jain, Sanjeev Sharma, BRaj, "Analysis of Triple Metal Surrounding Gate (TM-SG) III-V Nanowire MOSFET for Photosensing Application", Opto-electronics Journal, Elsevier, Volume 26, Issue 2, Pages 141-148, May 2018.

[48] Aakash Jain, Sanjeev Sharma, B Raj, "Design and Analysis of High Sensitivity Photosensor Using Cylindrical Surrounding Gate MOSFET For Low Power Sensor Applications", Engineering Science and Technology, an International Journal, Elsevier's, Volume 19, Issue 4, Pages 1864–1870, December 2016.

[49] Amandeep Singh, Mamta Khosla, B Raj, "Analysis of Electrostatic Doped Schottky Barrier Carbon Nanotube FET for Low Power Applications," Journal of Materials Science: Materials in Electronics, Springer, Vol.28, PP. 1762–1768, 2017.

[50] G. Saiphani Kumar, Amandeep Singh, BRaj, "Design and Analysis of Gate All Around CNTFET based SRAM cell Design", Journal of Computational Electronics, Springer, Volume 17, Issue 1, pp 138–145, March 2018.

[51] Gurinder pal Singh, B. S. Sohi, Balwinder Raj, "Material Properties Analysis of Graphene Base Transistor (GBT) for VLSI Analog Circuits, Indian Journal of Pure & Applied Physics (IJPAP), Vol. 55, pp. 896-902, Dec-2017.

[52] Amandeep Singh, Mamta Khosla, B Raj, "Comparative Analysis of Carbon Nanotube Field Effect Transistor and Nanowire Transistor for Low Power Circuit Design," Journal of Nanoelectronics and Optoelectronics, American Scientific Publishers, USA, Vol. 11, PP. 388-393, June 2016.

[53] Sunil Kumar and B Raj, "Estimation of Stability and Performance metric for Inward Access Transistor based 6T SRAM Cell Design using n-type/p-type DMDG-GDOV TFET" IEEE VLSI Circuits and Systems Letter, Volume3, Issue2, PP 25-39, June2017

[54] Shashikant Sharma, Anjan Kumar, Manisha Pattanaik and B Raj, "Forward Body Biased Multimode Multi-Threshold CMOS Technique for Ground Bounce Noise Reduction in Static CMOS Adders" International Journal of Information and Electronics Engineering, pp. 567-572, Vol. 3, issue 3, 2013.

[55] Hamendra Singh, Pankaj Kumar, Balwinder Raj, "Performance Analysis of Majority Gate SET Based 1-bit Full Adder", *International Journal of Computer and Communication Engineering* (IJCCE), IACSIT Press Singapore, ISSN: 2010-3743, Volume 2, No. 4, 2013.

[56] Anil Kumar Bhardwaj, Sumeet Gupta and B Raj, "Investigation of Parameters for Schottky Barrier (SB) Height for Schottky Barrier Based Carbon Nanotube Field Effect Transistor De-

vice", Journal of Nanoelectronics and Optoelectronics, ASP, Vol. 15, PP. 783–791, July 2020.

[57] Priya Bansal, BRaj, "MEMRISTOR: A VERSATILE NONLINEAR MODEL FOR DOPANT DRIFT AND BOUNDARY ISSUES," JCTN, American Scientific Publishers, Vol. 14 (5), PP. 2319-2325, May 2017.

[58] Neeraj Jain, BRaj, "An Analog and Digital Design Perspective Comprehensive Approach on Fin-FET (Fin-Field Effect transistor) Technology - A Review", Reviews in Advanced Sciences and Engineering (RASE), ASP, Vol. 5, PP. 1-14, 2016.

[59] Sanjeev Sharma, B Raj, Mamta Khosla, "Subthreshold Performance of In1-xGaxAs based Dual Metal with Gate Stack Cylindrical/Surrounding Gate Nanowire MOSFET for Low Power Analog Applications, Journal of Nanoelectronics and Optoelectronics, American Scientific Publishers, USA, Vol.12, PP. 171–176, 2017.

[60] Balwinder Raj, A. K. Saxena and S. Dasgupta, "Analytical Modeling for the Estimation of Leakage Current and Subthreshold Swing Factor of Nanoscale Double Gate FinFET Device" Microelectronics International, UK, vol. 26, pp. 53-63, 2009.

[61] Soniya, Shailendra Singh, Girish Wadhwa and Balwinder Raj, "An Analytical Modeling for Dual Source Vertical Tunnel Field Effect Transistor" submitted to 'International Journal of Recent Technology and Engineering (IJRTE)' at Volume-8 Issue-2, PP. 603-608, July 2019.

[62] Shailendra Singh, B Raj, "Design and analysis of hetrojunction Vertical T-shaped Tunnel Field Effect Transistor", Journal of Electronics Material, Springer, Volume 48, Issue 10, pp 6253–6260, October 2019.

[63] Candy Goyal, Jagpal Singh Ubhi and B Raj, "A Low Leakage CNTFET based Inexact Full Adder for Low Power Image Processing Applications", International Journal of Circuit Theory and Applications, Wiley, Volume47, Issue9, Pages 1446-1458, September 2019.

[64] B Raj, A. K. Saxena and S. Dasgupta, "A Compact Drain Current and Threshold Voltage Quantum Mechanical Analytical Modeling for FinFETs" Journal of Nanoelectronics and Optoelectronics (JNO), USA, vol. 3, no. 2, pp. 163-170, 2008.

[65] Girish Wadhwa, B Raj, "An Analytical Modeling of Charge Plasma based Tunnel Field Effect Transistor with Impacts of

Gate underlap Region" Superlattices and Microstructures, Elsevier, Vol. 142, PP.106512, June 2020.

[66] Shailendra Singh, B Raj, "Modeling and Simulation analysis of SiGe hetrojunction Double GateVertical t-shaped Tunnel FET", Superlattices and Microstructures, Elsevier Volume 142, PP. 106496, June 2020.

[67] Amandeep Singh, Dinesh Kumar Saini, Dinesh Agarwal, Sajal Aggarwal, Mamta Khosla, B Raj, "Modeling and Simulation of Carbon Nanotube Field Effect Transistor and its Circuit Application," Journal of Semiconductors (JoS), IOP Science, Vol.37, PP. 074001-6, July 2016.

[68] Neeraj Jain, Balwinder Raj, "Device and Circuit Co-Design Perspective Comprehensive Approach on FinFET Technology - A Review, Journal of Electron Devices, Vol. 23, Num. 1, pp 1890-1901, 2016.

[69] Sunil Kumar and B Raj, "Analysis of I_{ON} and Ambipolar Current for Dual-Material Gate-drain Overlapped DG-TFET," Journal of Nanoelectronics and Optoelectronics, American Scientific Publishers, USA, Vol.11, PP. 323-333, June 2016.

[70] Naveed Anjum, Tarun Bali and B Raj, Design And Simulation Of Handwritten Multiscript Character Recognition, International Journal of Advanced Research in Computer and Communication Engineering, Vol.2 (7), PP. 2544-2549, July 2013.

[71] Sanjeev Sharma, B Raj, Mamta Khosla, "A Gaussian Approach for Analytical Subthreshold Current Model of Cylindrical Nanowire FET with Quantum Mechanical Effects, Microelectronics Journal, Elsevier, Vol. 53, PP. 65-72, April 2016.

[72] Karmjit Singh, B Raj, "Performance and analysis of temperature dependent Multi-walled Carbon Nanotubes as Global Interconnects at different technology nodes," Journal of Computational Electronics, Springer, Volume 14 Issue 2, pp. 469-476, June 2015.

[73] Sunil Kumar and B Raj, "Compact channel potential analytical Modeling of DG-TFET based on Evanescent–mode Approach," Journal of Computational Electronics, Springer Vol. 14, No. 2, pp.820-827, July 2015.

[74] Karmjit Singh, B Raj, "Temperature Dependent Modeling and Performance Evaluation of Multi-Walled CNT and Single-Walled CNT as Global Interconnects," Journal of Electronic Materials, Springer, Vol. 44, Issue 12, pp 4825-4835, Dec 2015.

[75] V.K. Sharma, M. Pattanaik and B. Raj, "INDEP approach for leakage reduction in nanoscale CMOS circuits", International Journal of Electronics, Taylor & Francis, Vol.102, No.2, pages 200-215, 2014.

[76] Karmjit Singh, BRaj, "Influence of Temperature on MWCNT bundle, SWCNT bundle and Copper interconnects for Nanoscaled Technology nodes," Journal of Materials Science: Materials in Electronics. Springer, Vol. 26, No. 8, pp. 6134-6142, 2015.

[77] Naveed Anjum, Tarun Bali and B Raj, "Design and Simulation of Handwritten Gurumukhi and Devanagri Numerical Recognition, International Journal of Computer Applications, Published by Foundation of Computer Science, New York, USA, Vol. 73(12), pp. 16-21, 2013.

[78] Khandelwal, S., Gupta, V., BRaj, and Gupta, R D., Process Variability Aware Low Leakage Reliable Nano Scale DG-FinFET SRAM Cell Design Technique, Journal of Nanoelectronics and Optoelectronics, Volume 10, Number 6, pp. 810-817, December 2015.

[79] V.K. Sharma, M. Pattanaik and B Raj, "ONOFIC Approach: Low Power High Speed Nanoscale VLSI Circuits Design", International Journal of Electronics, Taylor & Francis, Vol. 101, issue 1, pages 61-73, 2014.

[80] Khandelwal, S., Balwinder Raj, and Gupta, R D., FinFET based 6T SRAM Cell Design: Analysis of Performance Metric, Process Variation and Temperature Effect. Journal of Computational and Theoretical Nanoscience, ASP, USA, Vol. 12, PP. 2500-2506, 2015.

[81] Sumit Singh; Yadav, Shekhar; Rahul, Jagdeep; Srivastava, Anurag; BRaj, Impact of HfO_2 in Graded Channel Dual Insulator Double Gate MOSFET, Journal of Computational and Theoretical Nanoscience, American Scientific Publishers, Volume 12, Number 6, pp. 950-953, April 2015.

[82] Vijay Kumar Sharma, Manisha Pattanaik, BRaj, PVT variations aware low leakage INDEP approach for nanoscale CMOS Circuits, Microelectronics Reliability, Elsevier, Vol. 54, pp 90-99, 2014.

[83] B Raj, A. K. Saxena and S. Dasgupta, "Quantum Mechanical Analytical Modeling of Nanoscale DG FinFET: Evaluation of Potential, Threshold Voltage and Source/Drain Resistance",

Elsevier's Journal of Material Science in Semiconductor Processing, Elsevier, Vol. 16, issue 4, pp. 1131- 1137, 2013.

[84] Maisagalla Gopal, Siva Sankar D Prasad and Balwinder Raj, 8T SRAM Cell Design for Dynamic and Leakage Power Reduction, International Journal of Computer Applications, Published by Foundation of Computer Science, New York, USA, Vol. 71(9), pp. 43-48, June 2013.

[85] Manisha Pattanaik, B Raj, Shashikant Sharma and Anjan Kumar, "Diode Based Trimode Multi-Threshold CMOS Technique for Ground Bounce Noise Reduction in Static CMOS Adders", Advanced Materials Research, Trans Tech Publications, Switzerland, vol. 548, pp 885-889, 2012.

[86] Balwinder Raj, A. K. Saxena and S. Dasgupta, "Nanoscale FinFET Based SRAM Cell Design: Analysis of Performance metric, Process variation, Underlapped FinFET and Temperature effect" IEEE Circuits and System Magazine, vol. 11, issue 2, pp. 38- 50, 2011.

[87] V.K. Sharma, M. Pattanaik & Balwinder Raj, "Leakage Current ONOFIC Approach for Deep Submicron VLSI Circuit Design", International Journal of Electrical, Computer, Electronics and Communication Engineering, World Academy of Sciences, Engineering and Technology, Vol. 7, issue 4, pages 239-244, 2013.

[88] Tulika Chawla, Mamta Khosla, BRaj, Design and simulation of Triple metal Double-gate Germanium on Insulator Vertical Tunnel Field Effect Transistor, Microelectronics Journal, Elsevier; Vol. 114, PP. 105125, August 2021.

[89] Parminder Kaur, Sandeep Singh Gill and B Raj, "Comparative Analysis of OFETs Materials and Devices for Sensor Applications", Journal of Silicon, Springer, Vol 14, PP. 4463–4471, 2022.

[90] Sanjeev Kumar Sharma, Parveen Kumar, Balwant Raj and BRaj, "$In_{1-x}Ga_xAs$ Double Metal Gate-Stacking Cylindrical Nanowire MOSFET for highly sensitive Photo detector", *Journal of Silicon, Springer Vol.*14, PP. 3535–3541, 2022.

[91] B Raj, A. K. Saxena and S. Dasgupta, "Analytical Modeling of Quasi Planar Nanoscale Double Gate FinFET with Source/Drain Resistance and field dependent carrier mobility: A quantum mechanical study" Journal of Computer (JCP), Academy Publisher, FINLAND, vol. 4, issue 9, PP.1-8, 2009.

[92] Bhushan, S., Khandelwal, S. and Raj, B., (2012). Analyzing Different Mode FinFET based Memory Cell at Different Power Supply for Leakage Reduction, Seventh International Conference on Bio-Inspired Computing: Theories and Application, (BIC-TA 2012) Advances in Intelligent Systems and Computing, Volume 202, p89-100 (2013).

[93] Jeetendra Singh, BRaj, "Temperature Dependent Analytical Modeling and Simulations of Nanoscale Memristor, Journal: Engineering Science and Technology, an International Journal, Elsevier's, Vol. 21, PP. 862-868, Oct 2018.

[94] Shradhya Singh, Shashi Bala, Balwant Raj, B Raj, "Improved Sensitivity of Dielectric Modulated Junctionless Transistor for Nanoscale Biosensor Design", Sensor Letter, ASP, Vol. 18, PP. 328–333, Apl 2020.

[95] Vivek Kumar, Santosh Kumar Vishvakarma, and B Raj, "Design and Performance Analysis of ASIC for IoT Applications" Sensor Letter ASP, Vol. 18, PP. 31–38, Jan 2020.

[96] Akanksha Jaiswal, R. K. Sarin, B Raj and Shikha Sukhija, "A Novel Circular Slotted Microstrip-fed Patch Antenna with Three Triangle Shape Defected Ground Structure for Multiband Applications", Advanced Electromagnetic (AEM), Vol. 7, Issue 3, PP. 56-63, Aug 2018.

[97] Girish Wadhwa, B Raj, "Label Free Detection of Biomolecules using Charge-Plasma-Based Gate Underlap Dielectric Modulated Junctionless TFET" Journal of Electronic Materials (JEMS), Springer, Volume 47, Issue 8, pp 4683–4693, August 2018.

[98] Gurmohan Singh, R. K. Sarin and B Raj, "Design and Performance Analysis of a New Efficient Coplanar Quantum-Dot Cellular Automata Adder", Indian Journal of Pure & Applied Physics (IJPAP), Vol. 55, pp. 97-103, February 2017.

[99] Amandeep Singh, Mamta Khosla, Balwinder Raj, "Design and Analysis of Electrostatic Doped Schottky Barrier CNTFET Based Low Power SRAM," International Journal of Electronics and Communications, (AEÜ), Elsevier, Vol. 80, PP.67–72, 2017.

[100] Parminder Kaur, Vikas Pandey and B Raj, "Comparative Study of Efficient Design, Control and Monitoring of Solar Power using IoT", Sensor Letter, ASP Vol. 18, PP. 419–426, May 2020.

[101] Anil Kumar Bhardwaj, Sumeet Gupta and B Raj, Development & Analysis of compact model for Double Gate Schottky Barrier

CNTFET, Journal of Nanoelectronics and Optoelectronics, ASP, Vol. 15, pp. 1199–1208, Aug 2020.

[102] Girish Wadhwa, Priyanka Kamboj; Jeetendra Singh, B Raj, "Design and Investigation of Junctionless DGTFET for Biological Molecule Recognition" Transactions on Electrical and Electronic Materials, Springer, 8 Aug, 2020.

[103] Tulika Chawla, Mamta Khosla, B Raj, "Optimization of Double-gate Dual material GeOI-Vertical TFET for VLSI Circuit Design, IEEE VLSI Circuits and Systems Letter, Vol. 6, issue-2, PP.13-25, Aug 2020.

[104] Sachin Kumar Verma, Shailendra Singh, Girish Wadhwa and B Raj, "Detection of Biomolecules using Charge-Plasma based Gate Underlap Dielectric Modulated Dopingless TFET" Transactions on Electrical and Electronic Materials (TEEM), Springer, Vol. 21, PP. 528-535, June 2020.

[105] Neeraj Jain, BRaj, "Impact of underlap spacer region variation on Electrostatic and analog/RF performance of symmetrical high-k SOI FinFET at 20 nm channel length", Journal of Semiconductors (JoS), IOP Science, Vol.38, Issue 12, PP- 122002, Dec 2017.

[106] Shailendra Singh, B Raj," Analytical Modeling and Simulation analysis of T-shaped III-V heterojunction Vertical T-FET", Superlattices and Microstructures, Elsevier, Vol. 147, PP. 106717, Nov 2020.

[107] Gurmohan Singh, R. K. Sarin and B Raj, "Design and Analysis of Area Efficient QCA Based Reversible Logic Gates", Journal of Microprocessors and Microsystems, Elsevier, Vol. 52, PP.59-68, May 2017.

[108] Amandeep Singh, Mamta Khosla, BRaj, "Compact Model for Ballistic Single Wall CNTFET under Quantum Capacitance Limit," Journal of Semiconductors (JoS), IOP Science, Vol.37, PP. 104001-8, Oct 2016.

[109] Sonal Singh; Mamta Khosla, Girish Wadhwa; B Raj, "Design and analysis of double-gate junctionless vertical TFET for gas sensing applications", Applied Physics A, Springer, Vol.127, No.16, 2 Jan 2021.

[110] Inderjit Singh, B Raj, Mamta Khosla, and Brajesh Kumar Kaushik, Potential MRAM Technologies for Low Power SoCs, SPIN World Scientific Publisher, SCIE; Vol. 10, No. 04, PP.2050027, Dec 2020.

[111] Shailendra Singh, B Raj,"Parametric variation analysis on hetero-junction Vertical t-shape TFET for Supressing ambipolar conduction", Indian Journal of Pure and Applied Physics, Vol. 58, pp. 478-485, June 2020.

[112] Soniya, Shailendra Singh, Girish Wadhwa and Balwinder Raj, Design and Analysis of Dual Source Vertical Tunnel Field Effect Transistor for high performance, Transactions on Electrical and Electronics Materials, Springer, Vol. 21, Pages 74–82, Oct 2019.

[113] Manjit Kaur; Neena Gupta; Sanjeev Kumar; B Raj; Arun Kumar Singh, "RF Performance Analysis of Intercalated Graphene Nanoribbon Based Global Level Interconnects" Journal of Computational Electronics, Springer, Vol. 19, PP.1002–1013, June 2020

[114] Girish Wadhwa, B Raj, "Design and Performance Analysis of Junctionless TFET Biosensor for high sensitivity" IEEE Nanotechnology, Vol.18, PP. 567 – 574, 2019.

[115] Jeetendra Singh, B Raj, "Enhanced Nonlinear Memristor Model Encapsulating Stochastic Dopant Drift, JNO, ASP 14, 958–963 (2019).

2

Design and Analysis of Charge Plasma based Vertical TFET

*Shailendra Singh[1,2]*and Balwinder Raj[1,3]*
[1]Department of Electronics and Communication Engineering,
[2]PSIT Kanpur
[3]NIT Jalandhar
**shailendras.nitj@gmail.com*

Abstract: This study supported the use of TFETs for better ON-current and steeper sub-threshold (SS) swings by utilising charge plasma (CP) and negative capacitance (NC). Recent research has shown that every metal oxide semiconductor is experiencing an increase in power dissipation (CMOS). Therefore, it is imperative to create new technology that can solve these problems. Then negative capacitance enters the picture, which can reduce the supply voltage and subthreshold slope while also minimizing the threshold voltage and subthreshold slope to some extent. In this book chapter, the performance of the device (CP-Ferro-VTEFT), is examined using a ferroelectric capacitor, which causes a negative capacitance effect as the device begins to operate. Device performance is assessed both with and without the use of negative capacitance. A comparison with different dioxide material (SiO_2 and Si_3N_4) and has been used to engineering the presence the effect of negative capacitance. The use of charge plasma and doping less silicon nanowire is improving electronics engineering. As a result, the device's CP and NC both produce better electrical properties, a reduced thermal budget, and process variation immunity. Investigations are also conducted into how differences in ferroelectric density affect device conducts such polarised charge and threshold voltage.

2.1 Introduction

The tunnel field-effect transistor (TFET), also known as a gated p-i-n diode, functions under a reverse bias condition; in contrast to MOSFETs, which employ thermal injection for carrier injection,

TFETs use band-to-band tunnelling as the source carrier injection mechanism [1]. By scaling the supply voltage, tunnel field effect transistors are intended to reduce energy dissipation [2]. The TFET may operate effectively below 1V and, by providing steep subthreshold swing devices (SS), one can reduce leakage current. The application of Moore's law is growing more challenging as technology advances, though.

In recent years, innovative channel materials and MOSFET structures that can control physical constraints have been used to control power consumption and operating voltages [3]. Tunnel FETs, nano-electromechanical FETs, and impact ionisation FETs are a few of the proposed structures on it. Other technological advancements, including as meta surfaces, near-zero-index materials, metamaterials, graphene, etc., can aid in improving the efficiency of semiconductor devices. Low refractive index, quantum wells, and high mobility are three qualities that could contribute to the new era of integrated circuits made possible by these technical developments [4-5].

In order to address a number of issues with conventional MOSFET device, such as high-power losses, leakage concerns, and a steep subthreshold slope, the new notion of negative capacitance was discovered. By varying the gate voltage, the negative capacitance offers differential potential amplifications and can function as a step-up transformer. Negative capacitance is subject to two opposing theories. Some claim that ferroelectrics can access their inherent negative capacitance, while others contend that negative capacitance is a temporary effect [6-7]. Even if the ferroelectric inside the negative capacitance is erratic, adding a series dielectric capacitor stabilises it. A device with ferroelectric material that uses negative capacitance to enhance performance is referred to as a negative capacitance FET [8-9].

In order to achieve the same surface potential, the total capacitance is made large by using the negative capacitance as a gate stack. The negative capacitance effect produced by the Ferro material used in (CP-Ferro-VTEFT) depends on the state variation of the Fe material. TFETs are well recognised for low power applications due to their lowest subthreshold properties, which also contribute to their low power dissipation and lower supply voltage [10-12]. After extensive investigation, it was discovered that the Landau-Khalatnikov (L-K) model, subthreshold slope, and numerical simulation are integrated because of the favourable effects of ferroelectric material's negative capacitance on various devices [13-14]. The vertical structure favours the scaling part, to integrate a greater number of transistors with same confined area [15].

Recently, researchers have begun utilising the charge plasma-based dopingless tunnel FET (CP-VTFET). Additionally, in this case, the source (p-type) and drain (n-type) regions are formed using work-function engineering without silicon doping. RDFs and thermal budget are no longer a problem thanks to CPTFET, but there is still a need to find a solution for the low ON-state current issue [16-17]. Therefore, a novel device known as a (CP-Ferro-VTEFT) is presented to address the ON-state current issue to a specific degree by evenly doping the drain and source [18-20]. CP-Ferro-VTEFT performs better than CP-VTFET in creating ON-state current when their respective performances are compared [21-24].

In this study, we employed CP-VTFET, which combines the benefits of CP-TFET with negative capacitance. It has been found that CP-Ferro-VTFET can achieve enhanced subthreshold slope values and decreased threshold voltage when there is a negative capacitance effect present.

2.2 Device Layout and Description

Table 2.1 gives a description of the CP-VTFET device with different dielectric constant of Gate stack SiO_2, Si_3N_4 and Ferro, Figure 2.1(a) & Figure 2.1(b) show the device's two-dimensional schematic without negative capacitance whereas Figure 2.1(c) shows the device's two-dimensional schematic with negative capacitance. Because it is based on a P-I-N structure, the device is designed with a p-type source, n-type drain, and slightly p-type gate [25-30]. The device's foundation is a silicon nanotube that accumulates N+ source/drain regions in doping-less transistors using the charge plasma concept [31-36].

Table 2.1 *Performance evaluation of charge plasma vertical tunnel FETs with negative capacitance*

Device Parameters	CP-SiO_2-VTFET	CP-Si_3N_4-VTFET	CP-Ferro-VTFET
Channel length	20 nm	20 nm	20 nm
Gate Oxide thickness	3 nm	3 nm	3 nm
Device Thickness	10 nm	10 nm	10 nm
Doping concentration	1×10^{16} cm^{-3}	1×10^{16} cm^{-3}	1×10^{16} cm^{-3}
Dielectric Material	SiO_2	Si_3N_4	Ferro
Work Function -source side	3.9 eV	3.9 eV	3.9 eV
Work Function - channel	5.1 eV	5.1 eV	5.1 eV
Work Function - Drain side	4.2 eV	4.2 eV	4.2 eV
Operating Voltage Vds/Vgs	1V/1V	1V/1V	1V/1V

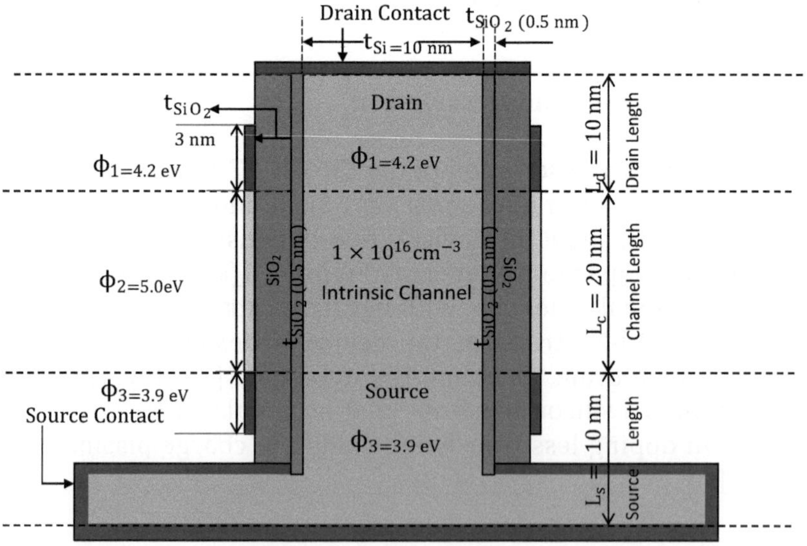

(a)

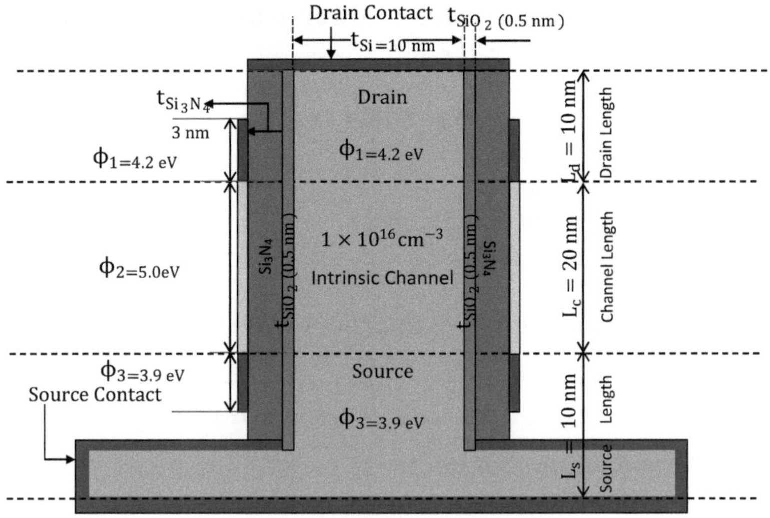

(b)

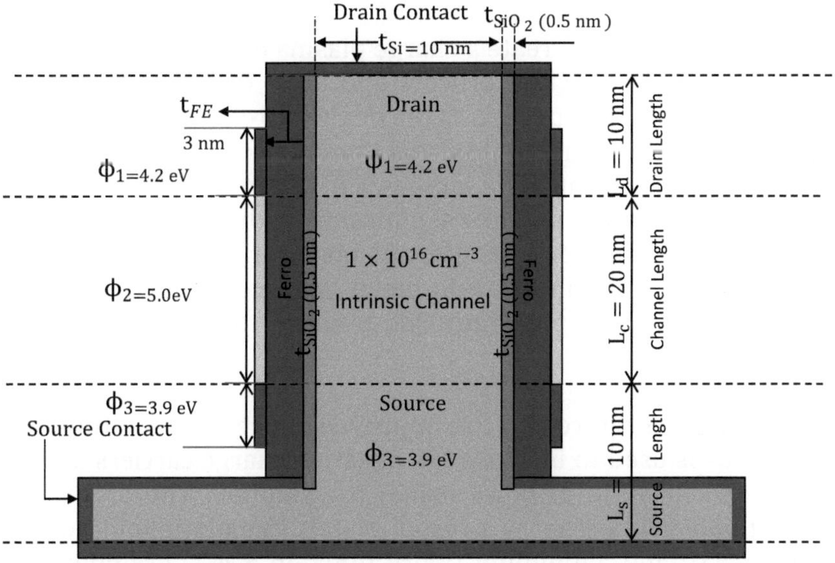

(c)

Figure 2.1 *(a) The CP-schematic Vertical TFET's 2D cross-sectional construction with SiO_2 (b) The CP-schematic Vertical TFET's 2D cross-sectional construction with Gate stack Si_3N_4. (c) Schematic 2D cross-sectional design of the negative capacitance CP-Ferro-VTFET.*

The idea of charge plasma reduces the overall metal-semiconductor contact resistance and eliminates the potential for silicide development. There are a few prerequisites for these ideas as source/drain metallic contact should have a lower work function than silicon [23].

$$\phi_m < \chi S_i + (E_G/2q) \quad (1)$$

q = elementary charge,
χS_i = electron affinity of bulk silicon,
E_G = band-gap of bulk silicon

The silicon body should have thickness < the Debye length i.e.,[24]

$$L_D = \sqrt{(\epsilon_{s_i} v_T / q.N)} \quad (2)$$

ϵ_{s_i} = *dielectric constant of silicon,*
v_T = *thermal voltage,*
N = *carrier concentration of body*

In order to perhaps create a charge plasma effect that would further improve the production process and area, the source metal is inserted into the core portion of the device[25-27]. Due to the entry of positive charge carriers from the core source metal's charge plasma, the hole concentration increases in the source region. The negative charge carriers are now significantly reduced at the source end. Additionally, the concentration of electron in the channel is decreased due to an increase in metal gate work function, but it remains constant on the drain side because to uniform doping [28-29].

To transpose the region for improved charge carrier tunneling, a gate bias is used to increase the negative charge carriers and lower the concentration of holes inside the channel. The drain current eventually gets better as a result. Metals like platinum (work function = 5.0) and aluminium (work function = 4.2) are employed as the core source metal to produce the shell gate. [29] In this study, a second ferroelectric capacitor is connected to the CP-shell VTFET's gate in order to provide the negative capacitance effect. Here, CP-Ferro-VTFET has been employed as the ferroelectric dielectric material [37-42]. When evaluating the device's performance, the intrinsic gate capacitance and ferroelectric capacitance should have similar absolute values (wielding the NC effect) [43-47].

Additionally, because hysteresis is created as a result of oscillations brought on by the total capacitance's negative value. Therefore, the series combination of C_{FE} and C_{INT} must be positive in the whole operating zone to prevent this and generate the NC effect. For simpler numerical simulation and calculation, the CP-Ferro-VTFET is kept at an absolute temperature of 300K.

$$V_f = a_0Q + b_0Q^3 + c_0Q^5 \tag{3}$$

The charge voltage (Q - V_f) properties of the ferroelectric material are described by the aforementioned L-K equation, were,Q = the ferroelectric material's charge per unit area, a_0, b_0, and c_0 = These coefficients correspond to the ferroelectric material's Landau Parameters a, b, and c, respectively.

$$a_0 = 2t_f a,\ b_0 = 4t_f b, \text{ and } c_0 = 6t_f c \tag{4}$$

Numerous models, including the Shockley-Read-Hall (SRH), Lombardi mobility model, local band-to-band tunnelling (BTBT) for tunneling the carriers, and auger recombination model, are used in the design of the CP-VTFET. Using a Wentzel-Kramer-Brillouin approximation, the numerical tunneling probability in this case has been evaluated. Additionally, the device's operating system is BTBT. The Silvaco atlas tool and Landau-Khalatnikov model are used to simulate and develop the device. [31-32].

2.3 Result Analysis

Cgg, C_{FE}, and C_T stand for the device's gate-oxide capacitance, ferroelectric-oxide capacitance, and total-capacitance, respectively. Where C_T is the outcome of the series of combinations of Cgg and C_{FE}, and Cgg is the result of adding (Cgs+Cgd). Itreflects the improvement in the device's net capacitance at low bias as a result of the ferroelectric oxide's negative differential capacitance compensating for the device's positive capacitance, which also improves the device's gross capacitance (C_T=C_{gg}+C_{FE}) and raises the surface potential at low bias [33-34]. For the goal of avoiding negative capacitance, the values of Cgg and C_{FE} are roughly maintained close to one another throughout the operation, and both capacitances should maintain positive values to avoid uncertainty [48-52].

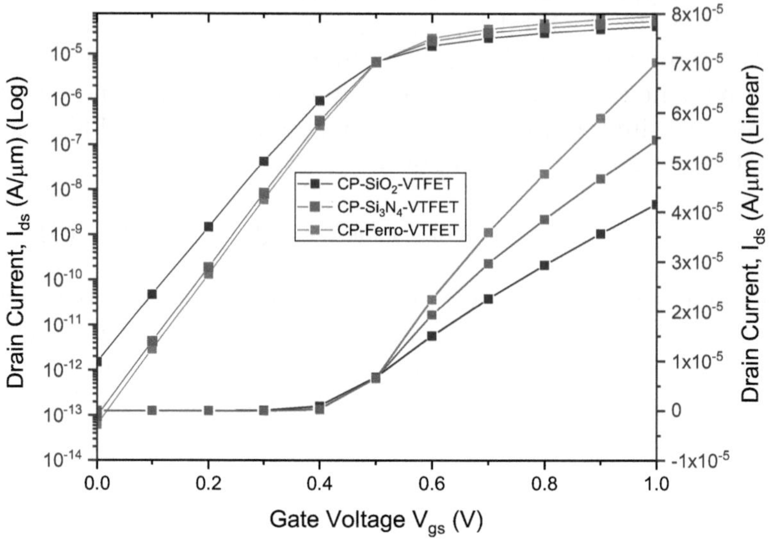

Figure 2.2 *Linear and Log drain current characteristics of CP-SiO2-VTFET, CP-Si3N4-VTFET and CP-Ferro-VTFET configuration with respect to Gate Voltage at V_{ds} 1.0 V and V_{gs} 1.0 V.*

Figure 2.2 represent the comparison graph drain Characteristics of Charge Plasma Vertical TFET for SiO_2, Si_3N_4 and Ferro material using gate stack technique. With this graph it can be observed that, the CP-Ferro-VTFET comes out with more improved graph in comparison to the CP-SiO_2-VTFET and CP-Si_3N_4-VTFET [53-57]. This happens due to addition of the negative capacitance to the device which significantly improve the Subthreshold slope (SS) by increasing the gate oxide capacitance. Due to reduction in the steep sub-threshold, the leakage current in exponentially reduce by given expression.

$$I_{Leakage} \approx \left(I_d @ V_{gs} = V_{th}\right) \times 10^{-V_{th}/SS} \qquad (5)$$

where, Vgs is gate-source Voltage, Vth is threshold voltage and I_d is device drain current respectively. Total capacitance will decide by the given equation 6, after adding the ferro material to the oxide.

$$\frac{1}{C_T} = \frac{1}{C_{ox}} + \frac{1}{C_{FE}} \qquad (6)$$

$$C_T = \frac{C_{ox} \cdot C_{FE}}{C_{ox} + C_{FE}} = \frac{Q}{V_{gs} - \psi_S} \qquad (7)$$

where, C_T, is the total capacitance, Cox is the oxide capacitance, C_{FE} is the ferro oxide capacitance. The term, $V_{gs} - \psi_S$, is the effective voltage which lied across the total capacitance respectively.

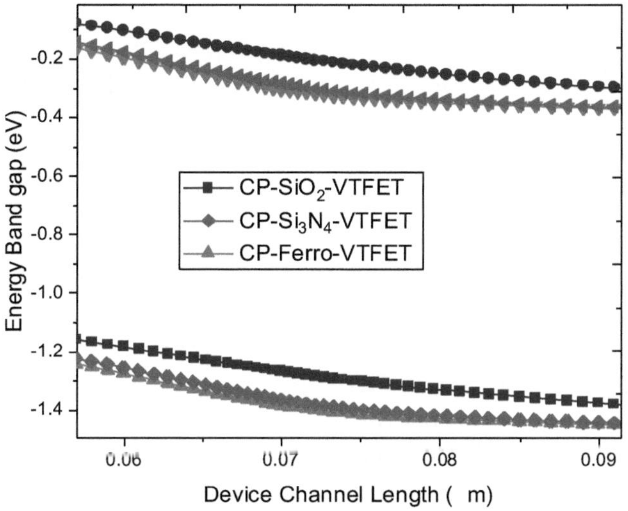

Figure 2.3 *Energy band diagram characteristics of CP-SiO_2-VTFET, CP-Si_3N_4-VTFET and CP-Ferro-VTFET configuration with respect to device length at V_{ds} 1.0 V and V_{gs} 1.0 V.*

Figure 2.3 depicts the energy band diagram of CP-SiO_2-VTFET, CP-Si_3N_4-VTFET and CP-Ferro-VTFET. Form the graph it can be observed that, the level of the three graphs of valance band and conduction band have been stepped up accordingly. The Ferro, which include the negative capacitance shown the highest value among all three configurations. Similarly, Figures 2.4, 2.5, 2.6 and 2.7 depict graphs that were computed both with and without the use of the negative capacitance effect.

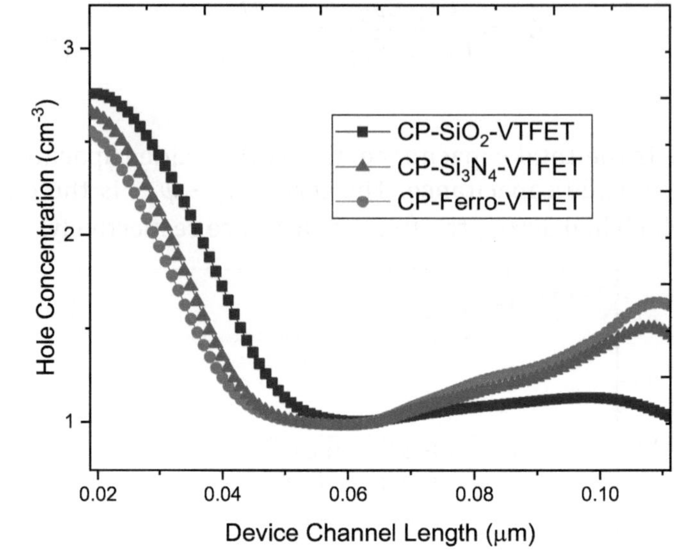

(a)

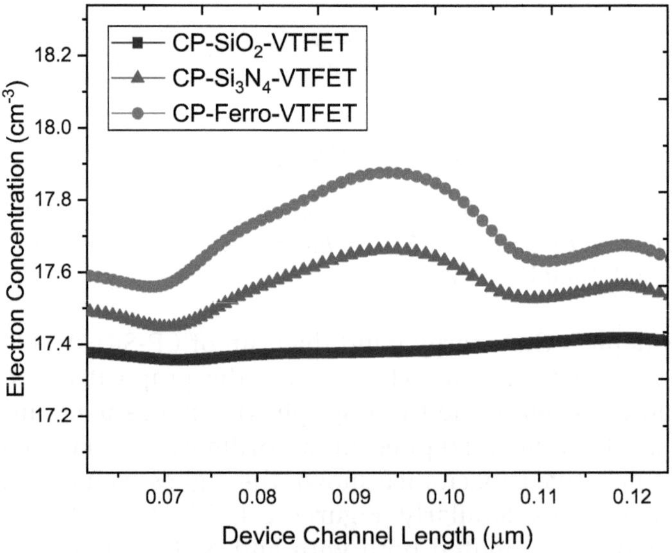

(b)

Figure 2.4 *(a) Charge carriers (electron) and (b) minority charge carriers (hole)of CP-SiO_2-VTFET, CP-Si_3N_4-VTFET and CP-Ferro-VTFET configuration with respect todevice length at V_{ds} 1.0 V and V_{gs} 1.0 V.*

Therefore, Figure 2.4(a) and 2.4(b) show a graph of the potential of a CP-VTFET with the majority charge carriers (electron) and minority charge carriers (hole) for all three configurations [58-63]. Due to negative capacitance, which increases electron concentration and amplifies internal voltage, electricity is created at low bias. As a result, there is an increase in the concentration of electrons, the majority of charge carriers, and a decrease in the concentration of positive charge [64-68]. The process of internal voltage inflation (in the ON state) resulted in a steeper graph of CP-Ferro- VTFET surface potential, as shown in Figure 2.5.

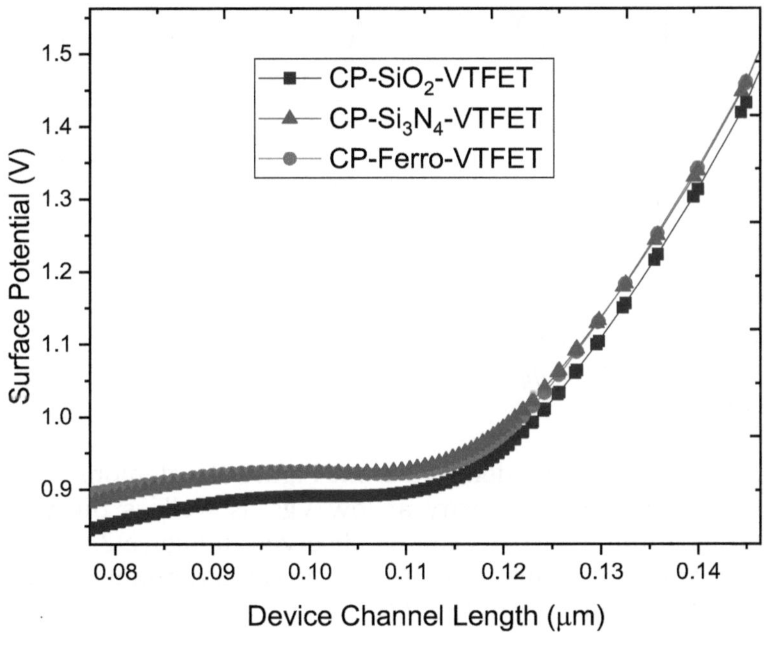

Figure 2.5 *Surface potential Characteristics of CP-SiO_2-VTFET, CP-Si_3N_4-VTFET and CP-Ferro-VTFET configuration with respect to device length at V_{ds} 1.0 V and V_{gs} 1.0 V.*

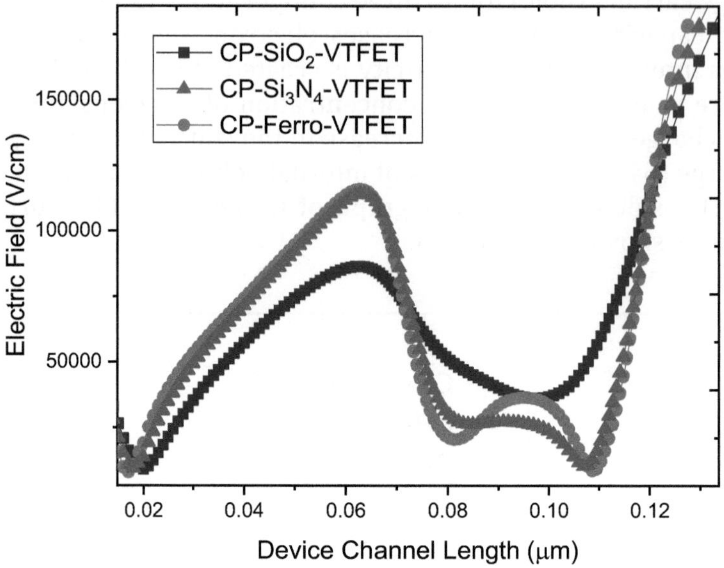

Figure 2.6 *Electric Field Characteristics of CP-SiO_2-VTFET, CP-Si_3N_4-VTFET and CP-Ferro-VTFET configuration with respect todevice length at V_{ds} 1.0 V and V_{gs} 1.0 V.*

The outcomes that are deduced from Figure 2.6 demonstrate the optimal electric field intensity at low Vgs as well as the various properties of the device. As can be observed in Figure 2.6, an optimum electric field results in more charge carriers tunneling, which increases the drain current [69-73].

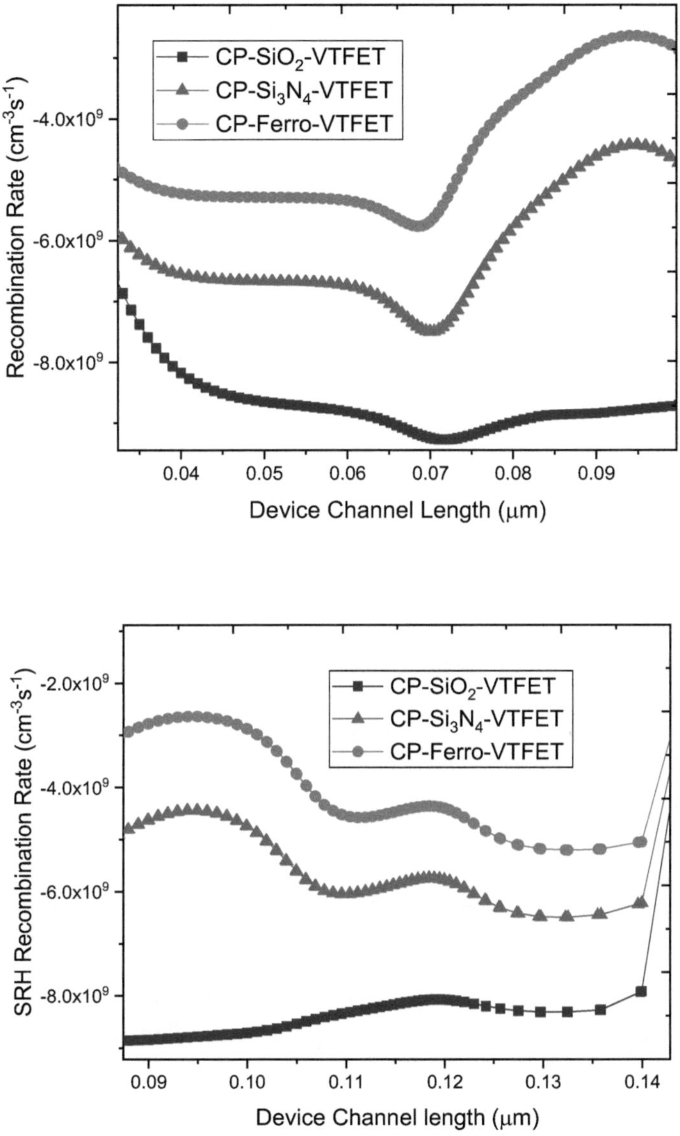

Figure 2.7 *(a) Recombination rate and (b) SRHRecombination rate Characteristics of CP-SiO_2-VTFET, CP-Si_3N_4-VTFET and CP-Ferro-VTFET configuration with respect to device length at V_{ds} 1.0 V and V_{gs} 1.0 V.*

Figure 2.7(a) and 2.7(b) also illustrates how the recombination rate depends on the strength of the electric field along the tunneling channel. Because momentum and quantized energy are maintained throughout the recombination process, the majority charge carriers in the valence band undergo less recombination with minority charge carriers in the conduction band [74-80].

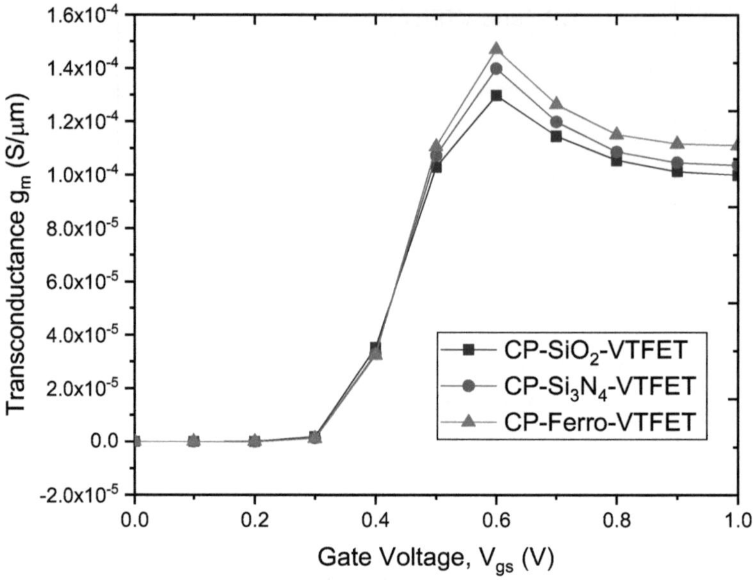

Figure 2.8 *Transconductance (g_m) Characteristics of CP-SiO_2-VTFET, CP-Si_3N_4-VTFET and CP-Ferro-VTFET configuration with respect to device length at V_{ds} 1.0 V and V_{gs} 1.0 V.*

The performance of any device, including a FET or a bipolar junction, is demonstrated in Figure 2.8 by the transconductance behaviour of that device. The device's amplification is increased by the bigger transconductance [81-85]. It has been noted that improved transconductance characteristics are displayed in the devices as a result of a sudden increase in total capacitance. The transconductance calculation formula is

$$\text{Transconductance } (g_m) = dI_D/dV_{GS} \text{ at const. } V_{DS} \tag{8}$$

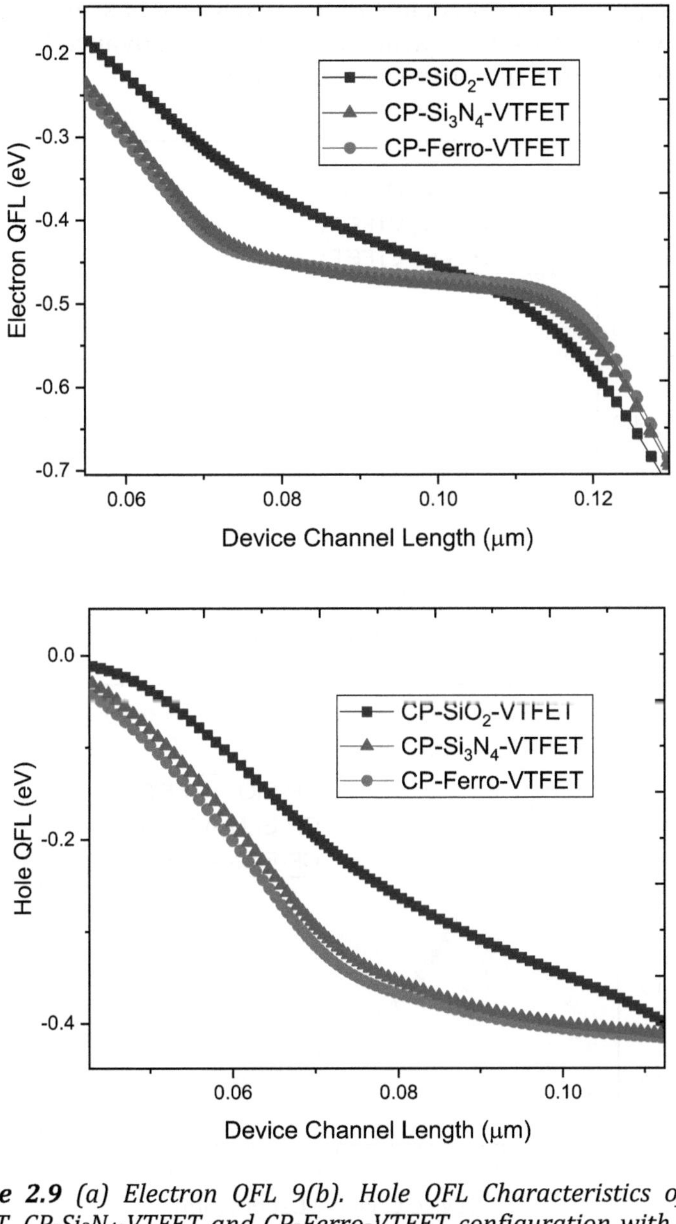

Figure 2.9 *(a) Electron QFL 9(b). Hole QFL Characteristics of CP-SiO_2-VTFET, CP-Si_3N_4-VTFET and CP-Ferro-VTFET configuration with respect to device length at V_{ds} 1.0 V and V_{gs} 1.0 V.*

Figure 2.9(a) and 2.9(b) shows the quasi-Fermi level of electrons and holes. The term "quasi-Fermi level" refers to the fermi level that depicts the concentration of electrons and holes at equilibrium [86-

90]. As was previously noted, negative capacitance has an impact on threshold voltage variation, which results in the advanced value of the quasi-Fermi level of both carriers at low bias.

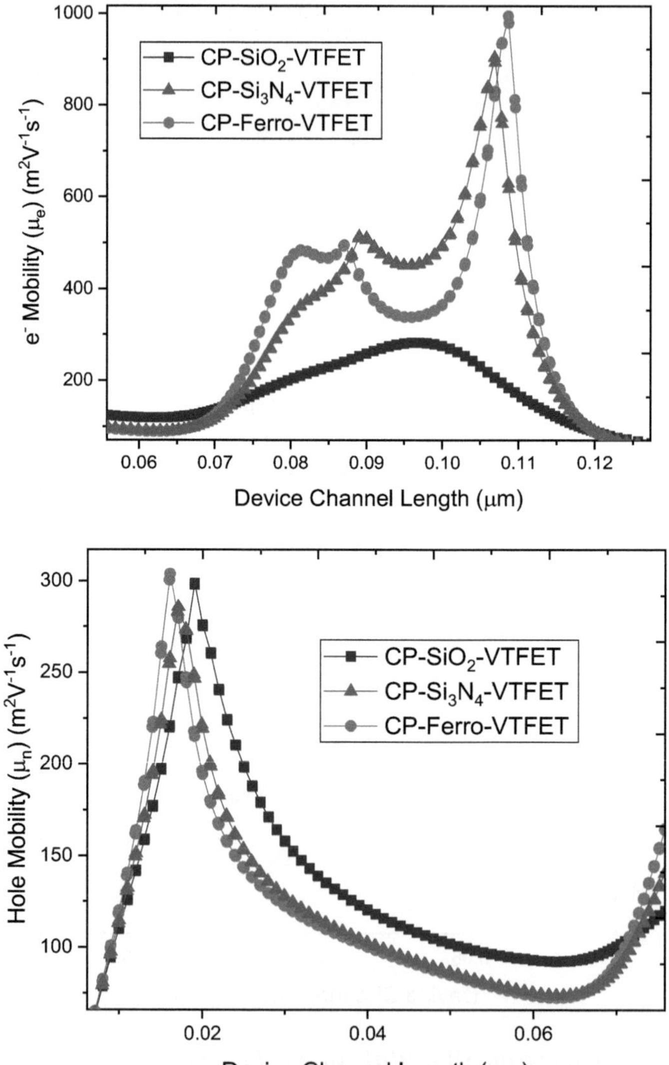

Figure 2.10 (a) *Electron Mobility 10(b). Hole Mobility Characteristics of CP-SiO_2-VTFET, CP-Si_3N_4-VTFET and CP-Ferro-VTFET configuration with respect to device length at V_{ds} 1.0 V and V_{gs} 1.0 V.*

Figure 2.10 (a) and 2.10(b) reflect the mobility factor of electrons and holes. The term mobility means the flow of the charge carries form source to destination. The flow of electron has always been shown to higher to the holes as can be seen form the Figure 2.10 (a) and 2.10 (b), However, among all three CP-Ferro-VTFET is shown majority for the electron and minority for the holes [91-94]. The is due to holes minority will mitigate the ambipolarity conduction.

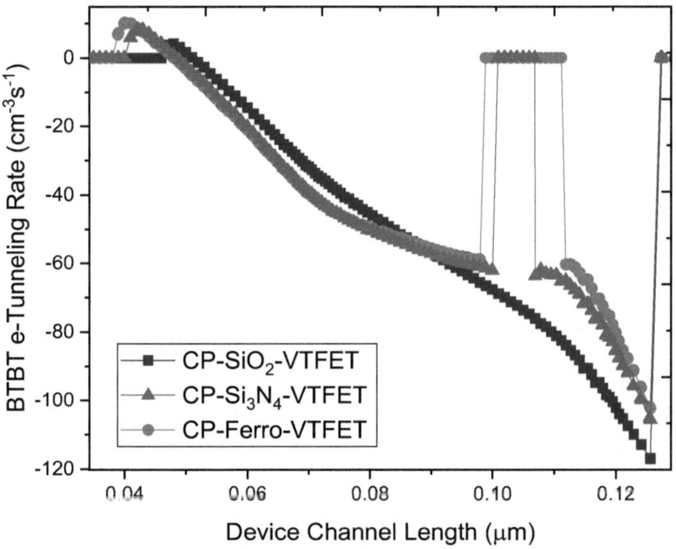

Figure 2.11 *Electron BTBT Tunneling Rate Characteristics of CP-SiO_2-VTFET, CP-Si_3N_4-VTFET and CP-Ferro-VTFET configuration with respect to to device length at V_{ds} 1.0 V and V_{gs} 1.0 V.*

Figure 2.11 shows the variation of the e-BTBT Tunneling with respect to device channel length for all the three configuration CP-SiO_2-VTFET, CP-Si_3N_4-VTFET and CP-Ferro-VTFET. With the incorporation of electron Band to band tunneling to all the proposed device one can easily predict form the graph that, for ferro configuration of the CP-VTFT, the graph is shown to be higher among all configuration [95-98]. This happens due to more capacitance generated due to involvement of the negative capacitance to the device which increase the barrier controlling capability and stop the leakage current.

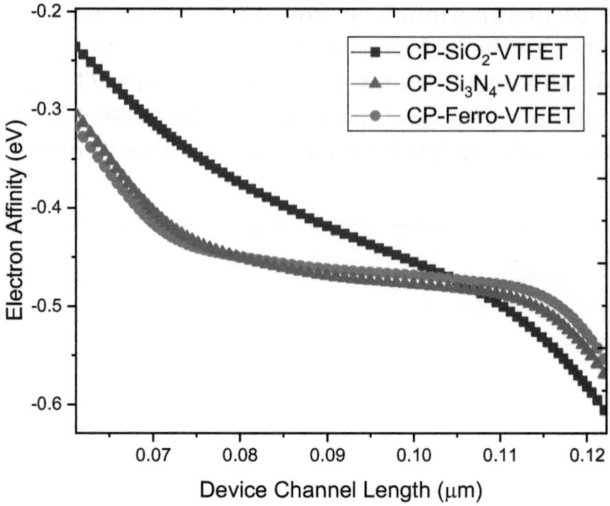

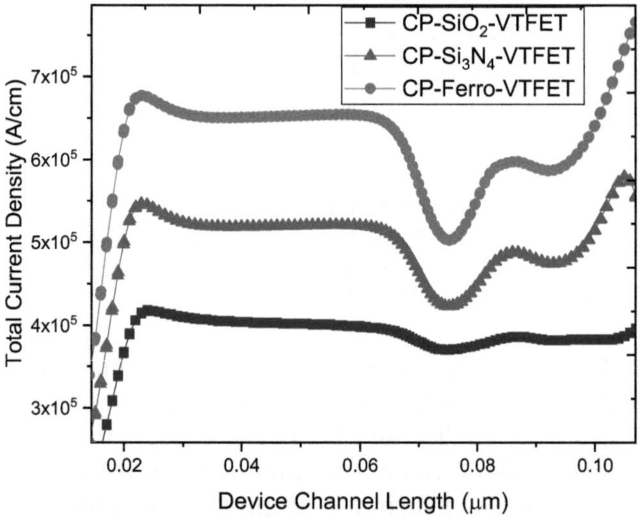

Figure 2.12 *(a) Electron Affinity 12, (b) Total Current Density characteristics of CP-SiO_2-VTFET, CP-Si_3N_4-VTFET and CP-Ferro-VTFET configuration with respect to device length at V_{ds} 1.0 V and V_{gs} 1.0 V.*

Figure 2.12 (a) and 2.12 (b) will show the graph of Electron Affinity and Total current density with respect to device channel length for all the three configuration CP-SiO_2-VTFET, CP-Si_3N_4-VTFET and CP-Ferro-VTFET. From the Figure 2.12 (a), electron affinity has varied along the channel length, which means that amount of energy released due attach of the electron to the neutral atom, which means that, after a certain duration the more electron will be varies in the CP-Ferro-VTFET configuration and highest among all [99-102]. Figure 2.12(b) shows the variation of the total current density, which is highest for the CP-Ferro-VTFET. This happens due to increases in the BTBT rate for ferro material configuration due to negative capacitance effect. Summarizing the graphs (Figure 2.2 to Figure 2.12) reveals that the negative capacitance impact enhances threshold and subthreshold slope at low bias without affecting any other characteristics.

Since the graph of thickness and threshold voltage exhibits an inversely proportional relationship, higher threshold voltagesare needed for thinner values of thickness (t_{FE}). While the subthreshold voltage, often referred to as Boltzmann tyranny, exhibits the switching rate of a FET transistor [103-107]. Calculating the subthreshold slope is as follows:

$$SS = (\delta VGK_BT)ln10/\delta\Phi q \qquad (9)$$

Since the factor K_BT/q in the equation above is constant (60 mV/decade at 300 K), the negative capacitance is applied to the remaining terms to further improve the subthreshold slope for higher values of t_{FE}. It shows how the negative capacitance's effect on internal voltage amplifies it while also increasing the surface potential at low bias.

The effect of altering the ferroelectric dielectric's thickness on the device's performance is depicted in this section [108-113]. The importance of fine-tuning the ferroelectric thickness is covered in this section. The graphs have been simulated using both negative and non-negative capacitances, and in both cases, the same level of **electron concentration** has been achieved. It is also important to keep in mind that for negative capacitance, the threshold voltage lowers as ferroelectric density rises [36].

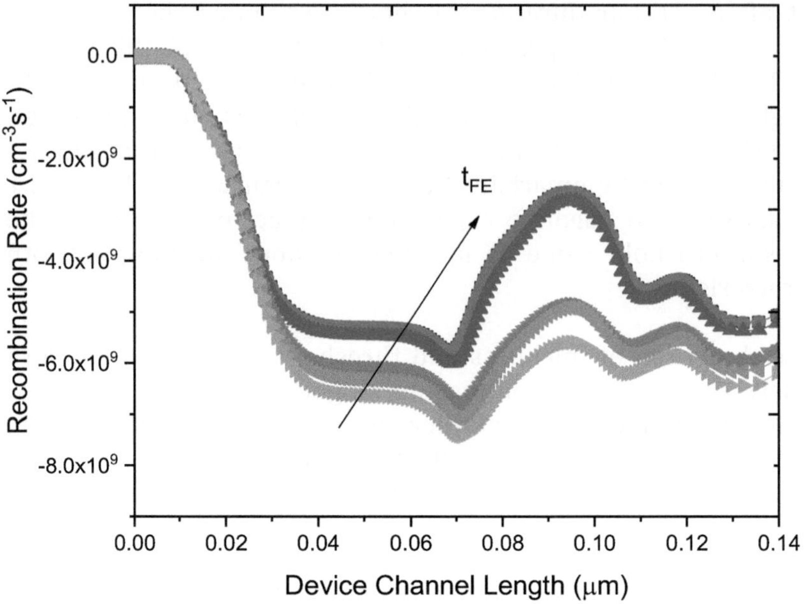

Figure 2.13 *Recombination rate characteristics of CP-Ferro-VTFET by the variation of ferro-oxide thickness.*

According to Figure 2.13, the **higher recombination rate** can be achieved by varying the thickness of the ferroelectric dielectric. In other words, thickness and transconductance are directly related to one another, meaning that higher t_{FE} results in larger transconductance [114-120]. As thickness and transconductance are directly proportional to one another while thickness and subthreshold voltage shows the inverse relationship, Figure 2.13 illustrates the combined effect of three different features and how the value of one parameter affects the other parameter. By combining these relationships, it can be seen that a higher ferroelectric dielectric thickness results in lower subthreshold slope and enhanced transconductance.

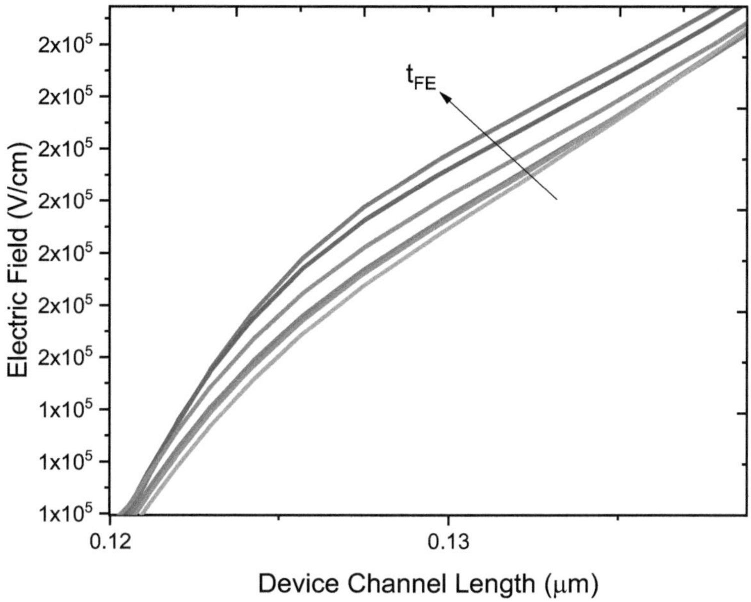

Figure 2.14 *Electric Field characteristics of CP-Ferro-VTFET by the variation of ferro-oxide thickness.*

The **electric field** likewise follows the same logic, and the greater value of t_{FE} leads the threshold voltage to shift to the left, producing practically the same electron concentration at a lower gate voltage as shown in Figure 2.14. These two graphs support the graph of derived transfer characteristics.

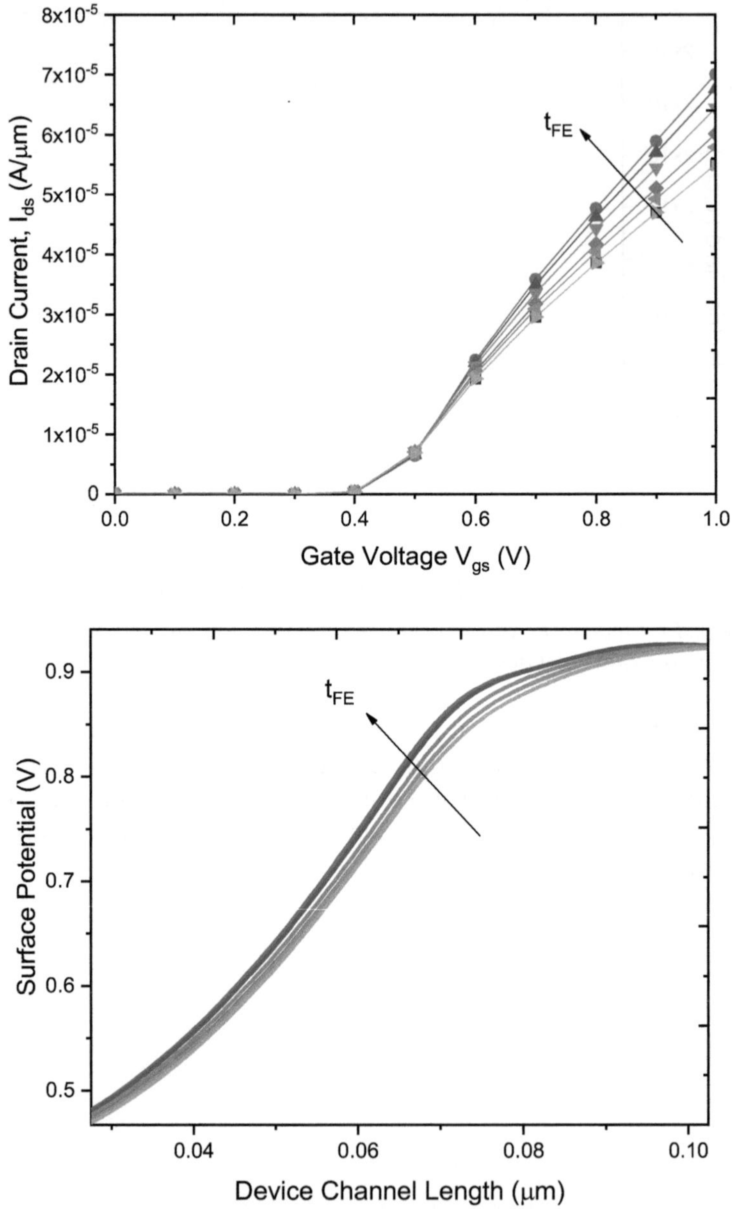

Figure 2.15 *(a) Drain current and (b) Surface Potential characteristics of CP-Ferro-VTFET by the variation of ferro-oxide thickness.*

The **drain current** is increased even at low bias because the higher ferroelectric thickness leads to a lower threshold voltage that creates an electric field and an electron concentration at low bias (refer to the relevant graphs shown above).It demonstrates that the **Surface potential** nearly maintains constant up to a thickness before dropping suddenly below that number [121-126].

The conclusion drawn from the results is that even at low threshold voltage, transconductance and recombination rate values will be optimised the higher the value of t_{FE}. In contrast, the device needs a higher gate voltage to accomplish the same recombination rate with a lower value of t_{FE}. The reduced threshold voltage and subthreshold slope caused by the higher ferroelectric density (in negative capacitance) lead to improved device characteristics. Another charge transfer in the inner electrode is accelerated during the phase transition due to a sharp increase in the charges connected by polarisation. As a result, the device's subthreshold slope is improved by the capacitor's series connection to the shell gate [127-130].

2.4 Conclusion

In this book chapter, the performance of a CP-Ferro-Vertical TFET is compared both with and without the influence of a negative capacitance. The proposed structure has the following benefits, reduced noise without compromising the device's ON current, reduced area, reduced cost, and reduced stray charges (due to the core part of the nanotube). In this study, the effect of varying ferroelectric density on device performance is noted and investigated. The simulated graphs show that the ferroelectric dielectric's negative capacitance property lowers the device's subthreshold slope (>60 mV/decade) and threshold voltage without influencing the drain current. Additionally, the voltage value decreases as t_{FE} rises. This study demonstrates improved device parameters for the negative capacitance effect, including potential, quasi-Fermi level at low bias, and carrier concentration. The thickness of the ferroelectric dielectric should be carefully controlled for better device performance analysis, according to the simulated graph. It means that the ferroelectric dielectric width should be the right thickness—that is, neither too thin nor too thick. A CP-Ferro-VTFET (with a negative capacitance effect) can be used for sensing and other low-power applications, according to observations.

References

[1] Frank, David J., Robert H. Dennard, Edward Nowak, Paul M. Solomon, Yuan Taur, and Hon-Sum Philip Wong. "Device scaling limits of Si MOSFETs and their application dependencies." *Proceedings of the IEEE* 89, no. 3 (2001): 259-288.10.1109/5.915374.

[2] Koswatta, Siyuranga O., Mark S. Lundstrom, and Dmitri E. Nikonov. "Performance comparison between pin tunneling transistors and conventional MOSFETs." *IEEE Transactions on Electron Devices* 56.3 (2009): 456-465.DOI: 10.1109/TED.2008.2011934.

[3] Thompson, Scott E., and Srivatsan Parthasarathy. "Moore's law: the future of Si microelectronics." Materials today 9, no. 6 (2006): 20-25.

[4] Koswatta, Siyuranga O., Steven J. Koester, and Wilfried Haensch. "On the possibility of obtaining MOSFET-like performance and sub-60-mV/dec swing in 1-D broken-gap tunnel transistors." *IEEE Transactions on electron devices* 57, no. 12 (2010): 3222-3230.

[5] Cheung, Kin P. "On the 60 mV/dec@ 300 K limit for MOSFET subthreshold swing." In *Proceedings of 2010 International Symposium on VLSI Technology, System and Application*, pp. 72-73. IEEE, 2010.

[6] Schaller, Robert R. "Technological innovation in the semiconductor industry: a case study of the International Technology Roadmap for Semiconductors (ITRS)." PhD diss., George Mason University, 2004.

[7] Baxter, Jason, Zhixi Bian, Gang Chen, David Danielson, Mildred S. Dresselhaus, Andrei G. Fedorov, Timothy S. Fisher et al. "Nanoscale design to enable the revolution in renewable energy." Energy & Environmental Science 2, no. 6 (2009): 559-588.

[8] Yang, Ming, Keqin Cao, Lang Sui, Ying Qi, Jian Zhu, Anthony Waas, Ellen M. Arruda, John Kieffer, M. D. Thouless, and Nicholas A. Kotov. "Dispersions of aramid nanofibers: a new nanoscale building block." Acs Nano 5, no. 9 (2011): 6945-6954.

[9] Roy, Kaushik, Saibal Mukhopadhyay, and Hamid Mahmoodi-Meimand. "Leakage current mechanisms and leakage reduction techniques in deep-submicrometer CMOS circuits." Proceedings of the IEEE 91, no. 2 (2003): 305-327.

[10] Choi, Woo Young, Byung-Gook Park, Jong Duk Lee, and Tsu-Jae King Liu. "Tunneling field-effect transistors (TFETs) with subthreshold swing (SS) less than 60 mV/dec." *IEEE Electron Device Letters* 28, no. 8 (2007): 743-745.DOI: 10.1109/LED.2007.901273.

[11] Yasin Khatami, Kaustav Banerjee, "Steep Subthreshold Slope n- and p-Type Tunnel-FET Devices for Low-Power and Energy- Efficient Digital Circuits," IEEE TRANSACTIONS ON ELECTRON DEVICES, VOL. 56, NO. 11, pp.2752- 2760, NOVEMBER2009.DOI:10.1109/TED.2009.2030831.

[12] Raad, B., K. Nigam, D. Sharma, and P. Kondekar. "Dielectric and work function engineered TFET for ambipolar suppression and RF performance enhancement." Electronics Letters 52, no. 9 (2016): 770-772.

[13] Kim, Chang-Hee, In-Tak Cho, Jong-Min Shin, Kyu-Bong Choi, Jung-Kyu Lee, and Jong-Ho Lee. "A new gas sensor based on MOSFET having a horizontal floating-gate." *IEEE Electron Device Letters* 35, no. 2 (2013): 265-267.

[14] Hong, Yoonki, Chang-Hee Kim, Jongmin Shin, Kyoung Yeon Kim, Jun Shik Kim, Cheol Seong Hwang, and Jong-Ho Lee. "Highly selective ZnO gas sensor based on MOSFET having a horizontal floating-gate." Sensors and Actuators B: Chemical 232 (2016): 653-659.

[15] Som, Debapriya, Budhaditya Majumdar, Souvik Kundu, and Sayan Kanungo. "Investigation of Charge Plasma-Enhanced Tunnel Field-Effect Transistor for Hydrogen Gas Sensing Application." IEEE Sensors Letters 4, no. 6 (2020): 1-4.

[16] Nigam, Kaushal, Pravin Kondekar, and Dheeraj Sharma. "High frequency performance of dual metal gate vertical tunnel field effect transistor based on work function engineering." *Micro & Nano Letters* 11, no. 6 (2016): 319-322.

[17] Chong, Chen, Hongxia Liu, Shulong Wang, and Shupeng Chen. "Simulation and Performance Analysis of Dielectric Modulated Dual Source Trench Gate TFET Biosensor." Nanoscale Research Letters 16, no. 1 (2021): 1-9.

[18] Shin, Wonjun, Gyuweon Jung, Seongbin Hong, Yujeong Jeong, Jinwoo Park, Dongkyu Jang, Byung-Gook Park, and Jong-Ho Lee. "Low frequency noise characteristics of resistor-and Si MOSFET-type gas sensors fabricated on the same Si wafer with In2O3 sensing layer." Sensors and Actuators B: Chemical 318 (2020): 128087.

[19] Seo, H., T. Endoh, H. Fukuda, and S. Nomura. "Highly sensitive MOSFET gas sensors with porous platinum gate electrode." Electronics letters 33, no. 6 (1997): 535-536.

[20] Damrongplasit, Nattapol, Sung Hwan Kim, and Tsu-Jae King Liu. "Study of random dopant fluctuation induced variability in the raised-Ge-source TFET." IEEE electron device letters 34, no. 2 (2013): 184-186.

[21] Damrongplasit, Nattapol, Changhwan Shin, Sung Hwan Kim, Reinaldo A. Vega, and Tsu-Jae King Liu. "Study of random dopant fluctuation effects in germanium-source tunnel FETs." IEEE transactions on electron devices 58, no. 10 (2011): 3541-3548.

[22] Yoon, Jun-Sik, and Rock-Hyun Baek. "Study on random dopant fluctuation in core–shell tunneling field-effect transistors." IEEE Transactions on Electron Devices 65, no. 8 (2018): 3131-3135.

[23] Manual, Atlas Users. "Device simulation software, Silvaco Int." Santa Clara, CA, Version 5, no. 0 (2010).

[24] Vandooren, Anne, Daniele Leonelli, R. Rooyackers, Andriy Hikavyy, Katia Devriendt, Marc Demand, Roger Loo, Guido Groeseneken, and Cedric Huyghebaert. "Analysis of trap-assisted tunneling in vertical Si homo-junction and SiGe hetero-junction tunnel-FETs." Solid-State Electronics 83 (2013): 50-55.

[25] Madan, Jaya, Rahul Pandey, and Rishu Chaujar. "Conducting Polymer Based Gas Sensor Using PNIN-Gate All Around-Tunnel FET." Silicon (2020): 1-9.

[26] Anjana Bhardwaj, Pradeep Kumar, B. Raj, Sunny Anand, "Design and Performance Optimization of doping-less Vertical Nanowire TFET using Gate Stack Technique", Journal of Electronic Materials (JEMS) Springer, Vol. 41, Issuse 7, PP. 4005-4013, 2022.

[27] Jeetendra Singh, BRaj, "Tunnel Current Model of Asymmetric MIM Structure Levying Various Image Forces to Analyze the Characteristics of Filamentary Memristor", Applied Physics A, Springer, Vol. 125, No.3, pp. 203.1 to 203.11 Feb 2019.

[28] Candy Goyal, Jagpal Singh Ubhi and B Raj, Low Leakage Zero Ground Noise Nanoscale Full Adder using Source Biasing Technique, "Journal of Nanoelectronics and Optoelectronics", American Scientific Publishers, Vol. 14, PP. 360–370, March 2019.

[29] Gurmohan Singh, R. K. Sarin and B Raj, "A Novel Robust Exclusive-OR Function Implementation in QCA Nanotechnology with Energy Dissipation Analysis", Journal of Computational Electronics, Springer, Volume 15, Issue 2, PP. 455-465, June 2016.

[30] Tanu Wadhera, Deepti Kakkar, Girish wadhwa, B Raj, " Recent Advances and Progress in Development of the Field Effect Transistor Biosensor: A Review" Journal of ELECTRONIC MATERIALS, Springer, Volume 48, Issue 12, pp 7635–7646, December 2019.

[31] Girish Wadhwa, Priyanka Kamboj, Balwinder Raj" Design optimisation of junctionless TFET biosensor for high sensitivity" Advances in Natural Sciences: Nanoscience and Nanotechnology, Vol. 10, PP. 045001 (7), (2019).

[32] Priya Bansal, BRaj, "MEMRISTOR MODELING AND ANALYSIS FOR LINEAR DOPANT DRIFT KINETICS," Journal of Nanoengineering and Nanomanufacturing, American Scientific Publishers, Vol. 6, pp. 1–7, 2016.

[33] Amandeep Singh, Mamta Khosla, BRaj, "Circuit Compatible Model for Electrostatic Doped Schottky Barrier CNTFET,"Journal of Electronic Materials, Springer, Vol. 45, Issue 12, pp 4825-4835, 2016.

[34] Ashima, D Vaithiyanathan, B Raj," Performance Analysis of Charge Plasma induced Graded Channel Si nanotube", **Journal of Engineering Research** (JER), EMSME Special Issue pp. 146-154 Aug 2021.

[35] Abhishek Singh Tomar, Vijay Kumar Magraiya, and B Raj, Scaling of Access and Data Transistor for High Performance DRAM Cell Design, Quantum Matter, Vol. 2, PP. 412-416 Oct. 2013.

[36] Neeraj Jain, BRaj, "Parasitic Capacitance and Resistance Model Development and Optimization of Raised Source/Drain SOI FinFET Structure for Analog Circuit Applications, Journal of Nanoelectronics and Optoelectronins, ASP, USA, Vol. 13, PP. 531–539, Apl 2018.

[37] Shradhya Singh, S. K. Vishvakarma and B Raj, "Analytical Modeling of Split-Gate Junction-Less Transistor for a Biosensor Application", Sensing and Bio-sensing, Elsevier, Vol. 18, Pages 31-36, April 2018.

[38] Maisagalla Gopal, and Balwinder Raj, Low Power 8T SRAM Cell Design for High Stability Video Applications, ITSI Transaction on

Electrical and Electronics Engineering, Vol. 1(5), pp. 91-97, 2013.
[39] Balwinder Raj, Jatin Mitra, Deepak Kumar Bihani, Rangharajan V, A.K. Saxena and S. Dasgupta, "Analysis of Noise Margin, Power and Process Variation for 32 nm FinFET Based 6T SRAM Cell" Journal of Computer (JCP), Academy Publisher, FINLAND, vol. 5, issue 6, PP.1-8, 2010.
[40] Divya Sharma, Rajesh Mehra and B Raj, "Comparative Analysis of Photovoltaic Technologies for High Efficiency Solar Cell Design", Superlattices and Microstructures, Elsevier, Volume 153, PP. 106861, May 2021.
[41] Pawandeep Kaur, Avtar Singh Buttar, B Raj, "A comprehensive Analysis of Nanoscale Transistor based Biosensor: A Review, Indian Journal of Pure and Applied Physics, Vol. 59, pp. 304-318, April 2021.
[42] Divya Yadav, Balwant Raj and B Raj, "Design and Simulation of Low Power Microcontroller for IoT Applications", Journal of Sensor Letters, ASP, Vol. 18, PP. 401–409, May 2020.
[43] Shailendra Singh, B Raj,"A 2-D Analytical Surface Potential and Drain current Modeling of Double-Gate Vertical t-shaped Tunnel FET", Journal of Computational Electronics, Springer, Vol. 19, PP.1154–1163, Apl 2020.
[44] Jeetendra Singh, B Raj, "An Accurate and Generic Window function for Non-linear Memristor Model" Journal of Computational Electronics, Springer, Volume 18 Issue 2, PP. 640-647 June 2019.
[45] Manjit Kaur; Neena Gupta; Sanjeev Kumar; BRaj; Arun Kumar Singh, "Comparative RF and Crosstalk Analysis of Carbon Based Nano Interconnects" IET Circuits, Devices & Systems, Vol.15(6), PP. 493-503, Feb 2021.
[46] Nehru Kandasamy, Firdous Ahmad, D. Ajitha, B Raj, and Nagarjuna Telagam "Quantum Dot Cellular Automata based Scan Flip Flop and Boundary Scan Register, IETE Journal of Research, Volume 66, PP. 535-548, 2020.
[47] Sharma, S. K.., Raj, B., Khosla, M., "Enhanced Photosensivity of Highly Spectrum Selective Cylindrical Gate In1-xGaxAs Nanowire MOSFET Photodetector, Modern Physics letter-B, Vol. 33, No. 12, PP. 1950144 (2019).
[48] Jeetendra Singh, BRaj, "Design and Investigation of 7T2M NVSARM with Enhanced Stability and Temperature Impact on Store/Restore Energy", IEEE Transactions on Very Large

Scale Integration Systems, Vol. 27, Issure 6, PP. 1322 - 1328 June 2019.

[49] Anil Kumar Bhardwaj, Sumeet Gupta, BRaj, Amandeep Singh, "Impact of Double Gate Geometry on the Performance of Carbon Nanotube Field Effect Transistor Structures for Low Power Digital Design", Computational and Theoretical Nanoscience, ASP, Vol. 16, PP. 1813–1820, 2019.

[50] Neeraj Jain, BRaj, "Thermal Stability Analysis and Performance Exploration of Asymmetrical Dual-k underlap Spacer (ADKUS) SOI FinFET for Security and Privacy Applications", Indian Journal of Pure & Applied Physics (IJPAP), Vol. 57, pp. 352-360, May 2019.

[51] Amandeep Singh, Mamta Khosla, BRaj, "Design and Analysis of Dynamically Configurable Electrostatic Doped Carbon Nanotube Tunnel FET"," Microelectronics Journal, Elesvier, Volume 85, Pages 17-24, March 2019.

[52] Neeraj Jain and Balwinder Raj, "Dual-k Spacer Region Variation at the Drain Side of Asymmetric SOI FinFET Structure: Performance Analysis towards the Analog/RF Design Applications", communicated in "Journal of Nanoelectronics and Optoelectronics", American Scientific Publishers, Vol. 14, PP. 349–359, March 2019.

[53] Jeetendra Singh, Sanjeev Sharma, B Raj, Mamta Khosla Analysis of barrier layer thickness on performance of In1-xGaxAs based Gate Stack Cylindrical Gate Nanowire MOSFET, " JNO, ASP, Vol. 13, PP 1473-1477, Oct 2018.

[54] Neeraj Jain, B Raj, "Analysis and Performance Exploration of High-k SOI FinFETs Over the Conventional Low-k SOI FinFET toward Analog/RF Design", Journal of Semiconductors (JoS), IOP Science, Vol.39, Issue 12, PP- 124002-1-7, Dec 2018.

[55] Candy Goyal, Jagpal Singh Ubhi and BRaj, A reliable leakage reduction technique for approximate full adder with reduced ground bounce noise, Journal of Mathematical Problems in Engineering, Hindawi, Volume 2018, Article ID 3501041, 16 pages, 15 Oct 2018.

[56] Anuradha, Jeetendra Singh, B Raj and Mamta Khosla, "Design and Performance Analysis of Nano-scale Memristor-based Nonvolatile SRAM" Journal of Sensor Letter", American Scientific Publishers, Vol. 16, PP 798–805, Oct 2018.

[57] Girish Wadhwa, BRaj, "Parametric Variation Analysis of Charge-Plasma-based Dielectric Modulated JLTFET for Bio-

sensor Application" IEEE Sensor Journal, VOL. 18, NO. 15, AUGUST 1, PP. 6070 – 6077, 2018 (SCI)

[58] Jeetendra Singh, BRaj, "Comparative Analysis of Memristor Models for Memories Design", JoS, IoP, Vol. 39, No. 7, PP. 074006-1-12, July 2018.

[59] Divya Yadav, Shailesh Singh Chouhan, Santosh Kumar Vishvakarma and BRaj, "Application Specific Microcontroller Design for IoT based WSN", Sensor Letter, ASP, Vol. 16, PP. 374–385, May 2018.

[60] Gurmohan Singh, R. K. Sarin and BRaj, "Fault-Tolerant Design and Analysis of Quantum-Dot Cellular Automata Based Circuits", IEEE/IET Circuits, Devices & Systems, Vol. 12, PP. 638 – 64, 2018.

[61] Jeetendra Singh, BRaj, "Modeling of Mean Barrier Height Levying Various Image Forces of Metal Insulator Metal Structure to Enhance the Performance of Conductive Filament Based Memristor Model", IEEE Nanotechnology, Vol. 17, No. 2, PP. 268-267, March 2018.

[62] Aakash Jain, Sanjeev Sharma, BRaj, "Analysis of Triple Metal Surrounding Gate (TM-SG) III-V Nanowire MOSFET for Photosensing Application", Opto-electronics Journal, Elsevier, Volume 26, Issue 2, Pages 141-148, May 2018.

[63] Aakash Jain, Sanjeev Sharma, B Raj, "Design and Analysis of High Sensitivity Photosensor Using Cylindrical Surrounding Gate MOSFET For Low Power Sensor Applications", Engineering Science and Technology, an International Journal, Elsevier's, Volume 19, Issue 4, Pages 1864–1870, December 2016.

[64] Amandeep Singh, Mamta Khosla, B Raj, "Analysis of Electrostatic Doped Schottky Barrier Carbon Nanotube FET for Low Power Applications," Journal of Materials Science: Materials in Electronics, Springer, Vol.28, PP. 1762–1768, 2017.

[65] G. Saiphani Kumar, Amandeep Singh, BRaj, "Design and Analysis of Gate All Around CNTFET based SRAM cell Design", Journal of Computational Electronics, Springer, Volume 17, Issue 1, pp 138–145, March 2018.

[66] Gurinder pal Singh, B. S. Sohi, Balwinder Raj, "Material Properties Analysis of Graphene Base Transistor (GBT) for VLSI Analog Circuits, Indian Journal of Pure & Applied Physics (IJPAP), Vol. 55, pp. 896-902, Dec-2017.

[67] Amandeep Singh, Mamta Khosla, B Raj, "Comparative Analysis of Carbon Nanotube Field Effect Transistor and Nanowire

Transistor for Low Power Circuit Design," Journal of Nanoelectronics and Optoelectronics, American Scientific Publishers, USA, Vol. 11, PP. 388-393, June 2016.

[68] Sunil Kumar and B Raj, "Estimation of Stability and Performance metric for Inward Access Transistor based 6T SRAM Cell Design using n-type/p-type DMDG-GDOV TFET" IEEE VLSI Circuits and Systems Letter, Volume3, Issue2, PP 25-39, June2017

[69] Shashikant Sharma, Anjan Kumar, Manisha Pattanaik and B Raj, "Forward Body Biased Multimode Multi-Threshold CMOS Technique for Ground Bounce Noise Reduction in Static CMOS Adders" International Journal of Information and Electronics Engineering, pp. 567-572, Vol. 3, issue 3, 2013.

[70] Hamendra Singh, Pankaj Kumar, Balwinder Raj, "Performance Analysis of Majority Gate SET Based 1-bit Full Adder", *International Journal of Computer and Communication Engineering* (IJCCE), IACSIT Press Singapore, ISSN: 2010-3743, Volume 2, No. 4, 2013.

[71] Anil Kumar Bhardwaj, Sumeet Gupta and B Raj, "Investigation of Parameters for Schottky Barrier (SB) Height for Schottky Barrier Based Carbon Nanotube Field Effect Transistor Device", Journal of Nanoelectronics and Optoelectronics, ASP, Vol. 15, PP. 783–791, July 2020.

[72] Priya Bansal, BRaj, "MEMRISTOR: A VERSATILE NONLINEAR MODEL FOR DOPANT DRIFT AND BOUNDARY ISSUES," JCTN, American Scientific Publishers, Vol. 14 (5), PP. 2319-2325, May 2017.

[73] Neeraj Jain, BRaj, "An Analog and Digital Design Perspective Comprehensive Approach on Fin-FET (Fin-Field Effect transistor) Technology - A Review", Reviews in Advanced Sciences and Engineering (RASE), ASP, Vol. 5, PP. 1-14, 2016.

[74] Sanjeev Sharma, B Raj, Mamta Khosla, "Subthreshold Performance of In1-xGaxAs based Dual Metal with Gate Stack Cylindrical/Surrounding Gate Nanowire MOSFET for Low Power Analog Applications, Journal of Nanoelectronics and Optoelectronics, American Scientific Publishers, USA, Vol.12, PP. 171–176, 2017.

[75] Balwinder Raj, A. K. Saxena and S. Dasgupta, "Analytical Modeling for the Estimation of Leakage Current and Subthreshold Swing Factor of Nanoscale Double Gate FinFET Device" Microelectronics International, UK, vol. 26, pp. 53-63, 2009.

[76] Soniya, Shailendra Singh, Girish Wadhwa and Balwinder Raj, "An Analytical Modeling for Dual Source Vertical Tunnel Field Effect Transistor" submitted to 'International Journal of Recent Technology and Engineering (IJRTE)' at Volume-8 Issue-2, PP. 603-608, July 2019.

[77] Shailendra Singh, B Raj, "Design and analysis of hetrojunction Vertical T-shaped Tunnel Field Effect Transistor", Journal of Electronics Material, Springer, Volume 48, Issue 10, pp 6253–6260, October 2019.

[78] Candy Goyal, Jagpal Singh Ubhi and B Raj, "A Low Leakage CNTFET based Inexact Full Adder for Low Power Image Processing Applications", International Journal of Circuit Theory and Applications, Wiley, Volume47, Issue9, Pages 1446-1458, September 2019.

[79] B Raj, A. K. Saxena and S. Dasgupta, "A Compact Drain Current and Threshold Voltage Quantum Mechanical Analytical Modeling for FinFETs" Journal of Nanoelectronics and Optoelectronics (JNO), USA, vol. 3, no. 2, pp. 163-170, 2008.

[80] Girish Wadhwa, B Raj, "An Analytical Modeling of Charge Plasma based Tunnel Field Effect Transistor with Impacts of Gate underlap Region" Superlattices and Microstructures, Elsevier, Vol. 142, PP.106512, June 2020.

[81] Shailendra Singh, B Raj,"Modeling and Simulation analysis of SiGe hetrojunction Double GateVertical t-shaped Tunnel FET", Superlattices and Microstructures, Elsevier Volume 142, PP. 106496, June 2020.

[82] Amandeep Singh, Dinesh Kumar Saini, Dinesh Agarwal, Sajal Aggarwal, Mamta Khosla, BRaj, "Modeling and Simulation of Carbon Nanotube Field Effect Transistor and its Circuit Application ," Journal of Semiconductors (JoS), IOP Science, Vol.37, PP. 074001-6, July 2016.

[83] Neeraj Jain, Balwinder Raj, "Device and Circuit Co-Design Perspective Comprehensive Approach on FinFET Technology - A Review, Journal of Electron Devices, Vol. 23, Num. 1, pp 1890-1901, 2016.

[84] Sunil Kumar and BRaj, "Analysis of I_{ON} and Ambipolar Current for Dual-Material Gate-drain Overlapped DG-TFET," Journal of Nanoelectronics and Optoelectronics, American Scientific Publishers, USA, Vol.11, PP. 323-333, June 2016.

[85] Naveed Anjum, Tarun Bali and B Raj, Design And Simulation Of Handwritten Multiscript Character Recognition, Interna-

tional Journal of Advanced Research in Computer and Communication Engineering, Vol.2 (7), PP. 2544-2549, July 2013.

[86] Sanjeev Sharma, BRaj, Mamta Khosla, "A Gaussian Approach for Analytical Subthreshold Current Model of Cylindrical Nanowire FET with Quantum Mechanical Effects, Microelectronics Journal, Elsevier, Vol. 53, PP. 65-72, April 2016.

[87] Karmjit Singh, BRaj, "Performance and analysis of temperature dependent Multi-walled Carbon Nanotubes as Global Interconnects at different technology nodes," Journal of Computational Electronics, Springer, Volume 14 Issue 2, pp. 469-476, June 2015.

[88] Sunil Kumar and BRaj, "Compact channel potential analytical Modeling of DG-TFET based on Evanescent–mode Approach," Journal of Computational Electronics, Springer Vol. 14, No. 2, pp.820-827, July 2015.

[89] Karmjit Singh, BRaj, "Temperature Dependent Modeling and Performance Evaluation of Multi-Walled CNT and Single-Walled CNT as Global Interconnects," Journal of Electronic Materials, Springer, Vol. 44, Issue 12, pp 4825-4835, Dec 2015.

[90] V.K. Sharma, M. Pattanaik and B.Raj, "INDEP approach for leakage reduction in nanoscale CMOS circuits", International Journal of Electronics, Taylor & Francis, Vol.102, No.2, pages 200-215, 2014.

[91] Karmjit Singh, BRaj, "Influence of Temperature on MWCNT bundle, SWCNT bundle and Copper interconnects for Nanoscaled Technology nodes," Journal of Materials Science: Materials in Electronics. Springer, Vol. 26, No. 8, pp. 6134-6142, 2015.

[92] Naveed Anjum, Tarun Bali and B Raj, "Design and Simulation of Handwritten Gurumukhi and Devanagri Numerical Recognition, International Journal of Computer Applications, Published by Foundation of Computer Science, New York, USA, Vol. 73(12), pp. 16-21, 2013.

[93] Khandelwal, S., Gupta, V., BRaj, and Gupta, R D., Process Variability Aware Low Leakage Reliable Nano Scale DG-FinFET SRAM Cell Design Technique, Journal of Nanoelectronics and Optoelectronics, Volume 10, Number 6, pp. 810-817, December 2015.

[94] V.K. Sharma, M. Pattanaik and B Raj, "ONOFIC Approach: Low Power High Speed Nanoscale VLSI Circuits Design", Interna-

tional Journal of Electronics, Taylor & Francis, Vol. 101, issue 1, pages 61-73, 2014.

[95] Khandelwal, S., Balwinder Raj, and Gupta, R D., FinFET based 6T SRAM Cell Design: Analysis of Performance Metric, Process Variation and Temperature Effect. Journal of Computational and Theoretical Nanoscience, ASP, USA, Vol. 12, PP. 2500-2506, 2015.

[96] Sumit Singh; Yadav, Shekhar; Rahul, Jagdeep; Srivastava, Anurag; BRaj, Impact of HfO_2 in Graded Channel Dual Insulator Double Gate MOSFET, Journal of Computational and Theoretical Nanoscience, American Scientific Publishers, Volume 12, Number 6, pp. 950-953, April 2015.

[97] Vijay Kumar Sharma, Manisha Pattanaik, BRaj, PVT variations aware low leakage INDEP approach for nanoscale CMOS Circuits, Microelectronics Reliability, Elsevier, Vol. 54, pp 90-99, 2014.

[98] B Raj, A. K. Saxena and S. Dasgupta, "Quantum Mechanical Analytical Modeling of Nanoscale DG FinFET: Evaluation of Potential, Threshold Voltage and Source/Drain Resistance ", Elsevier's Journal of Material Science in Semiconductor Processing, Elsevier, Vol. 16, issue 4, pp. 1131- 1137, 2013.

[99] Maisagalla Gopal, Siva Sankar D Prasad and Balwinder Raj, 8T SRAM Cell Design for Dynamic and Leakage Power Reduction, International Journal of Computer Applications, Published by Foundation of Computer Science, New York, USA, Vol. 71(9), pp. 43-48, June 2013.

[100] Manisha Pattanaik, B Raj, Shashikant Sharma and Anjan Kumar, "Diode Based Trimode Multi-Threshold CMOS Technique for Ground Bounce Noise Reduction in Static CMOS Adders", Advanced Materials Research, Trans Tech Publications, Switzerland, vol. 548, pp 885-889, 2012.

[101] Balwinder Raj, A. K. Saxena and S. Dasgupta, "Nanoscale FinFET Based SRAM Cell Design: Analysis of Performance metric, Process variation, Underlapped FinFET and Temperature effect" IEEE Circuits and System Magazine, vol. 11, issue 2, pp. 38- 50, 2011.

[102] V.K. Sharma, M. Pattanaik & Balwinder Raj, "Leakage Current ONOFIC Approach for Deep Submicron VLSI Circuit Design", International Journal of Electrical, Computer, Electronics and Communication Engineering, World Academy of Sciences, En-

gineering and Technology, Vol. 7, issue 4, pages 239-244, 2013.

[103] Tulika Chawla, Mamta Khosla, BRaj, Design and simulation of Triple metal Double-gate Germanium on Insulator Vertical Tunnel Field Effect Transistor, Microelectronics Journal, Elsevier; Vol. 114, PP. 105125, August 2021.

[104] Parminder Kaur, Sandeep Singh Gill and B Raj, "Comparative Analysis of OFETs Materials and Devices for Sensor Applications", Journal of Silicon, Springer, Vol 14, PP. 4463–4471, 2022.

[105] Sanjeev Kumar Sharma, Parveen Kumar, Balwant Raj and BRaj, "$In_{1-x}Ga_xAs$ Double Metal Gate-Stacking Cylindrical Nanowire MOSFET for highly sensitive Photo detector", *Journal of Silicon, Springer Vol.*14, PP. 3535–3541, 2022.

[106] B Raj, A. K. Saxena and S. Dasgupta, "Analytical Modeling of Quasi Planar Nanoscale Double Gate FinFET with Source/Drain Resistance and field dependent carrier mobility: A quantum mechanical study" Journal of Computer (JCP),Academy Publisher, FINLAND, vol. 4, issue 9, PP.1-8, 2009.

[107] Bhushan, S., Khandelwal, S. and Raj, B., (2012). Analyzing Different Mode FinFET based Memory Cell at Different Power Supply for Leakage Reduction, Seventh International Conference on Bio-Inspired Computing: Theories and Application, (BIC-TA 2012) Advances in Intelligent Systems and Computing, Volume 202, p89-100 (2013).

[108] Jeetendra Singh, BRaj, "Temperature Dependent Analytical Modeling and Simulations of Nanoscale Memristor, Journal: Engineering Science and Technology, an International Journal, Elsevier's, Vol. 21, PP- 862-868, Oct 2018.

[109] Shradhya Singh, Shashi Bala, Balwant Raj, B Raj, "Improved Sensitivity of Dielectric Modulated Junctionless Transistor for Nanoscale Biosensor Design", Sensor Letter, ASP, Vol.18, PP.328–333, Apl 2020.

[110] Vivek Kumar, Santosh Kumar Vishvakarma, and B Raj, "Design and Performance Analysis of ASIC for IoT Applications" Sensor Letter ASP, Vol. 18, PP. 31–38, Jan 2020.

[111] Akanksha Jaiswal, R. K. Sarin, B Raj and Shikha Sukhija, "A Novel Circular Slotted Microstrip-fed Patch Antenna with Three Triangle Shape Defected Ground Structure for Multi-

band Applications", Advanced Electromagnetic (AEM), Vol. 7, Issue 3, PP. 56-63, Aug 2018.
[112] Girish Wadhwa, B Raj, "Label Free Detection of Biomolecules using Charge-Plasma-Based Gate Underlap Dielectric Modulated Junctionless TFET" Journal of Electronic Materials (JEMS), Springer, Volume 47, Issue 8, pp 4683–4693, August 2018.
[113] Gurmohan Singh, R. K. Sarin and B Raj, "**Design and Performance Analysis of a New Efficient Coplanar Quantum-Dot Cellular Automata Adder**", Indian Journal of Pure & Applied Physics (IJPAP), Vol. 55, pp. 97-103, February 2017.
[114] Amandeep Singh, Mamta Khosla, Balwinder Raj, "Design and Analysis of Electrostatic Doped Schottky Barrier CNTFET Based Low Power SRAM," International Journal of Electronics and Communications, (AEÜ), Elsevier, Vol. 80, PP.67–72, 2017.
[115] Parminder Kaur, Vikas Pandey and B Raj, "Comparative Study of Efficient Design, Control and Monitoring of Solar Power using IoT", Sensor Letter, ASP Vol. 18, PP. 419–426, May 2020.
[116] Anil Kumar Bhardwaj, Sumeet Gupta and B Raj, Development & Analysis of compact model for Double Gate Schottky Barrier CNTFET, Journal of Nanoelectronics and Optoelectronics, ASP, Vol. 15, pp. 1199–1208, Aug 2020.
[117] Girish Wadhwa, Priyanka Kamboj; Jeetendra Singh, B Raj, "Design and Investigation of Junctionless DGTFET for Biological Molecule Recognition" Transactions on Electrical and Electronic Materials, Springer, Vol. , PP. 8 Aug, 2020.
[118] Tulika Chawla, Mamta Khosla, B Raj, "Optimization of Double-gate Dual material GeOI-Vertical TFET for VLSI Circuit Design, IEEE VLSI Circuits and Systems Letter, Vol. 6, issue-2, PP.13-25, Aug 2020.
[119] Sachin Kumar Verma, Shailendra Singh, Girish Wadhwa and B Raj, "Detection of Biomolecules using Charge-Plasma based Gate Underlap Dielectric Modulated Dopingless TFET" Transactions on Electrical and Electronic Materials (TEEM), Springer, Vol. 21, PP. 528-535, June 2020.
[120] Neeraj Jain, BRaj, "Impact of underlap spacer region variation on Electrostatic and analog/RF performance of symmetrical high-k SOI FinFET at 20 nm channel length", Journal of Semiconductors (JoS), IOP Science, Vol.38, Issue 12, PP- 122002, Dec 2017.

[121] Shailendra Singh, B Raj," Analytical Modeling and Simulation analysis of T-shaped III-V heterojunction Vertical T-FET", Superlattices and Microstructures, Elsevier, Vol. 147, PP. 106717, Nov 2020.

[122] Gurmohan Singh, R. K. Sarin and B Raj, "Design and Analysis of Area Efficient QCA Based Reversible Logic Gates", Journal of Microprocessors and Microsystems, Elsevier, Vol. 52, PP.59-68, May 2017.

[123] Amandeep Singh, Mamta Khosla, BRaj, "Compact Model for Ballistic Single Wall CNTFET under Quantum Capacitance Limit," Journal of Semiconductors (JoS), IOP Science, Vol.37, PP. 104001-8, Oct 2016.

[124] Sonal Singh; Mamta Khosla, Girish Wadhwa; B Raj, "Design and analysis of double-gate junctionless vertical TFET for gas sensing applications", Applied Physics A, Springer, Vol.127, No.16, 2 Jan 2021.

[125] Inderjit Singh, BRaj, Mamta Khosla, and Brajesh Kumar Kaushik, Potential MRAM Technologies for Low Power SoCs, SPIN World Scientific Publisher, SCIE; Vol. 10, No. 04, PP.2050027, Dec 2020.

[126] Shailendra Singh, B Raj,"Parametric variation analysis on hetero-junction Vertical t-shape TFET for Supressing ambipolar conduction", Indian Journal of Pure and Applied Physics, Vol. 58,pp. 478-485 June 2020.

[127] Soniya, Shailendra Singh, Girish Wadhwa and Balwinder Raj, Design and Analysis of Dual Source Vertical Tunnel Field Effect Transistor for high performance, Transactions on Electrical and Electronics Materials, Springer, Vol. 21, Pages 74–82, Oct 2019.

[128] Manjit Kaur; Neena Gupta; Sanjeev Kumar; BRaj; Arun Kumar Singh, "RF Performance Analysis of Intercalated Graphene Nanoribbon Based Global Level Interconnects" Journal of Computational Electronics, Springer, Vol. 19, PP.1002–1013, June 2020

[129] Girish Wadhwa, BRaj, "Design and Performance Analysis of Junctionless TFET Biosensor for high sensitivity" IEEE Nanotechnology, Vol.18, PP. 567 – 574, 2019.

[130] Jeetendra Singh, BRaj, "Enhanced Nonlinear Memristor Model Encapsulating Stochastic Dopant Drift, JNO, ASP 14, 958–963 (2019).

3

CNFET and Nano Tubes

Meenakshi Gupta[a], K. Deepa[a], Bhanu Pratap Chaudhary[a], Kumod Kumar Gupta[b] and Vijay Kumar Gill[a]
[a]ECE, Manav Rachna University, India
[b]Noida Institute of Engineering & Technology, India

Abstract: Carbon Nanotube Field-Effect Transistors (CNFETs) are brilliant possibility for building exceptionally energy-productive computerized frameworks. Notwithstanding, flaws intrinsic in carbon nanotubes (CARBON NANO TUBEs) present critical obstacles to acknowledging down to earth CNFET circuits. To accomplish CNFET VLSI frameworks within the sight of these innate flaws, cautious organization of plan and handling is expected: from gadget handling and circuit reconciliation, the entire way to enormous scope framework plan and streamlining. In this chapter, we sum up the key thoughts that empowered the primary exploratory exhibition of CNFET math and capacity components. We likewise present an outline of a probabilistic structure to examine the effect of different CNFET circuit plan procedures and CARBON NANO TUBE handling choices on framework level energy. We show the way that this structure can be utilized to further develop the energy-delay-item (EDP) of CNFET-based advanced frameworks.

3.1 Introduction

Parallel circuits for long been connected with the arrangement of cutting edge systems for performing estimations on two potential reasoning qualities (0 or substantial and 1 or counterfeit) in the Boolean space. In any case, the twofold planned circuit design encounters different issues, for example, complexity associated with the interconnection of the pin out issues related with internal and external connection of the circuit. Interconnects are the essential source of power dissipation in a VLSI chip and includes almost 70% of complete on-chip capacitance [1]. In case of coordinated circuits, these interconnects possess extra region, subsequently causing a few sign trustworthiness issues like commotion and postpone butterflies. The execution of energy productive computerized plans turns out to be very difficult while downsizing the component size into a nano-meter system, inferable from dependability and execution issues. Additionally, some genuine applications, for example, mechanical technology and interaction control, face hardships for precise execution just with the ordinary parallel rationale [2]. To move past the restrictions of double rationale, a rationale framework having multiple qualities related to the rationale levels, which is regularly termed as MULTI VALUED LOGIC, can be utilized to carry out a computerized framework plan. In present-day circuits MVL is an exceptionally viable and functional means to decide the issue of interconnects. Inferable from the way that the MULTI VALUED LOGIC uses multiple rationale levels, MULTI VALUED LOGIC based frameworks give the transmission of a more prominent measure of data when contrasted with double rationale frameworks over a solitary wire, which serves to diminish the on-chip and off-chip interconnections fundamentally. The plans using MULTI VALUED LOGIC for its execution offer benefits regarding expanded data thickness, decreased chip region and working on the computational ability of the incorporated circuits. In this manner, the utilization of MULTI VALUED LOGIC plans assists with diminishing the circuit intricacy and further develops energy productivity in computerized framework plan because of decreased circuit upward [3]. MULTI VALUED LOGIC is one of the new strategies, which builds the radix for the calculation to bring down the intricacy of the circuit. The principal motivation behind utilizing MULTI VALUED LOGIC over double rationale is to:

(i) lessen the chip region by the transmission of more data through each wire when contrasted with parallel;

(ii) Speed up data transmission as the data per unit time is expanded;

(iii) Decrease the circuit intricacy as each MULTI VALUED LOGIC component processes more data when contrasted with the twofold component. The acquaintance of extra levels with the current parallel rationale brings about MULTI VALUED LOGIC frameworks, with the run of the mill models being that of the ternary and quaternary rationale frameworks [4].

The ternary rationale is of impressive interest because of its simplicity of execution and energy productivity due to the diminished intricacy of interconnects and chip region [5]. Moreover, ternary circuits can assist with decreasing the issues of pin-out and increment in interconnection thickness in the VLSI frameworks. MULTI VALUED LOGIC plans can displace the current twofold advancements because they are exceptionally reliant upon the circuit acknowledgement strategies utilized to execute MULTI VALUED LOGIC frameworks. Various ways to deal with fabricating MULTI VALUED LOGIC based circuit is utilizing CMOS or joining parallel and multi-esteemed blocks in rationale circuits as shown [6]. With the advancement of nanotechnology, atomic gadgets are becoming promising options in contrast to the traditional silicon-based CMOS innovation [7]. The most encouraging alternative to silicon semiconductor to accomplish low power and superior execution is the Carbon nanotube field impact semiconductor (CARBON NANO TUBEFET) [8-10]. Carbon Nano Tubes (CARBON NANO TUBEs) have acquired critical interest, fundamentally because of their moderately little aspects and capability of execution in many existing as well as arising advances. CARBON NANO TUBEFET could be the most reasonable choice for supplanting regular MOSFET in light of silicon innovation, due to its likenesses as far as different attributes, for example, close ballistic vehicle activity because of its exceptional one-layered (1D) band structure, which gives the gadget a phenomenal driving capacity and less power dissemination [11]. The limit voltage of CARBON NANO TUBEFET can be constrained by changing the CARBON NANO TUBE breadth, which is extremely powerful in the plan of MULTI VALUED LOGIC based circuits. Moreover, CARBON NANO TUBEFETs (both p-type and n-type) have comparable versatility values, which is very huge for gadget measuring in complex circuits. CARBON NANO TUBEFETs face a few difficulties in chirality control, creation cycle, development, and the situating of single-divider CARBON NANO TUBEs. Ternary frameworks can likewise be executed while utilizing one more reasonable arising innovation, normally alluded to as the Resistive Random Access Memory (RRAM). This is credited to the way that the RRAM gadget can successfully execute numerous rationale values (at least two) without the requirement for additional equipment for the acknowledgment of advanced plans. Executing ternary rationale along with RRAM gives exceptional results. The circuit schematic of RRAM is very straightforward, having a two terminal metal-encasing metal (MIM) structure, with the outside voltage beat causing an obstruction change inside the gadget [12,13]. Its actual component depends on the development and demolition of a conductive fiber (CF) in the oxide (dielectric) interfacing the two metal terminals [14,15]. This opposition shift between the high obstruction state (HRS) and low opposition state (LRS) is successfully utilized for information capacity as "0" and "1" [16-18], individually. In light of the different appealing elements of RRAM, for example, higher on/off proportion, non-unpredictability, great versatility, and falling short ON-state obstruction [19,20], it is utilized for arranging rationale doors, including the field programmable entryway clusters [21] that are utilized in the plan of computerized frameworks. The RRAM is picked as the up-and-comer gadget here, as it has a straightforward design, empowers high reconciliation thickness, and can display exchanging in the low-power system [22]. Although utilizing RRAM for carrying out ternary rationale frameworks is not full-fledged use, different procedures for the execution of parallel rationale along with RRAM have proactively been proposed [23] . TUBEFET-RRAM ternary rationale doors for the plan of ternary number-crunching circuits to investigate the benefits as far as decrease of entryway count by using the invalidation of literals procedure. In this original copy, CARBON NANO TUBEFETRRAM ternary combinational rationale circuits are proposed and the reproduction results while utilizing the HSPICE test system exhibit benefits as far as power utilization and part count when contrasted with the multi-esteemed rationale plans existing in the writing. Rationale states can likewise be put away in the RRAM because of the expansion of RRAM as a circuit component. This will guarantee the combination of rationale calculation and information stockpiling into the proposed plans, which opens another area of chances for investigating clever calculation frameworks. The remainder of the paper is introduced, underneath.

3.2 Carbon Nanotube Field Effect Transistors (CNFET)

CNFETs are considered potential replacements for silicon MOSFET as their threshold voltage can be controlled by varying the chirality vector. Carbon Nanotubes were discovered by Japanese Physicist S. Lijima in late 1991[24]. Usage of Carbon Nanotubes along with silicon technology speeds up to THz or PHz range with a reduced threshold voltage. This behavior is utilized in CNFET for reduced static power consumption. Carbon Nanotubes are tubular structures with reduced diameters up to 1nm and their length varies in the range between a few nm to μm. These tubular structures are made with rolled graphene sheets which can be single-walled

(SWCNT) or multi-walled (MWCNT). A single wall structure is very simple in construction with one cylinder (Nanotube) whereas multi-walled is constructed with multiple cylinders with 0.34nm as their interlayer spacing. The single wall structure is considered to be the best alternative for MOSFETs [25]. The behavior of SWCNT depends on the chirality vector (angle arrangement of the atom) and is denoted by the integer pair (n, m). By considering the indices (n, m) the tube can be characterized as metallic or semiconducting. i.e. if the indices n and m are the same (n=m) or n-m=3i else it is semiconducting, where i is an integer [26]. The schematic of CARBON NANO TUBEFET is shown in Figure 3.1. A highly doped carbon Nanotube is sandwiched between source and gate electrodes. When the CARBON NANO TUBEFET is in the ON state the highly doped carbon Nanotubes offers low resistance. The undoped carbon Nanotubes serve as channels and are placed under the gate. The gate width of the CARBON NANO TUBEFET is calculated using equation (1) [27].

$$W_{gate} \approx \text{Max}\,(W_{min}, M \times P) \tag{1}$$

where M represents the tube number of CARBON NANOTUBE, Wmin represents the minimum gate width, and P is the distance between two adjacent **CARBON NANO TUBEFET** referred to as pitch value.

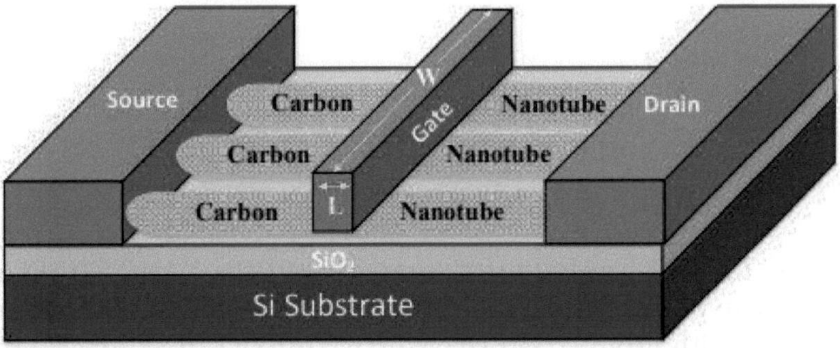

Figure 3.1 *Schematic of a CARBON NANO TUBEFET.*

The diameter across of the CARBON NANO TUBE is computed using Equation (2) in nanometers [28].

$$D_{CNT} = \frac{\sqrt[3]{a_0}}{\pi} * (n^2 + m^2 + nm)^{1/2} \quad (2)$$

In the above equation a_0 is the distance between two adjacent carbon atoms whose value is 0.142 nm. Equation (2) can be rewritten as

$$D_{CNT} = 0.0783 * (n^2 + m^2 + nm)^{1/2} \quad (3)$$

CARBON NANO TUBEFETs exhibit similar properties like the standard MOSFETs. The minimum voltage required to turn on the transistor is expressed using Equation (4)

$$V_{th} \approx \frac{E_g}{2e} = \frac{\sqrt{3}}{3}\frac{a}{e}\frac{V}{D_{CNT}} \quad (4)$$

In the above equation V_{th} is the threshold voltage, E_g is band gap, e is the electron charge, 'a' is the carbon atom distance whose value is 2.49 Å, V is the carbon π-π energy whose value is 3.033eV in the tight bonding model. From equations 3 and 4 it can be inferred that the threshold voltage of the CARBON NANO TUBEFETs changes with the chiral indices n and m. It is clear that to control the threshold voltage of CARBON NANO TUBEFETs either the diameter of CARBON NANO TUBE or the chirality vector must be changed [29].

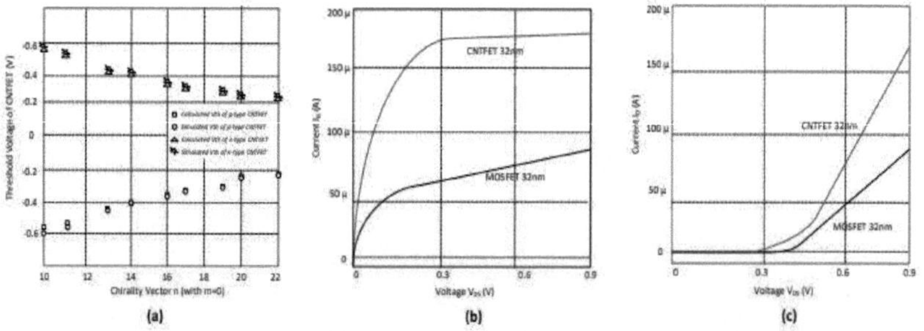

Figure 3.2 (a) Threshold variation of CARBON NANO TUBEFET for various n values with m=0. (b) I-V curve comparison of n-type MOSFET with n-type CARBON NANO TUBEFET in 32 nm. (c) ID-VGS Linear Scale.

Figure 3.2(a) depicts the threshold variation of CARBON NANO TUBEFET for various n values with m=0.calculated from Equation (4). It can be observed that the distribution of threshold voltage is same for both n and p type CARBON NANO TUBEFET with opposite sign.

Figure 3.2(b) depicts the comparison of VDS-ID characteristics of CARBON NANO TUBEFET and MOSFET in 32 nm technology. It can be observed that the performance of CARBON NANO TUBEFET is more desirable in comparison with MOSFET. From Figure 3.2(c) it can be further observed in the saturation region curve is quadric when compared to MOSFET.

3.3 Circuit Level Implementation of Ternary Logic

Using binary logic in terms of interconnect in VLSI technology has been a serious concern in terms of interconnect complexity and circuit density. Scaling down to a nanometer regime also results in severe issues in reliability and performance restricting the usage in the case of energy-efficient designs. In these cases, multi-valued logic levels can be adapted. Multi-valued logic uses more than two levels to facilitate the transmission and subsequently reduces the on and off interconnection issues [30,31]. Ternary logic is of radix 3 and is a subset of Multi-valued logic. Numbers represented in Multi-valued logic can be either in balanced representation or unbalanced representation. However, balanced representation is not widely used as it is difficult to represent negative voltages [48-53]. In the case of binary logic 0 and V_{dd} are the voltage level that represents '0' and'1' respectively whereas in ternary systems levels 0, and V_{dd} represent '0' '1' and '2' respectively. In the case of binary logic if A is the input there will be four functions (O=A, O=, O=0 and O=1).as given in Table 3.1. Table 3.2 describes the binary operators for corresponding inputs, outputs and operators.

Table 3.1 *Unary operations in binary logic*

I1	I0	Function	Description
0	0	O=0	Constant
0	1	O=A	True
1	0	O=$\overline{A}$	inversion
1	1	O=1	Constant

Table 3.2 *Input, output & operators in binary logic*

Input	Output	Function	Description
A	0	O=0	Constant
A	A	O=A	Buffer
A	$\overline{A}$	O=$\overline{A}$	inversion
A	1	O=1	Constant

Adapting Ternary Logic-based designs decreases the problem of large off-chip resistors but results in high hardware complexity. This is because of the technique implemented to convert ternary to binary logic. In general, decoders are implemented for ternary to binary conversion and then for computation binary gates are implemented and at last buffer/encoder circuits are implemented to convert back to ternary Logic [54-59]. Adapting Ternary Logic, the transmission channels can be better used due to their high storage density of the line. Let us assume logic '0' and '1', '2' represent three conditions false, undefined and true. The ternary function of any n variable can $\{N_1,N_2.......N_n)$ can be mapped from n to [32].

Equation (5) represents the basic NOT, AND and OR ternary logic operations. Inversion of OR and AND results in NOR and NAND operations. NAND and NOR have 3 statuses like NTN AND, NTNOR, PTNAND, PTNOR, STNAND AND STNOR respectively. In most cases, the ALU is designed through basic logical blocks and in miscellaneous cases, it is done as mentioned in Table 3.3.

$$\text{NOT: } \overline{N_i} = 2 - N_i$$
$$\text{AND: } N_i . N_j = \min\{N_1, N_2\}$$
$$\text{OR: } N_i + N_j = \max\{N_{1,}N_2\} \quad (5)$$

Table 3.3 describes the different unary operators for different ternary logic values. Three ternary inverters and permutative operations are defined as Positive ternary inverter (PTI), Negative ternary inverter (NTI), Standard ternary inverter (STI), single shift, double shift and self-shift [60-64].

Table 3.3 *Truth table for different unary operators*

N	NTI(N)	PTI(N)	STI(N)	Single shift(N(+1))	Dual shift(N(+2))	Self-shift(1 2)
0	2	2	2	1	2	0
1	0	2	1	2	0	2
2	0	0	0	0	1	1

In case of ternary logic there are 3^{3^n} modal functions and 3^n combinations. If n=1 there can be 27 modal functions and 3 combinations respectively [33]. As CARBON NANO TUBEFET technique is the best alternative for MOSFET where threshold voltage can be altered by adjusting physical dimension of the device, it is apt to choose ternary logic based circuits [65-71]. Ternary functions can be graphically represented using cube like structure [45] as shown in Figure 3.3.

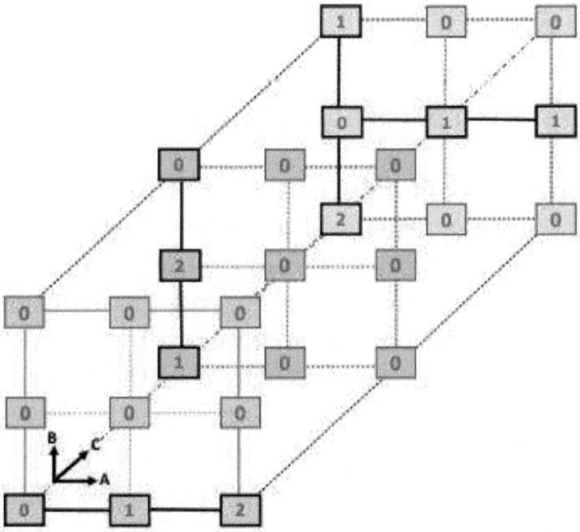

Figure 3.3 *Graphical representation of ternary functions.*

For three ternary variables there are 33 distinct members as depicted in Figure 3.3. In case of the ternary logic, the output of the function can be 0, 1 or 2. With the 'presence' of a term alone the function can't be characterized. Therefore, labels of 0, 1 and 2 represented in Figure 3.3. From the cube, the equation can be written as

$$O= 1 \cdot C_0B_1A_0 + 2 \cdot C_0B_2A_0 + 1 \cdot C_1B_0A_0 + 2 \cdot C_1B_0A_1 +2 \cdot C_2B_0A_0 + 1 \cdot C_2B_0A_2 + 1 \cdot C_2B_1A_1 + 1 \cdot C_2B_2A_1 \quad (6)$$

Equation (6) can be simplified as

$$O = 0 \cdot C_0B_0A_0 + 1 \cdot C_0B_1A_0 + 2 \cdot C_0B_2A_0 +1 \cdot C_1B_0A_0 + 2 \cdot C_1B_0A_1 + 0 \cdot C_1B_0A_2 +2 \cdot C_2B_0A_0 + 0 \cdot C_2B_0A_1 + 1 \cdot C_2B_0A_2 +2 \cdot C_2B_0A_0 + 1 \cdot C_2B_1A_1 + 1 \cdot C_2B_2A_1 \quad (7)$$

$$O = C_0A_0 \cdot (0 \cdot B_0 + 1 \cdot B_1 + 2 \cdot B_2) +C_1B_0 \cdot (1 \cdot A_0 + 2 \cdot A_1 + 0 \cdot A_2) +C_2B_0 \cdot (2 \cdot A_0 + 0 \cdot A_1 + 1 \cdot A_2) + C_2A_1 \cdot (0 \cdot B_0 + 1 \cdot B_1 + 1 \cdot B_2) \quad (8)$$

Equation (8) simplified as

$$O = C_0A_0 \cdot (B) + C_1B_0 \cdot (A_1) +C_2B_0 \cdot (A_2) +C_2A_1 \cdot (1 \cdot B_n) \quad (9)$$

$$O= C_0A_0 \cdot (B) + C_1B_0 \cdot (A_1) +C_2 \cdot (B_0 \cdot A_2 + A_1 \cdot (1 \cdot B_n)) \quad (10)$$

3.4 Preparation of carbon nanotubes

The properties of CARBON NANO TUBEs depend on the conditions where they are synthesised. CARBON NANO TUBEs exhibit properties like high mechanical strength, high stability, high thermal conductivity, low resistivity and narrow distribution size [1,2]. These properties depend on the structure of the CARBON NANO TUBEs [72-77]. In recent times CARBON NANO TUBEs are produced using three techniques namely arc discharge, chemical vapour deposition and laser ablation. In the case of arc discharge and laser ablation mixture of various forms of carbon is used to manufacture with low yield carbon nanotubes.

3.4.1 Arc discharge

This is one of the oldest techniques for nanotube synthesis. In this technique two electrodes are used and under the influence of DC current arc will be produced [46]. The Electrodes are kept in a vacuum chamber and inert gas at constant rate is passed into the chamber [78-83]. As a result of this carbon particles are deposited on the electrodes. Once the electric supply is ON and the positive electrode comes in contact with the negative electrode arc is produced. Once the electrodes become red hot plasma (atoms in vapour state at high temperature) is formed. This technique can be used for the synthesis of SWCNT and MWCNT, the former is synthesised by using appropriate catalyst and later is synthesised by maintaining constant pressure of the inert gas [84-92]. Using this technique nanotube from 2-20nm in diameter can be grown. Figure 3.4 describes the Arc discharge technique for nanotube synthesis [47]. The electrodes are separated by a distance of 1mm, pressure of 50 torr and a potential of 20-25V is applied inside the chamber.

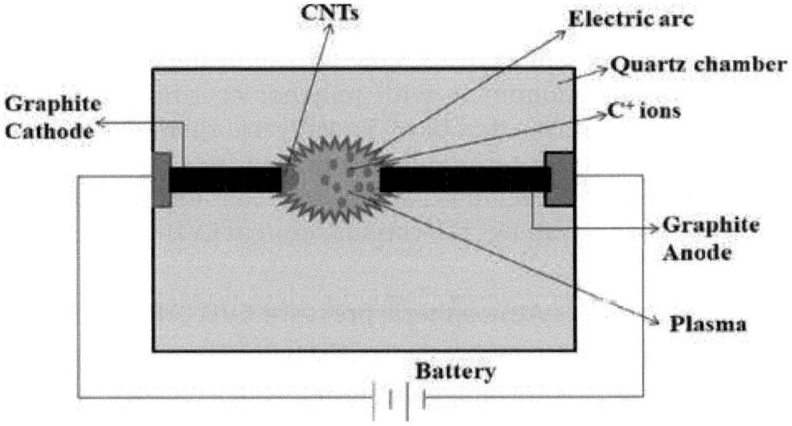

Figure 3.4 *Arc discharge technique for nanotube synthesis.*

3.4.2 Chemical vapour deposition

In this technique, the vaporised reactants chemical reaction forms a nanomaterial product which is deposited on the substrate as shown in Figure 3.5. Along with the desired deposit volatile by products are also formed which is removed by passing gas in the reaction chamber.

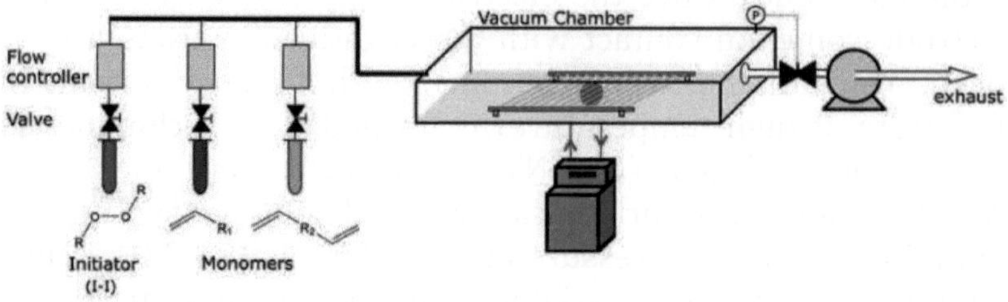

Figure 3.5 CVD process.

In general, the CVD process begins in large tanks with polymer coating carrying initiator materials which helps in speeding up the process and monomers which helps in polymer formation on the surface of the substrate. In Figure 3.5 the tank indicated in red contains the initiators and monomers with polymer coating are placed in tank indicated in purple and blue. To vaporize them either heating technique or reduction technique is followed [93-97]. Then the resultant is fed into the vacuum chamber that contains the material to be coated. Figure 3.6 shows the classification of CVD techniques.

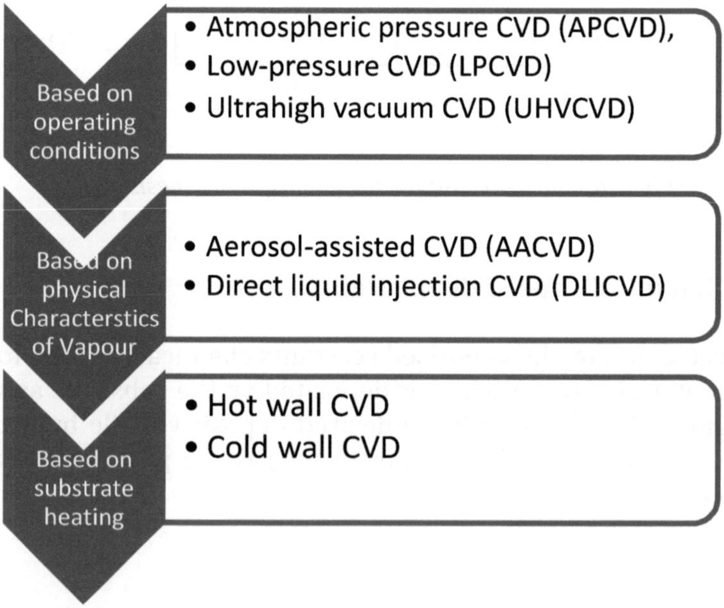

Figure 3.6 Classifications of CVD techniques

3.5 Carbon nanotube metrology

Wide commercialization of CARBON NANO TUBEs requires mass material assessments that are quick reliable, and perceivable to intensive and quantitative procedures. So as to upgrade the fabricating process the volume part of semiconducting to metallic mass material is considered engaging productive end-client applications like improvement of multi-work composites. Considering chirality and breadth of specific CARBON NANO TUBEs excellent quality applications can be built in (micro)electronics and optoelectronics. Also, this beginning part depicts NIST standards to spread out recommended practice guides and reference materials considering industry interest, and shows that collaboration of examiners, accomplices like CARBON NANO TUBE makers, NASA, and far edge clients is expected for outrageous advancement of CARBON NANO TUBEs accessible [98-103]. Single-walled carbon nanotubes (SWCARBON NANO TUBEs) offer different entryways for applications in the fields of equipment, optoelectronics, photonics, or photovoltaics. SWCARBON NANO TUBEs are absolutely essentially ideal models of unlimited π orbital arrangements, where the sp2 network gives exceptional charge carrier properties. Additionally, their real one-layered character enables unmistakable optical changes from excitonic states. While their Raman and photoluminescence characteristics have proactively been broadly thought of, the fundamental ingestion properties of SWCARBON NANO TUBE are still barely investigated on a quantitative reason, but especially these properties are of critical significance for applications. In all honesty, depicted ingestion assessments on SWCARBON NANO TUBE social occasions experience the evil impacts of the issue of test heterogeneity brand name for standard mix procedures [104-107]. The monstrous collection of nanotube chirality's and perspectives, alongside the occasion of contaminations from the association collaboration impede the assessment of the particular SWCARBON NANO TUBE chirality adding to an absorption signature [4].

3.6 Electrochemistry of carbon nanotubes

CARBON NANO TUBEs most certainly hang out considering their unprecedented morphology, nanosized angles, excellent physical-substance properties, and their different fields of use. As shown in Figure 3.1, CARBON NANO TUBEs could be imagined as wrapped graphene sheets (sp2 carbon coordinated in a honeycomb-like organization). As referred to the north, two get-togethers of carbon nanotubes are perceived, to be explicit multi-walled carbon nanotubes (MWNTs) and single-walled carbon nanotubes (SWNTs). Under MWNTs, these can be envisioned as concentric and closed graphite tubules with various graphene sheet layers describing an opening of typically 2-25 nm, disengaged by around 0.34 nm. SWNTs include a clear single-moved graphite sheet, forming an office of 1-2 nm width. Dependent upon the plan, generally on the distance across and helicity, CARBON NANO TUBEs can carry on like metals or semiconductors. Electrically, MWNTs show superconductivity with a to some degree high change temperature. Electron transport in nanotubes is ballistic; hence, the nanotubes' impediment doesn't depend upon its length, considering the way that the mean freeway λm is longer than the length of the real nanotube. For example, Berger et al. uncovered the mean freeway with λm = 30 μm, which was astonishingly longer than the nanotube used for the audit [108-112]. Progressing assessments of the alluring properties of nanotubes confirm that SWNT might be the seriously believed materials of choice that should be used as room-temperature superconductors.

3.7 Applications of carbon nanotubes

Different potential applications have been proposed recorded as a hard copy for CARBON NANO TUBEs, encompassing conductive and high-strength composites, sensors, energy limit and energy change devices, hydrogen limit media, field release introductions and radiation sources, and nanometre-sized semiconductor contraptions, tests, and interconnects [6]. A part of these applications was successfully recognized in things, while others were simply shown in lab-scale to display devices. Cost for nanotube status, poly dispersity in a given sort of nanotubes, and adolescent dealing with and get together procedures are at this point huge impediments for specific applications, especially by virtue of SWNTs [7].

3.7.1 Potential application of CARBON NANO TUBEs in vacuum microelectronics

Field radiation, a quantum influence, is a sharp source to make electrons interestingly, with a thermionic release. Right when reliant upon a satisfactorily high electric field, electrons near the Fermi level can beat the energy hindrance to getting conveyed to the vacuum level. The essential material study of electron radiation is at this point without a doubt known. The transmission current from metal is not settled by the Fowler-Nordheim condition: I = aV 2 exp (−bφ3/2/βV), where I, V, φ, β are the current, voltage applied, work capacity, and field redesign factor, independently. Electron field transmission materials have been widely analyzed for different mechanical purposes, for example, for level board shows, microwave speakers, or electron weapons in electron amplifying focal points. For imaginative applications, electron emissive materials should have low breaking point outpouring fields and should be consistent at high current thickness. A continuous thickness of 1-10 mA/cm2 is normal for shows and >500 mA/cm2 for a microwave intensifier. To restrict the electron surge edge field, makers with low work and an immense field improvement factor are great. The work is a characteristic material property. The field improvement factor chiefly depends upon the maker's computation and can be approximated by: β = 1/5r, where r is the breadth of the maker tip. Different taking care techniques to make different maker types have been spread out like Spin dt-type makers, with a sub-micron tip range. Be that as it may, this cycle is expensive, and the makers have recently confined the future. The breakdown is often achieved by molecule flood from the waiting gas species that dull the outpouring tips.

3.7.2 Energy storage

CARBON NANO TUBEs are similarly considered for energy creation and limit. Graphite, carbonaceous materials, and carbon fiber cathodes are used since seemingly forever in power modules, batteries, and different other electrochemical applications. Nanotubes are momentous due to their little viewpoints, the smooth surface geology, and awesome surface distinction, which relies upon the way that primary the basal graphite planes are uncovered in their plan. The electron move rate at carbon anodes finally manages the display of energy parts, which depends upon different factors, similar to game-plan and morphology of the carbon material used in the cathodes. A couple of tests have shown that, stood out from traditional carbon terminals, electron move energy are fastest on nanotubes, following ideal Nernstian lead. Nanotube microelectrodes have been manufactured using a cover; such devices have been successfully pursued for bio electrochemical reactions (e.g., oxidation of dopamine). They beat the spread out carbon cathodes to the extent that reaction rates and reversibility. Pure MWNTs and MWNTs put away with decent metal catalysts (Pd, Pt and Ag) have been applied for electro-catalysis of oxygen decline reactions, which are critical for energy units. A couple of examinations display that nanotubes contain fantastic decisions for standard carbon-based cathodes. Likewise, the updated selectivity of nanotube-based catalysts in heterogeneous catalysis was certified. In this particular circumstance, Ru-maintained nanotubes beat comparable metal on graphite and on other carbon materials in the liquid stage hydrogenation reaction of cinnamaldehyde. The properties of artificially evolved carbon nanofibers (which can mostly be envisioned as lacking nanotubes) turned out to be fitting for high power electrochemical capacitors.

3.7.3 Filled composites

The mechanical characteristics of CARBON NANO TUBEs are fortifying since they are seen as "a conclusive" carbon fiber anytime conveyed. Differentiated and steel, standard carbon strands have on numerous occasions the specific strength (strength/thickness), and contain momentous weight bearing strongholds when united in composites. Hence, nanotubes should be the contenders of choice for basic applications. Believe it or not, carbon strands have really been applied as strongholds in high strength, low thickness, and unrivaled execution composites. They are consistently found in a grouping of things going from exorbitant tennis rackets to plane and space mechanical assembly body parts. In this particular situation, NASA has actually made huge interests in making novel CARBON NANO TUBEs-based composites to be applied for, for example, the advanced Mars mission. Furthermore, carbon nanotubes were in like manner applied as filler materials to adjust the mechanical properties of films containing biodegradable, microbial bioplastics [8].

3.7.4 Nanoprobes and sensors

The little and steady parts of nanotubes prepare for a couple of extraordinarily intriguing applications. With infinitesimal sizes, high mechanical strength and flexibility, and high conductivity, nanotubes might conceivably finally become key nanoprobes for imaginative use. Such tests can be considered as being applied for various applications, similar to significant standard imaging, nanoelectrodes, nanolithography, drug transport systems, sensors, or field makers. The decision of making nanotube-based field emanating devices has recently been analyzed above in this part. Furthermore, moreover using a single MWNT associated with the uttermost furthest reaches of a checking test amplifying instrument method for imaging has recently been portrayed. Since MWNT tips are electrically conductive, they can be applied in STM and AFM devices as well as in other looking at test instruments, as in electrostatic power amplifying focal point. The upside of nanotube tips is their slimness and the probability to picture explicit plans (like infinitesimal, significant surface breaks), which are unquestionably difficult to test with greater, blunter-scratched Si or metal tips. Stood out from customary STM tips, similarly colossal biomolecules, similar to DNA or proteins, can be imaged with glorious objective using nanotube tips. MWNT and SWNT tips were really used in a tapping mode to picture biomolecules like amyloid-b-protofibrils (associated with Alzheimer's disease) with exceptional objective. Furthermore, considering the incredible adaptability of the nanotubes, the tips are not hurt by contact with the substrates. Impacts just objective fastening of the nanotube, which regularly is reversible on pulling out the tip from the substrate.

3.7.5 Templates

Since nanotubes have very straight and meager diverts in their centers, it was talked about all along that it very well may be feasible to fill various materials into these holes to create one-layered nanowires. Early presumptions suggested that articulated slender powers exist in nanotubes, adequately solid to keep down gases and fluids inside the cavities. In 1993, this was interestingly sealed tentatively by filling and setting liquid lead inside MWNT channels. By this procedure, wires just 1.2 nm in measurement were effectively produced inside nanotubes. These days, an enormous number of studies managing this subject exist, a considerable lot of them depicting the filling of nanotubes with metallic and earthenware materials. Accordingly, nanotubes comprise capable formats to make nanowires of various design and structure.

3.8 Conclusion

Hardly any materials can have been pretty much as seriously concentrated as carbon nanotubes. Because of the gigantic measure of work did on these designs starting around 1991, we presently have an abundance of data on their electronic, mechanical, warm, optical and other actual properties. We know how to connect things to the beyond tubes and to put things inside. We can manoeuvre nanotubes toward characterized game plans and integrate them into polymer, ceramic or metal grids. As of now, nonetheless, there are still somewhat not many nanotube-containing items available. The couple of business items that have arisen are in exceptionally concentrated regions, for example, AFM tips. Maybe the most economically huge use of nanotubes to date has been as parts of lithium particle batteries, albeit these utilize generally low quality chemically delivered MWNTs. It would be reasonable to say, then, at that point, that carbon nanotubes have not yet lived up to their true capacity. The primary justification behind this is very basic: they are still unreasonably costly. The cost of single-walled nanotubes is as of now around \$100 per gram, contrasted and about \$30 per gram for gold. Obviously, there is still a lot of work to be done in working on the quality and yield of nanotube creation.

In this last part, a brief and emotional outline is given of a portion of the features of carbon nanotube research. A few remarks are then gained on the areas where headway actually should be made.

References

1. Sharma, T.; Kumre, L. Energy-Efficient Ternary Arithmetic Logic Unit Design in CARBON NANO TUBEFET Technology. Circuits Syst. Signal Process. 2019, 39, 3265-3288.

2. Vudadha, C.; Katragadda, S.; Phaneendra, P.S. 2: 1 Multiplexer based plan for ternary rationale circuits. In Proceedings of the 2013 IEEE Asia Pacific Conference on Postgraduate Research in Microelectronics and Electronics (PrimeAsia), Visakhapatnam, India, 19-21 December 2013.

3. Murotiya, L.S.; Gupta, A. Plan of CARBON NANO TUBEFET-based 2-digit ternary ALU for nanoelectronics. Int. J. Electron. 2014, 101, 1244-1257.
4. Chen, L.J.L.; Han, J.; Lombardi, F. Plan and assessment of various esteemed rationale entryways utilizing pseudo N-type carbon nanotube FETs. IEEE Trans. Nanotechnol. 2014, 13, 695-708.
5. Lin, S.; Kim, Y.- B.; Lombardi, F. Plan of a ternary memory cell utilizing CARBON NANO TUBEFETs. IEEE Trans. Nanotechnol. 2012, 11, 1019-1025.
6. Levashenko, Z.E.V.; Lukyanchuk, I.; Rabcan, J.; Kvassay, M.; Rusnak, P. Utilization of Generalized Reed-Muller Expression for Development of Non-Binary Circuits. Gadgets 2020, 9, 12.
7. Moaiyeri, M.H.; Shamohammadi, M.; Sharifi, F.; Navi, K. Superior execution ternary rationale doors for nanoelectronics. Int. J. High Perform. Syst. Archit. 2015, 5, 209-215.
8. Prakash, P.; Sundaram, K.M.; Bennet, M.A. A survey on carbon nanotube field impact semiconductors (CARBON NANO TUBEFETs) for super low power applications. Reestablish. Support. Energy Rev. 2018, 89, 194-203.
9. Obite, F.; Ijeomah, G.; Bassi, J.S. Carbon nanotube field impact semiconductors: Toward future nanoscale gadgets. Int. J. Comput. Appl. 2019, 41, 149-164.
10. Hu, Z.; Tulevski, G.S.; Hannon, J.B.; Afzali, A.; Liehr, M.; Park, H. Changeability and unwavering quality examination in self-collected multichannel carbon nanotube field-impact semiconductors. Appl. Phys. Lett. 2015, 106, 243106.
11. Khurshid, T.; Fatima, S.; Khanday, F.A.; Bashir, F.; Zahoor, F.; Hussin, F.A. Carbon nanotube field impact semiconductor (CARBON NANO TUBEFET) functional transconductance enhancer (OTA) based plan of high recurrence memristor emulator. Int. J. Numer. Model. Electron. Netw. Gadgets Fields 2020, 2827.
12. Banerjee, W. Difficulties and Applications of Emerging Nonvolatile Memory Devices. Hardware 2020, 9, 1029.
13. Lim, E.W.; Ismai, R. Conduction Mechanism of Valence Change Resistive Switching Memory: A Survey. Hardware 2015, 4, 586-613.
14. Arumi, D.; Manich, S.; Gomez-Pau, A.; Rodriguez-Montanes, R.; Montilla, V.; Hernandez, D.; Gonzalez, M.B.; Campabadal, F. Effect of Laser Attacks on the Switching Behavior of RRAM Devices. Hardware 2020, 9, 200.
15. Perez E.; Ossorio, O.G.; Duenas, S.; Castan, H.; Garcia, H.; Wenger, C. Programming Pulse Width Assessment for Reliable and Low-

Energy Endurance Performance in Al:HfO2-Based RRAM Arrays. Hardware 2020, 9, 864.
16. Zahoor, F.; Zulkifli, T.Z.A.; Khanday, F.A. Resistive Random Access Memory (RRAM): An Overview of Materials, Switching Mechanism, Performance, Multilevel Cell (mlc) Storage, Modeling, and Applications. Nanoscale Res. Lett. 2020, 15, 90.
17. Zahoor, F.; Zulkifli, T.Z.A.; Khanday, F.A.; Murad, S.A.Z. Carbon Nanotube and Resistive Random Access Memory Based Unbalanced Ternary Logic Gates and Basic Arithmetic Circuits. IEEE Access 2020, 8, 104701-104717.
18. Shulaker, M.M.; Hills, G.; Park, R.S.; Howe, R.T.; Saraswat, K.; Wong, H.- S.P.; Mitra, S. Three-layered coordination of nanotechnologies for figuring and information stockpiling on a solitary chip. Nature 2017, 547, 74-78.
19. Zhang, H.; Zhang, Z.; Gao, M.; Luo, L.; Duan, S.; Dong, Z.; Lin, H. Execution of Unbalanced Ternary Logic Gates with the Combination of Spintronic Memristor and CMOS. Gadgets 2020, 9, 542.
20. Zahoor, F.; Zulkifli, T.Z.A.; Khanday, F.A.; Fida, A.A. Low-power RRAM Device based 1T1R Array Design with CARBON NANO TUBE-FET as Access Device. In Proceedings of the 2019 IEEE Student Conference on Research and Development (SCOReD), Bandar Seri Iskandar, Malaysia, 15-17 October 2019; pp. 280-283.
21. Strukov, D.; Kohlstedt, H. Resistive exchanging peculiarities in dainty movies: Materials, gadgets, and applications. MRS Bull. 2012, 37, 108-114.
22. Prabhu, N.L.; Jun, D.L.J.; Dananjaya, P.A.; Lew, W.S.; Toh, E.H.; Raghavan, N. Investigating the Impact of Variability in Resistance Distributions of RRAM on the Prediction Accuracy of Deep Learning Neural Networks. Hardware 2020, 9, 414.
23. Dish, F.; Gao, S.; Chen, C.; Song, C.; Zeng, F. Late advancement in resistive arbitrary access recollections: Materials, exchanging systems, and execution. Mater. Sci. Eng. R Rep. 2014, 83, 1-59.
24. Natsuki, T. Carbon Nanotube-Based Nanomechanical Sensor: Theoretical Analysis of Mechanical and Vibrational Properties. Gadgets 2017, 6, 56.
25. Appenzeller, J. Carbon Nanotubes for High-Performance Electronics — Progress and Prospect. Proc. IEEE 2008, 96, 201-211.
26. Sethi, D.; Kaur, M.; Singh, G. Plan and execution investigation of a CNFET-based TCAM cell with double chirality choice. J. Comput. Electron. 2017, 16, 106-114.

27. Moaiyeri, M.H.; Mirzaee, R.F.; Doostaregan, A.; Navi, K.; Hashemipour, O. A general technique for planning low-power carbon nanotube FET-based numerous esteemed rationale circuits. IET Comput. Digit. Tech. 2013, 7, 167-181.
28. Samadi, H.; Shahhoseini, A.; Aghaei-liavali, F. Another technique on planning and recreating CARBON NANO TUBEFET based ternary entryways and number-crunching circuits. Microelectron. J. 2017, 63, 41-48.
29. Jaber, R.A.; Kassem, A.; EL-Hajj, A.M.; EL-Nimri, L.A.; Haida, A.M. Elite execution and energy-effective CNFET-based plans for ternary rationale circuits. IEEE Access 2019, 7, 93871-93886.
30. Rokhani, F.Z.; Sobelman, G.E. Transport Energy Consumption for Multilevel Signals. IEEE Trans. Circuits Syst. Regul. Pap. 2009, 57, 64-71.
31. Soliman, N.; Fouda, M.E.; Alhurbi, A.G.; Said, L.A.; Madian, A.H.; Radwan, A.G. Ternary Functions Design Using Memristive Threshold Logic. IEEE Access 2019, 7, 48371-48381.
32. Lin, S.; Kim, Y.- B.; Lombardi, F. CARBON NANO TUBEFET-based plan of ternary rationale entryways and number-crunching circuits. IEEE Trans. Nanotechnol. 2011, 10, 217-225.
33. Zahoor, F., Hussin, F. A., Khanday, F. A., Ahmad, M. R., Mohd Nawi, I., Ooi, C. Y., & Rokhani, F. Z. (2021). Carbon nanotube field effect transistor (CNTFET) and resistive random access memory (RRAM) based ternary combinational logic circuits. *Electronics, 10*(1), 79.
34. Mohammed, M.U.; Vijjapuram, R.; Chowdhury, M.H. Novel CARBON NANO TUBEFET and Memristor based Unbalanced Ternary Logic Gate. In Proceedings of the 2018 IEEE 61st International Midwest Symposium on Circuits and Systems (MWSCAS), Windsor, ON, Canada, 5-8 August 2018; pp. 1106-1109.
35. Priest, M.D.; Hills, G.; Srimani, T.; Lau, C.; Murphy, D.; Fuller, S.; Humes, J.; Ratkovich, A.; Nelson, M.; Shulaker, M.M. Creation of carbon nanotube field-impact semiconductors in business silicon fabricating offices. Nat. Electron. 2020, 3, 492-501.
36. Stanford CNFET Model. Accessible on the web: https://nano.stanford.edu/stanford-cnfet-model (got to on 1 November 2020).
37. Stanford RRAM Model. Accessible on the web: https://nano.stanford.edu/stanford-rram-model (got to on 1 November 2020).

38. Shahrom, E.; Hosseini, S.A. A new low power multiplexer based ternary multiplier utilizing CARBON NANO TUBEFETs. AEU Int. J. Electron. Commun. 2018, 93, 191-207.
39. Das, S.; Bhattacharya, S.; Das, D. Plan of computerized rationale circuits utilizing carbon nanotube field impact semiconductors. Int. J. Delicate Comput. Eng. 2011, 1, 173-178.
40. Navi, K.; Momeni, A.; Sharifi, F.; Keshavarzian, P. Two novel ultra high velocity carbon nanotube Full-Adder cells. IEICE Electron. Express 2009, 6, 1395-1401.
41. Navi, K.; Rashtian, M.; Khatir, A.; Keshavarzian, P.; Hashemipour, O. Fast capacitor-inverter based carbon nanotube full snake . Nanoscale Res. Lett. 2010, 5, 859-862.
42. Jaber, R.A.; El-Hajj, M.; Kassem, A.; Nimri, L.A.; Haidar, A.M. CNFET-based plans of ternary half-snake utilizing a novel 'decoder-less' ternary multiplexer in view of unary administrators. Microelectron. J. 2019, 96, 104698.
43. Sandhie, Z.T.; Ahmed, F.U.; Chowdhury, M.H. Plan of Ternary Logic and Arithmetic Circuits utilizing GNRFET. IEEE Open J. Nanotechnology. 2020, 1, 77-87.
44. Meenakshi Gupta, Vipin Chauhan, CVD grown single walled carbon nanotubes in organic solvents, https://aip.scitation.org/doi/10.1063/1.3526242.
45. Hurst, S.L.: An extension of binary minimization techniques to ternary equations. Comput. J. 11(3), 277–286 (1968)
46. Sinha N, Ma J, Yeow JTW (2006) Carbon nanotube-based sensors. J Nanosci Nanotechnol 6:573–590
47. Jagadeesan, Aravind Kumar, Krithiga Thangavelu, and Venkatesan Dhananjeyan. "Carbon nanotubes: Synthesis, properties and applications." *21st Century Surface Science—A Handbook* (2020).
48. Anjana Bhardwaj, Pradeep Kumar, B. Raj, Sunny Anand, "Design and Performance Optimization of doping-less Vertical Nanowire TFET using Gate Stack Technique", Journal of Electronic Materials (JEMS) Springer, Vol. 41, Issuse 7, PP. 4005-4013, 2022.
49. Jeetendra Singh, BRaj, "Tunnel Current Model of Asymmetric MIM Structure Levying Various Image Forces to Analyze the Characteristics of Filamentary Memristor", Applied Physics A, Springer, Vol. 125, No.3, pp. 203.1 to 203.11 Feb 2019.
50. Candy Goyal, Jagpal Singh Ubhi and B Raj, Low Leakage Zero Ground Noise Nanoscale Full Adder using Source Biasing Technique, "Journal of Nanoelectronics and Optoelectronics", American Scientific Publishers, Vol. 14, PP. 360–370, March 2019.

51. Gurmohan Singh, R. K. Sarin and B Raj, "A Novel Robust Exclusive-OR Function Implementation in QCA Nanotechnology with Energy Dissipation Analysis", Journal of Computational Electronics, Springer, Volume 15, Issue 2, PP. 455-465, June 2016.
52. Tanu Wadhera, Deepti Kakkar, Girish wadhwa, B Raj, " Recent Advances and Progress in Development of the Field Effect Transistor Biosensor: A Review" Journal of ELECTRONIC MATERIALS, Springer, Volume 48, Issue 12, pp 7635–7646, December 2019.
53. Girish Wadhwa, Priyanka Kamboj, Balwinder Raj" Design optimisation of junctionless TFET biosensor for high sensitivity" Advances in Natural Sciences: Nanoscience and Nanotechnology, Vol. 10, PP. 045001 (7), (2019).
54. Priya Bansal, BRaj, "MEMRISTOR MODELING AND ANALYSIS FOR LINEAR DOPANT DRIFT KINETICS," Journal of Nanoengineering and Nanomanufacturing, American Scientific Publishers, Vol. 6, pp. 1–7, 2016.
55. Amandeep Singh, Mamta Khosla, BRaj, "Circuit Compatible Model for Electrostatic Doped Schottky Barrier CNTFET,"Journal of Electronic Materials, Springer, Vol. 45, Issue 12, pp 4825-4835, 2016.
56. Ashima, D Vaithiyanathan, B Raj," Performance Analysis of Charge Plasma induced Graded Channel Si nanotube", **Journal of Engineering Research** (JER), EMSME Special Issue pp. 146-154 Aug 2021.
57. Abhishek Singh Tomar, Vijay Kumar Magraiya, and B Raj, Scaling of Access and Data Transistor for High Performance DRAM Cell Design, Quantum Matter, Vol. 2, PP. 412-416 Oct. 2013.
58. Neeraj Jain, BRaj, "Parasitic Capacitance and Resistance Model Development and Optimization of Raised Source/Drain SOI FinFET Structure for Analog Circuit Applications, Journal of Nanoelectronics and Optoelectronins, ASP, USA, Vol. 13, PP. 531–539, Apl 2018.
59. Shradhya Singh, S. K. Vishvakarma and B Raj, "Analytical Modeling of Split-Gate Junction-Less Transistor for a Biosensor Application", Sensing and Bio-sensing, Elsevier, Vol. 18, Pages 31-36, April 2018.
60. Maisagalla Gopal, and Balwinder Raj, Low Power 8T SRAM Cell Design for High Stability Video Applications, ITSI Transaction on Electrical and Electronics Engineering, Vol. 1(5), pp. 91-97, 2013.

61. Balwinder Raj, Jatin Mitra, Deepak Kumar Bihani, Rangharajan V, A.K. Saxena and S. Dasgupta, "Analysis of Noise Margin, Power and Process Variation for 32 nm FinFET Based 6T SRAM Cell" Journal of Computer (JCP), Academy Publisher, FINLAND, vol. 5, issue 6, PP.1-8, 2010.
62. Divya Sharma, Rajesh Mehra and B Raj, "Comparative Analysis of Photovoltaic Technologies for High Efficiency Solar Cell Design", Superlattices and Microstructures, Elsevier, Volume 153, PP. 106861, May 2021.
63. Pawandeep Kaur, Avtar Singh Buttar, B Raj, "A comprehensive Analysis of Nanoscale Transistor based Biosensor: A Review, Indian Journal of Pure and Applied Physics, Vol. 59, pp. 304-318, April 2021.
64. Divya Yadav, Balwant Raj and B Raj, "Design and Simulation of Low Power Microcontroller for IoT Applications", Journal of Sensor Letters, ASP, Vol. 18, PP. 401–409, May 2020.
65. Shailendra Singh, B Raj,"A 2-D Analytical Surface Potential and Drain current Modeling of Double-Gate Vertical t-shaped Tunnel FET", Journal of Computational Electronics, Springer, Vol. 19, PP.1154–1163, Apl 2020.
66. Jeetendra Singh, B Raj, "An Accurate and Generic Window function for Non-linear Memristor Model" Journal of Computational Electronics, Springer, Volume 18 Issue 2, PP. 640-647 June 2019.
67. Manjit Kaur; Neena Gupta; Sanjeev Kumar; BRaj; Arun Kumar Singh, "Comparative RF and Crosstalk Analysis of Carbon Based Nano Interconnects" IET Circuits, Devices & Systems, Vol.15(6), PP. 493-503, Feb 2021.
68. Nehru Kandasamy, Firdous Ahmad, D. Ajitha, B Raj, and Nagarjuna Telagam "Quantum Dot Cellular Automata based Scan Flip Flop and Boundary Scan Register, IETE Journal of Research, Volume 66, PP. 535-548, 2020.
69. Sharma, S. K.., Raj, B., Khosla, M., "Enhanced Photosensivity of Highly Spectrum Selective Cylindrical Gate In1-xGaxAs Nanowire MOSFET Photodetector, Modern Physics letter-B, Vol. 33, No. 12, PP. 1950144 (2019).
70. Jeetendra Singh, BRaj, "Design and Investigation of 7T2M NVSARM with Enhanced Stability and Temperature Impact on Store/Restore Energy", IEEE Transactions on Very Large Scale Integration Systems, Vol. 27, Issure 6, PP. 1322 - 1328 June 2019.
71. Anil Kumar Bhardwaj, Sumeet Gupta, BRaj, Amandeep Singh, "Impact of Double Gate Geometry on the Performance of Car-

bon Nanotube Field Effect Transistor Structures for Low Power Digital Design", Computational and Theoretical Nanoscience, ASP, Vol. 16, PP. 1813–1820, 2019.

72. Neeraj Jain, BRaj, "Thermal Stability Analysis and Performance Exploration of Asymmetrical Dual-k underlap Spacer (ADKUS) SOI FinFET for Security and Privacy Applications", Indian Journal of Pure & Applied Physics (IJPAP), Vol. 57, pp. 352-360, May 2019.

73. Amandeep Singh, Mamta Khosla, BRaj, "Design and Analysis of Dynamically Configurable Electrostatic Doped Carbon Nanotube Tunnel FET"," Microelectronics Journal, Elesvier, Volume 85, Pages 17-24, March 2019.

74. Neeraj Jain and Balwinder Raj, "Dual-k Spacer Region Variation at the Drain Side of Asymmetric SOI FinFET Structure: Performance Analysis towards the Analog/RF Design Applications", communicated in "Journal of Nanoelectronics and Optoelectronics", American Scientific Publishers, Vol. 14, PP. 349–359, March 2019.

75. Jeetendra Singh, Sanjeev Sharma, B Raj, Mamta Khosla Analysis of barrier layer thickness on performance of In1-xGaxAs based Gate Stack Cylindrical Gate Nanowire MOSFET, " JNO, ASP, Vol. 13, PP 1473-1477, Oct 2018.

76. Neeraj Jain, B Raj, "Analysis and Performance Exploration of High-k SOI FinFETs Over the Conventional Low-k SOI FinFET toward Analog/RF Design", Journal of Semiconductors (JoS), IOP Science, Vol.39, Issue 12, PP- 124002-1-7, Dec 2018.

77. Candy Goyal, Jagpal Singh Ubhi and BRaj, A reliable leakage reduction technique for approximate full adder with reduced ground bounce noise, Journal of Mathematical Problems in Engineering, Hindawi, Volume 2018, Article ID 3501041, 16 pages, 15 Oct 2018.

78. Anuradha, Jeetendra Singh, B Raj and Mamta Khosla, "Design and Performance Analysis of Nano-scale Memristor-based Nonvolatile SRAM" Journal of Sensor Letter", American Scientific Publishers, Vol. 16, PP 798–805, Oct 2018.

79. Girish Wadhwa, BRaj, "Parametric Variation Analysis of Charge-Plasma-based Dielectric Modulated JLTFET for Biosensor Application" IEEE Sensor Journal, VOL. 18, NO. 15, AUGUST 1, PP. 6070 – 6077, 2018 (SCI)

80. Jeetendra Singh, BRaj, "Comparative Analysis of Memristor Models for Memories Design", JoS, IoP, Vol. 39, No. 7, PP. 074006-1-12, July 2018.

81. Divya Yadav, Shailesh Singh Chouhan, Santosh Kumar Vishvakarma and BRaj, "Application Specific Microcontroller Design for IoT based WSN", Sensor Letter, ASP, Vol. 16, PP. 374–385, May 2018..

82. Gurmohan Singh, R. K. Sarin and BRaj, "Fault-Tolerant Design and Analysis of Quantum-Dot Cellular Automata Based Circuits", IEEE/IET Circuits, Devices & Systems, Vol. 12, PP. 638 – 64, 2018.

83. Jeetendra Singh, BRaj, "Modeling of Mean Barrier Height Levying Various Image Forces of Metal Insulator Metal Structure to Enhance the Performance of Conductive Filament Based Memristor Model", IEEE Nanotechnology, Vol. 17, No. 2, PP. 268-267, March 2018.

84. Aakash Jain, Sanjeev Sharma, BRaj, "Analysis of Triple Metal Surrounding Gate (TM-SG) III-V Nanowire MOSFET for Photosensing Application", Opto-electronics Journal, Elsevier, Volume 26, Issue 2, Pages 141-148, May 2018.

85. Aakash Jain, Sanjeev Sharma, B Raj, "Design and Analysis of High Sensitivity Photosensor Using Cylindrical Surrounding Gate MOSFET For Low Power Sensor Applications", Engineering Science and Technology, an International Journal, Elsevier's, Volume 19, Issue 4, Pages 1864–1870, December 2016.

86. Amandeep Singh, Mamta Khosla, B Raj, "Analysis of Electrostatic Doped Schottky Barrier Carbon Nanotube FET for Low Power Applications," Journal of Materials Science: Materials in Electronics, Springer, Vol.28, PP. 1762–1768, 2017.

87. G. Saiphani Kumar, Amandeep Singh, BRaj, "Design and Analysis of Gate All Around CNTFET based SRAM cell Design", Journal of Computational Electronics, Springer, Volume 17, Issue 1, pp 138–145, March 2018.

88. Gurinder pal Singh, B. S. Sohi, Balwinder Raj, "Material Properties Analysis of Graphene Base Transistor (GBT) for VLSI Analog Circuits, Indian Journal of Pure & Applied Physics (IJPAP), Vol. 55, pp. 896-902, Dec-2017.

89. Amandeep Singh, Mamta Khosla, B Raj, "Comparative Analysis of Carbon Nanotube Field Effect Transistor and Nanowire Transistor for Low Power Circuit Design," Journal of Nanoelectronics and Optoelectronics, American Scientific Publishers, USA, Vol. 11, PP. 388-393, June 2016.

90. Sunil Kumar and B Raj, "Estimation of Stability and Performance metric for Inward Access Transistor based 6T SRAM Cell De-

sign using n-type/p-type DMDG-GDOV TFET" IEEE VLSI Circuits and Systems Letter, Volume3, Issue2, PP 25-39, June2017
91. Shashikant Sharma, Anjan Kumar, Manisha Pattanaik and B Raj, "Forward Body Biased Multimode Multi-Threshold CMOS Technique for Ground Bounce Noise Reduction in Static CMOS Adders" International Journal of Information and Electronics Engineering, pp. 567-572, Vol. 3, issue 3, 2013.
92. Hamendra Singh, Pankaj Kumar, Balwinder Raj, "Performance Analysis of Majority Gate SET Based 1-bit Full Adder", *International Journal of Computer and Communication Engineering* (IJCCE), IACSIT Press Singapore, ISSN: 2010-3743, Volume 2, No. 4, 2013.
93. Anil Kumar Bhardwaj, Sumeet Gupta and B Raj, "Investigation of Parameters for Schottky Barrier (SB) Height for Schottky Barrier Based Carbon Nanotube Field Effect Transistor Device", Journal of Nanoelectronics and Optoelectronics, ASP, Vol. 15, PP. 783–791, July 2020.
94. Priya Bansal, BRaj, "MEMRISTOR: A VERSATILE NONLINEAR MODEL FOR DOPANT DRIFT AND BOUNDARY ISSUES," JCTN, American Scientific Publishers, Vol. 14 (5), PP. 2319-2325, May 2017.
95. Neeraj Jain, BRaj, "An Analog and Digital Design Perspective Comprehensive Approach on Fin-FET (Fin-Field Effect transistor) Technology - A Review", Reviews in Advanced Sciences and Engineering (RASE), ASP, Vol. 5, PP. 1-14, 2016.
96. Sanjeev Sharma, B Raj, Mamta Khosla, "Subthreshold Performance of In1-xGaxAs based Dual Metal with Gate Stack Cylindrical/Surrounding Gate Nanowire MOSFET for Low Power Analog Applications, Journal of Nanoelectronics and Optoelectronics, American Scientific Publishers, USA, Vol.12, PP. 171–176, 2017.
97. Balwinder Raj, A. K. Saxena and S. Dasgupta, "Analytical Modeling for the Estimation of Leakage Current and Subthreshold Swing Factor of Nanoscale Double Gate FinFET Device" Microelectronics International, UK, vol. 26, pp. 53-63, 2009.
98. Soniya, Shailendra Singh, Girish Wadhwa and Balwinder Raj, "An Analytical Modeling for Dual Source Vertical Tunnel Field Effect Transistor" submitted to 'International Journal of Recent Technology and Engineering (IJRTE)' at Volume-8 Issue-2, PP. 603-608, July 2019.
99. Shailendra Singh, B Raj, "Design and analysis of hetrojunction Vertical T-shaped Tunnel Field Effect Transistor", Journal of

Electronics Material, Springer, Volume 48, Issue 10, pp 6253-6260, October 2019.

100. Candy Goyal, Jagpal Singh Ubhi and B Raj, "A Low Leakage CNTFET based Inexact Full Adder for Low Power Image Processing Applications", International Journal of Circuit Theory and Applications, Wiley, Volume47, Issue9, Pages 1446-1458, September 2019.

101. B Raj, A. K. Saxena and S. Dasgupta, "A Compact Drain Current and Threshold Voltage Quantum Mechanical Analytical Modeling for FinFETs" Journal of Nanoelectronics and Optoelectronics (JNO), USA, vol. 3, no. 2, pp. 163-170, 2008.

102. Girish Wadhwa, B Raj, "An Analytical Modeling of Charge Plasma based Tunnel Field Effect Transistor with Impacts of Gate underlap Region" Superlattices and Microstructures, Elsevier, Vol. 142, PP.106512, June 2020.

103. Shailendra Singh, B Raj,"Modeling and Simulation analysis of SiGe hetrojunction Double GateVertical t-shaped Tunnel FET", Superlattices and Microstructures, Elsevier Volume 142, PP. 106496, June 2020.

104. Amandeep Singh, Dinesh Kumar Saini, Dinesh Agarwal, Sajal Aggarwal, Mamta Khosla, BRaj, "Modeling and Simulation of Carbon Nanotube Field Effect Transistor and its Circuit Application ," Journal of Semiconductors (JoS), IOP Science, Vol.37, PP. 074001-6, July 2016.

105. Neeraj Jain, Balwinder Raj, "Device and Circuit Co-Design Perspective Comprehensive Approach on FinFET Technology - A Review, Journal of Electron Devices, Vol. 23, Num. 1, pp 1890-1901, 2016.

106. Sunil Kumar and BRaj, "Analysis of I_{ON} and Ambipolar Current for Dual-Material Gate-drain Overlapped DG-TFET," Journal of Nanoelectronics and Optoelectronics, American Scientific Publishers, USA, Vol.11, PP. 323-333, June 2016.

107. Naveed Anjum, Tarun Bali and B Raj, Design And Simulation Of Handwritten Multiscript Character Recognition, International Journal of Advanced Research in Computer and Communication Engineering, Vol.2 (7), PP. 2544-2549, July 2013.

108. Sanjeev Sharma, BRaj, Mamta Khosla, "A Gaussian Approach for Analytical Subthreshold Current Model of Cylindrical Nanowire FET with Quantum Mechanical Effects, Microelectronics Journal, Elsevier, Vol. 53, PP. 65-72, April 2016.

109. Karmjit Singh, BRaj, "Performance and analysis of temperature dependent Multi-walled Carbon Nanotubes as Global Intercon-

nects at different technology nodes," Journal of Computational Electronics, Springer, Volume 14 Issue 2, pp. 469-476, June 2015.

110. Sunil Kumar and BRaj, "Compact channel potential analytical Modeling of DG-TFET based on Evanescent–mode Approach," Journal of Computational Electronics, Springer Vol. 14, No. 2, pp.820-827, July 2015.

111. Karmjit Singh, BRaj, "Temperature Dependent Modeling and Performance Evaluation of Multi-Walled CNT and Single-Walled CNT as Global Interconnects," Journal of Electronic Materials, Springer, Vol. 44, Issue 12, pp 4825-4835, Dec 2015.

112. V.K. Sharma, M. Pattanaik and B.Raj, "INDEP approach for leakage reduction in nanoscale CMOS circuits", International Journal of Electronics, Taylor & Francis, Vol.102, No.2, pages 200-215, 2014.

4

Study and Analysis of Low Power Dynamic Sequential Circuits

Preeti Verma, Ajay K Sharma, Swatantrata Shukla, Alok K Mishra, D. Vaithiyanathan and Baljit Kaur
National Institute of Technology Delhi, India

4.1 Introduction

The electronic industry has accomplished remarkable growth since last few decades. This is essentially because of prompt expeditious developments in Integrated Circuit (IC) technologies and also in design of large-scale systems. There is no doubt that our routine lives are remarkably dependent on microelectronics and related technologies. This is verifiable on the domestic scene, in our professional exercise, in the workplace and in leisure actions. Today, electronics is characterized by low power dissipation, low cost, high reliability, incorporating with the ability to survive comfortably with high degree of refinement and complexity [1] [2] [3]. Integrated circuits have made feasible the design of robust and workable processors which present extremely intelligent and conformable devices for the users. Silicon CMOS technology has been considered superior from the past several years, for cost effective and high-performance systems. The requirement of image processing, quality improvement in multimedia content and future generation multimedia, together with the exposure of new and complex optical and photonics technologies being driven by microelectronics, put heavy demands on current CMOS technology integrated systems. Despite the fact that the technology is progressively developing to compose smaller systems with reduced power dissipation, the IC industry is facing crucial challenges due to limitations on power density and high dynamic and static power dissipation. The clue to vanquish these challenges reside in the refinement of design, material and manufacturing processes [4]. The main issue faced by the VLSI designer is to find the effective technique for balancing the requirement of performance along with power dissipation [5].

Primary interest for complex, VLSI circuits is low-power designing. Technologies below 130nm are experiencing contemporary set of design issues in relation to the chip's power usage. During present days due to the use of tens of millions of gates, power density and total power consumption are at the verge that can be supported by available packaging, cooling and other infrastructure. Leakage current is increasing greatly as the technology has reduced to 90nm and the leakage power even becoming approximately as high as dynamic power consumption for below 65 nm technologies.

4.2 Design Methodologies intended for Low Power

Low-power approaches essentially be used all over the design procedure, from the system level to the process level, for reducing power-consumption of digital systems though keeping in mind that performance is still crucial. Power distribution within processor must be known during optimization. So that the block consuming maximum amount of power can be identified.

Various design levels of an integrated circuit can be represented in following sequence:

System: Algorithm: Architecture: Circuit/ Logic: Device / Process. The process technology can be controlled by the device/ process designer while the circuit designer has control over all other levels. There are several ways to reduce power consumption.

4.2.1 Power Reduction Through Process Technology

Power dissipation can be effectively reduced by lowering the supply voltage. Nonetheless, reducing the supply voltage increases delay significantly, especially if supply voltage value is equivalent to threshold voltage. To overcome this issue proper scaling of the device is required. Through appropriate scaling following achievement can be seen:

i. Increases current drive capabilities improves the device characteristics for low voltage operation;
ii. Reduction in capacitance due to small geometries
iii. Enhanced interconnect technology
iv. Active and standby power consumption can be managed with the use of several and variable threshold devices

v. High level of integration on a single chip provides power saving.

4.2.2 Lowering of Power dissipation by way of circuit/ logic design

Many procedures can be used to minimize the power-dissipation at circuit and logic level such as:

i. Emphasising use of static logic instead of dynamic logic
ii. Logic optimization to achieve reduced switching activity
iii. By optimizing clock and bus loading
iv. Use of circuit technique that requires a smaller number of transistors and also have smaller internal swing
v. Custom design can work for reduction of power consumption at an expense of increased cost
vi. Supply voltage reduction in non-critical paths and appropriate transistor sizing;
vii. Multi-threshold logic can be used

4.2.3 Lowering Power Consumption Through Architectural Design

The following strategies can be used for architecture-level design

i. By shutting down the unused blocks
ii. Using parallelism and pipelining for low power architectures
iii. Partitioning of memory with specifically enabled blocks
iv. By reducing count of global buses

4.2.4 Choosing an Algorithm to Cut Down Power Consumption

Power consumption can be minimized at algorithm level also through the following ways:

i. Number of hardware can be minimized by minimizing the number of operations
ii. Data coding can be done for less number of switching activity

The entire process of power optimization also takes into consideration the system level. Several methods include:

- Use a system with slow clocks. On-chip phase locked loops are used to create higher frequencies and
- Extremely high integration. Integrate other ICs, such as digital and analogue peripherals, and off-chip memory (ROM, RAM, etc.).

Classification of digital CMOS circuits, based on their operation can be done as static, and dynamic circuits. Static complementary metal oxide semiconductor design is improved with respect to energy, and robustness, but it has a higher delay so this is inefficient for use in critical paths and substantial designs. Dynamic logics are having smaller delay but are more energy consuming and are less robust imposing a lot of problem to the designer. So, an efficient dynamic logic design style is required that utilizes the high-speed property of their family with an improvement in robustness in performance [6]. However, current corrective methods increase the performance of dynamic circuits at a significant expense to one or more other crucial design criteria like power-consumption, speed and area etc. The fact is that when the criteria for low power and noise tolerance are raised coupled with ongoing process technology downscaling, the amount of overhead increases significantly. The benefit of dynamic logic and whether it will contrast with existing techniques in which convincing circuits are created using static logic are two general questions related to it. To bring more testimony the first trial to answer the question. For instance, the first Alpha chip [7] had several dynamic latches and few static gates were replaced by dynamic logic gates. The development releases [8] [9] shows that the results obtained by using dynamic logic were used as a part of design strategy from the very beginning are far better when compared to the performance of designs using only static logic.

Another way to entrust that the dynamic logic has acquired from a structural viewpoint. For example, the circuit if made from only static logic gates, will become one viable design within the search space of this type of circuit's breed. To the contrary, the same circuit may acquire a possible solution from a different search space, if designed using dynamic logic. Both static and dynamic logic circuits can be merged by an overlapping set of rules, to enhance the province of possible solutions. Although it does not guarantee a superior design, this increases the chance of finding a standout design. Further, the proof is found from a measurement of the performance of absolute static and dynamic logic gates. Performance of individual gate depicts that the dynamic techniques are better performer as compared to static counterpart [10] but restricted to circumstances. However, a designer must have knowledge of the relationship between one type of gate with the other type of gates. The following promising characteristics of dynamic logic make them fascinating design option in addition to the static logic gate.

1. Pre-charged dynamic logic has only one logic evaluation block rather than two. Faster charging and discharging of node capacitances are allowed due to reduced load.
2. These are compact circuits, designed by combining complex latch and logic function [7][11]. Due to this, the data has to propagate through less number of stages.
3. Static logic gates may end up getting up to 60% transient short circuit current of total power [12]. This does not exist in dynamic logic circuits.
4. Complex logic can be included within the dynamic latch; this minimizes the number of devices, and hence reduced the dynamic power-consumption.
5. Dynamic latches are less complex and use a lesser number of devices thana conventional design. Single voltage transition is propagated during evaluation, the power consuming glitches are reduced [12].

4.3 Dynamic CMOS Logic Design Overview

High capacitive load, extensive interconnects, and complex chip clock distribution presented some compelling challenges for high-speed digital CMOS in the early 1980s. Due to the sluggish transitions and clock skew produced by this, it is frequently challenging to distribute several high-speed clock phases evenly throughout the circuit. On the other hand, dynamic logic allowed for simpler, faster, and smaller circuits, which was already known. There are various dynamic logic design approaches, each with pros and cons.

Clocked CMOS (C2MOS) logic design style was proposed in 1973 [13]. In this, more complex latches were replaced by four transistors dynamic implementation. Nevertheless, it required two non-overlapping phases of clock. So, the clock generation and application had to manage clock skew along with the organization of loss of timing generated because of non-overlapping intervals. Henceforth, a single-phase-clock approach was presented by merging with the dynamic latches and the logic being realized by N-metal oxide field effect transistor or P-metal oxide field effect transistor devices at alternate stages. But this design had race condition impediment during the evaluation of second stage, while the first stage pre-charge. The race conditions can result in a spurious intermediate logic level. For resolving this problem either the clock at the second stage may be delayed or an inverter may be placed at the output of each stage.

Domino logic [14] was proposed to overcome the issue of race condition. It uses single-phase-clock and is a non-inverting circuit, restricting some logical functions. The static inverter placed at the output stage consumes power while perceiving no particularly useful function. Later on, NORA technique was put forward by including the dynamic logic within the inverter [15] but at the cost of two-clock-phases. Both Domino and NORA designs faced the charge sharing problem. The problem is less significant in Domino due to the makeover contributed by the inverter. Complementing this, the C2MOSis out of charge sharing issue. Advantages of using Single phase clocking scheme made it more winning over two phase or multiphase clocking schemes. Dynamic Single-Phase Clock Logic Dynamic CMOS logic gates cannot be cascaded easily in conventional non-inverting single-phase clock dynamic logic due to a cascading problem that prevents undesirable charge leakage from pre-charged nodes. To overcome this issue a combination of dynamic logic with a static complementary metal oxide semiconductor inverter at the second stage is used and this design was named as Domino logic style. However, this design style was having a high delay due to static inverter used at second stage.

Later, an alternative concept known as True Single-Phase Clock (TSPC) logic [18] emerged with promising qualities. It has a dynamic inverter at second stage rather than a static inverter (as in case of Domino logic) [19]. The Alpha microprocessor from Digital Equipment Corporation reportedly uses TSPC logic [7]. Further, the idea was used in design of an accumulator its potential was demonstrated in a design of prescaler. Domino and the TSPC logic gates stand in the class of circuits identified as dynamic logic styles using single phase clock scheme. At times, such techniques are reported as an equivalent solution to one another. The complexity of dynamic logic is accessed proficiently by several techniques.

The above-mentioned group of circuits comprise several techniques and ideas underlying in relation to these gates and in multiple cases, their fundamental concepts overlap. In fact, enhanced gates and circuit design methodologies make it possible to gain the long-lasting knowledge needed to create a dynamic logic circuit.

4.4 Key Issues and Challenges

Dynamic logic style retaliates into high performance systems as compared to static logic style. At the same time, it suffers from following issues and challenges that may deteriorate its performance. For a candid performance, these should be taken care of.

1. If the clock period is too lengthy, transistor leakages from source-drain junctions and subthreshold conduction contaminate the stored states.
At high temperatures, this problem often gets worse, necessitating careful simulations.
2. Noise immunity in dynamic circuits is reduced.
3. Use of single-phase clock may cause race condition, i.e., when the first stage precharges, the second stage starts to evaluate. Thereby, creating glitches at few nodes due to a slow clock. This issue can be resolved by taking up the rise-time and fall-time of clock signal as three to five times smaller than that of the propagation-delay of the circuit [16]. In bigger circuits, we can easily fulfil this condition but for smaller circuits we need to take proper care of delay, frequency, etc. to wave off the race condition.
4. Dynamic circuits are more prone to process and environment-related variations such as temperature, voltage, etc.

4.5 State of the art

Low-power consumption and low noise are the chief constituents of performance characterising parameters for any dynamic circuit. Even yet, the evolution of CMOS only gave high performance systems a temporary reprieve in terms of power-consumption. Device density on an IC with high frequencies is being enhanced in consideration to satisfy the market demands for high performance systems. Due to the constant rise in power-consumption and power density, controlling the on-die temperature for optimum IC operation is getting more challenging. Dynamic power dissipation dominates the overall power-consumption of high-performance ICs while they are operating in active mode. With the development of technology, naturally there is an increase in leakage power. Further, the growth of CMOS technology, leakage has moreover come to light as a fresh and significant cause of on-chip noise. Dynamic complementary metal oxide semiconductor circuits are less noise indulgent due to their high-power consumption and noise.

Over the past 20 years, numerous design methodologies have been created in an effort to strengthen this weak point. However, current corrective methods increase the performance of dynamic circuits at a significant expense to one or more other crucial design criteria like area, speed, and power dissipation, etc. Matter is this, when the criteria for low power and noise tolerance are raised coupled with ongoing process technology downscaling, the amount of overhead increases significantly. Efficient low-power, low noise circuit design strategies that have small impact on silicon area and speed of the circuit are therefore in high demand.

In past years researchers have tried to improve these factors by developing different techniques. In this section significant single-phase clock dynamic circuit design techniques are discussed. Specifically, some useful Domino logic techniques and substantial TSPC logic techniques are discussed. All previous reported research works are divided into three categories based on the following performance parameters:

1. Power dissipation, delay and Power delay product (PDP).
2. Noise minimization/ Noise tolerance.
3. Robustness in terms of parameter variations such as temperature, voltage, size and process corners.

PMOS transistor named as keeper is placed between the supply voltage and switching node was the basic design to change the Domino technique as suggested [14] and shown in Figure 4.1. The transistor's size was minimised, and the gate is always grounded. Its function is to maintain the logic high voltage at dynamic node as and when required. Noise tolerance of the circuit is improved by this technique. Feedback keeper method for circuit designing is presented in [17]. As compared to previous modification this technique was capable of reducing DC power consumption due to its always ON keeper transistor.

Without employing an inverted clock, two different single phase clock dynamic circuit design methods are presented [18] and the second stage of this circuit is a dynamic inverter. Limitation of first design is charge redistribution effect that may result in wrong output logic that is removed in second design by making changes so that the input signal changes occur only when N block evaluates. It is depicted in Figure 4.2, where a dynamic stage is followed by another dynamic latch. In [19], the experimental performance of previous theoretical work is presented. Using this technique, speed can be increased with simple clock distribution. The single-phase clock is unaffected by rise and fall time of the clock.

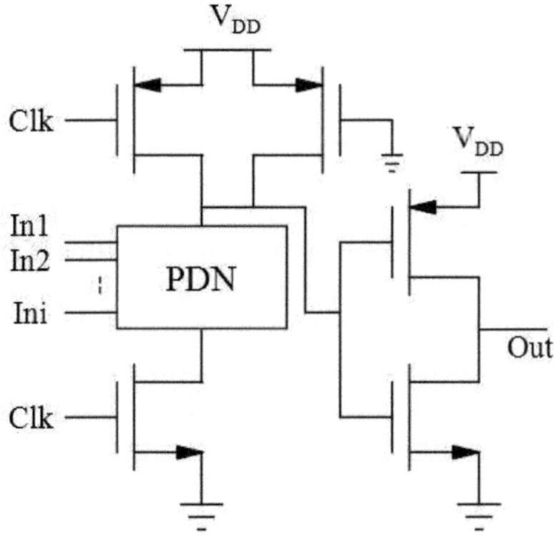

***Figure 4.1** Domino logic with Keeper Transistor (DKT) [14].*

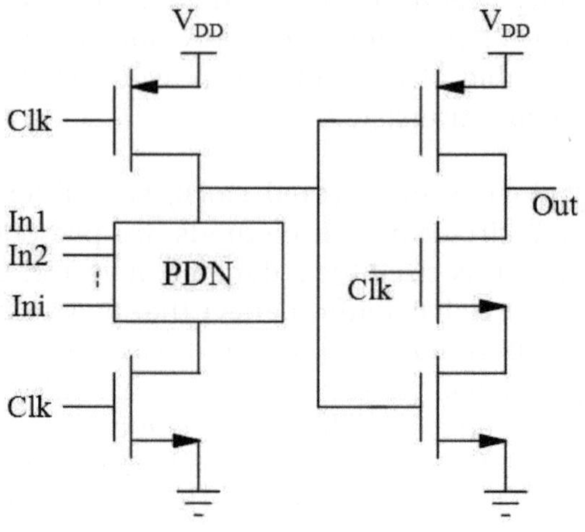

Figure 4.2 *TSPC logic [18].*

Previously proposed design [18] was modified by connecting source of lower NMOS of latch to the drain of evaluation transistor [11]. Therefore, the latch is not directly connected to ground. Performance of the circuit is increased as compared to the previous design. Further, a pipeline encoder was designed that provides each comparator with a load capacitance not higher than that of a unit inverter [20]. Error possibility of most significant bit in this design is less than its least significant bit. A simple design with high speed is achieved with this technique a consolidated true single-phase clocking system is presented in [21]. Signals are divided into two different classes; one is data signal and another is clock signal. Clock signal has a restriction on its reference edge, it should occur only when specified. But the gating of clock signal results in clock skew problem.

A gate array architecture for pipelined circuits is presented in [22]. TSPC technique is used to design the core cells to remove clock skew problem and parallel connected transistors are used for fast operation. Frequency of operation is 200 MHz. Circuits designed using this technique are faster, smaller in size but consume more power. The number of transistors in an ICpackage is restricted by power dissipation and clock frequency limits of the design. TSPC full adder circuit, having six clock transistors is designed [23]. It consumes less power and occupies smaller area. Improvement is attained by designing a half cycle full-adder and avoiding pre-charge transistors.

Buffered single phase clock logic was introduced in [24]. In this technique only N blocks are used and clock distribution is done by straight line method. Accordingly, data flow clock distribution may be in forward or backward directions. This removes the requirement of high delay PMOS transistors. This design style can be used for high speed circuits.

The back-gate bias technique is presented in [25]. It is integrated with width/length ration of MOSFET. A Non-full voltage Swing TSPClogic circuit for dynamic design was introduced [26]. This technique works for low voltage and high-speed applications. Two additional transistors are required for implementation. This technique results in faster operation but consumes more power than conventional counterpart. Consequently, an overall improvement in PDP is obtained with enhance operating frequency. Moreover, low power, high speed non full voltage swing carry look ahead adder is analysed in [27]. It utilizes a dynamic logic with dynamic latch. Due to low swing at internal nodes power dissipation is reduced and speed is improved. A noise tolerant technique is proposed by utilizing an NMOS transistor and a delay element [28]. This technique makes the circuit more noise immune at an expense of increased power consumption, delay and increase area due to delay circuitry. The delay circuitry can be used for simultaneously for several stages. To illustrate the value of the back-gate forward substrate bias approach in N-well complementary metal oxide semiconductor technology, a fast 4-bit ALU has been developed for 1V operation. Four arithmetic and four logical operations can be carried out by the 4-bit ALU, which uses a ripple-carry-adder [29]. Concept of disconnecting the keeper for short time during switching is used in this technique. Researchers have proposed a variable threshold keeper design that reduces power consumption and delay [30]. Threshold voltage of keeper transistor is kept variable to control the current drive of the transistor. This technique is depicted in Figure 4.3. Contention current is reduced, whereas the noise-immunity is not altered. Dynamic body bias generator produces two different voltage levels at its output. This technique produced a high speed, low power consuming dynamic circuit. Use of body bias generator makes the size too large that increases area. Authors show two innovative dynamic body-bias Schmitt trigger circuits for low voltage applications. The first circuit is intended to run at one volt. The second circuit, which is a derivative of the first circuit, is intended for use at 0.4 V [31].

In [32] forward body biased keeper transistor is used to enhance noise-immunity of Domino-logic circuit. Current drive of keeper transistor is increased without changing its size. Enhancement in noise immunity is greatly dependent on temperature and the power consumption is increased in this design. A diode connected NMOS is used for leakage reduction in domino circuits [33]. Implementation of diode transistor reduces leakage but increases delay. To overcome this delay degradation a mirror transistor is paced in evaluation network as shown in Figure 4.4. So that the evaluation current is increased and delay is reduced. Dynamic comparator and multiplexers are designed using this technique and found to consume less power and are immune to noise. One NMOS in pull up network and one PMOS in pull down network is used to reduce the voltage swing [34]. At the dynamic node both lower and upper range of swing are modified. For high-speed operation full voltage swing is maintained at the input and output nodes. Power delay product is reduced along with the improvement in noise immunity.

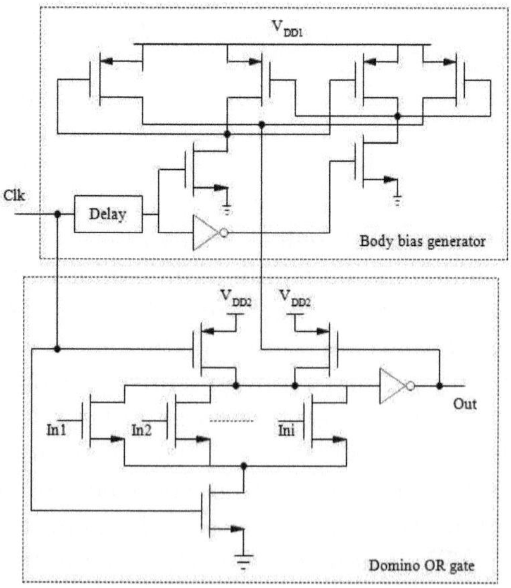

Figure 4.3 *Domino OR gate with body biased keeper [31]*

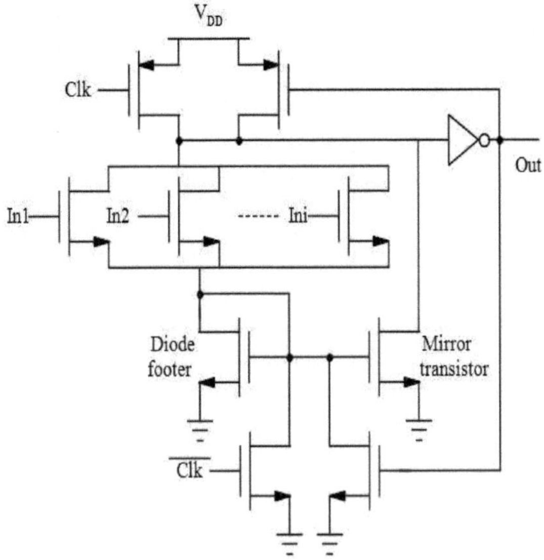

Figure 4.4 *Wide OR gate with diode footed technique [34].*

Body biasing with true single-phase logic is utilized, to have the advantages of both [35] design techniques. Transistors are kept at high threshold during OFF state to reduce leakage power consumption. Power is further reduced by charge recovery technique of substrate. Using this technique power consumption is reduced and the speed is enhanced. This design technique is shown in Figure 4.5. The authors of [36] have suggested an approach for improving noise tolerance. This method is equally effective for TSPC and Domino logic design approaches. In Figure 4.6, it is shown. The benefit is keeping the input capacitive load constant. The trade-off between performance and noise immunity is calculated using transparency window width. The approach is applied to a one bit carry look ahead complete adder circuit in addition to the gate, and it is tested at a clock frequency of 200 MHz. Two strategies used for improvement are precharging of internal nodes and isolation of inputs from dynamic nodes in pull down network. Noise metric analysis is done using unity noise gain. Dynamic circuit design technique for low power consumption is introduced [37]. For case study booth multiplier circuit is taken as reference circuit. In this technique, pull down network (pull up network) of gates are slitted into evaluation sub networks. This reduces the number of series connected devices in pull down network (pull up network) thus the gate capacitance is minimized. Consequently, energy dissipation and delay are reduced. Domino logic is designed using TSPC logic design style for low power applications [38]. Number of clock transistors are reduced from three to two, as a result of this clock load capacitance is reduced, and the power consumption is reduced [57-61].

A twist coupled pair of N-channel MOS and P-channel MOS is used to create small swing dynamic logic [39]. As shown in Figure 4.7, there is one pair in individual path between the supply node and the ground node. The dimensions of the additional transistors and output capacitance value can be adjusted to change the output's swing range.

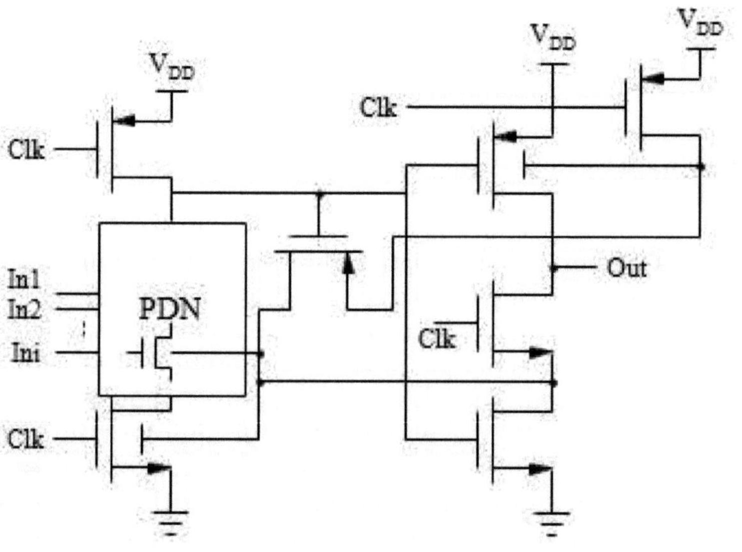

Figure 4.5 *Body biased TSPC design technique [35].*

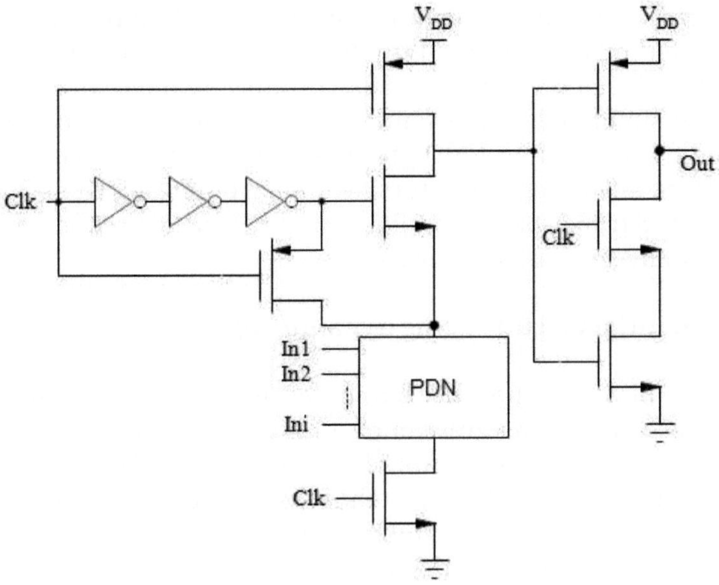

Figure 4.6 *Noise reducing Dynamic circuit [38].*

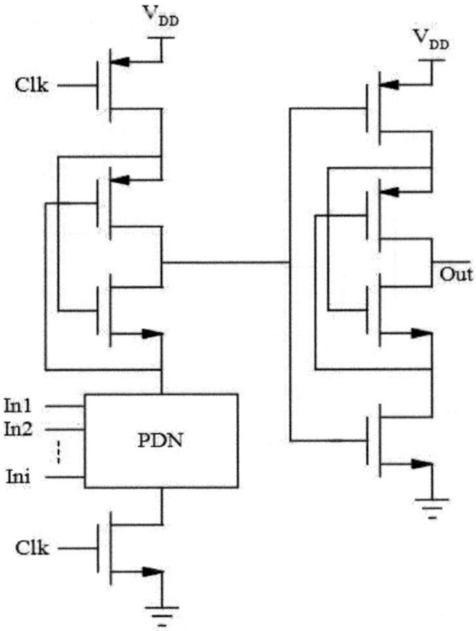

Figure 4.7 *Small Swing Technique (SST).*

By altering the typical Domino logic design style, a dynamic circuit design method is suggested [40]. In Figure 4.8, this design aesthetic is displayed. For a low power delay product, two possible modifications are simulated and examined [62-69]. One alteration involves moving the evaluation NMOS transistor in the typical circuit without adding any more transistors, while the other involves adding an extra NMOS transistor in parallel with the logic transistor. Later one does not have evaluation transistor.

4.6 Sequential Circuits using dynamic logic

Flip flops play a significant role in designing of sequential circuits [41]. These are having great impact on various parameters such as speed and power consumption. Flip-flops can be dynamic or static [42]. The two inverter-based latches that serve as the foundation of static FF typically have a significant influence on area, and latency. However, compared to static FF, those designed with dynamic logic have superiority in terms of area and smaller operating delay [70-75]. The majority of FF need both the clock signal and its reversal, which makes clock tree synthesis difficult. A single clock signal is all that is required for the dynamic flipflop known as the TSPC flipflop.

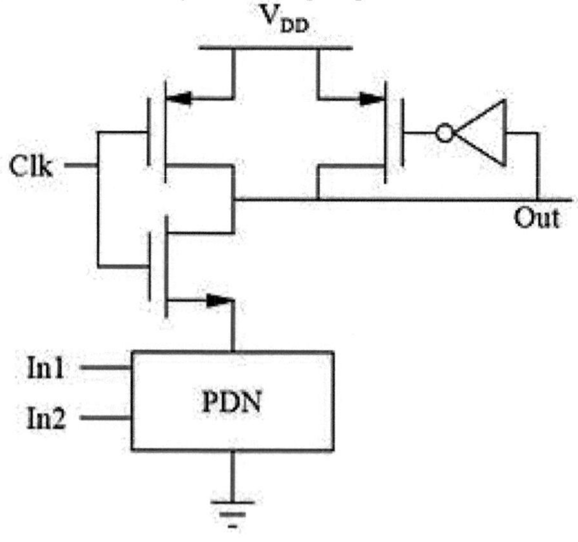

Figure 4.8 *Up-Footed Technique (UFT).*

Figure 4.9 shows the conventional TSPC flip flop. Without a complementing clock signal, this flip flop (FF) can only be driven by a single-phase clock signal. As a result, it operates quickly and has a basic architecture. When the clock signal is high and the signal D is low, node an is in the floating state. Dynamic logic and a single-phase clock enable not just a more compressed design but also a quicker reaction [76-81]. Moreover, the TSPC flip-flop has the unique ability to reduce set-up time and delay overall by including logic gates into the design [42]. The clock system alone accounts for 20% to 40% of the integrated circuit's overall power consumption in digital hardware circuits (IC) [43]. The traditional flip-flop is the Set Reset flip-flop (SRFF), and the most extensively used industry standard flip-flop is the TGFF [44], as seen in Figure 4.10.

TGFF contains twelve transistors and uses a clock signal as an input, more power is lost. It is fairly difficult to directly lower transistor counts in this circuit that are related to the clock. One reason is that the clock driver cannot be removed since TG requires a dual-phase clock signal. To prevent the deterioration of data transfer characteristics brought on by the use of single-channel MOS, P-MOS or and N-MOS should both be used to make transmission gates.

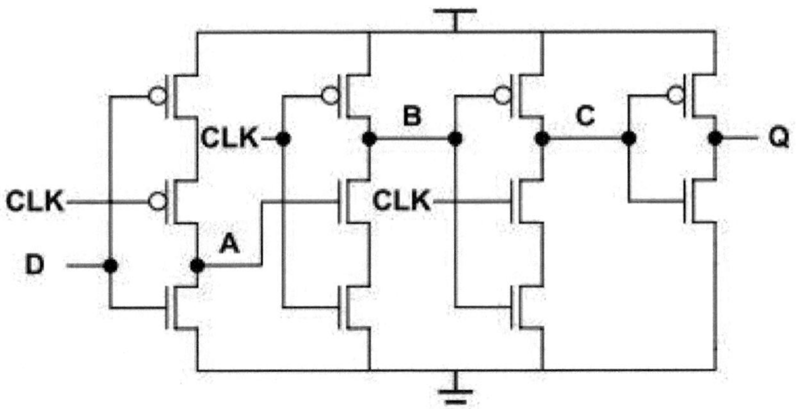

Figure 4.9 *Conventional TSPC flip flop.*

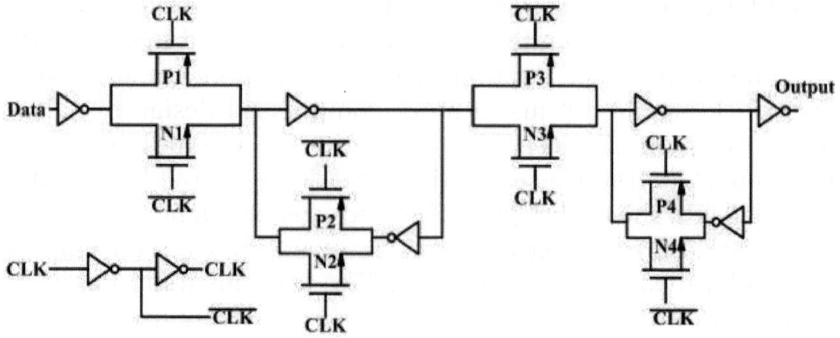

Figure 4.10 *CMOS circuit of TGFF [44].*

TGFF devices show worst power dissipation performance due to a larger clock tree loading than other true-single-phase clocking-based FF systems. There is more static power utilisation because TGFF switches at CLK and inverted CLK even when data is stable [82-87].

Clock overloading is the main issue with these designs [45] [46]. CLK and complementary CLK, two clock signals make up these flip-flop logic structures. These signals increase the number of transistors in addition to the stress placed on the clock signal. Due to the additional circuitry needed to generate the CLK and inverter CLK signals, in the overall design structure [41],[47]. As a result, even with minimal data switching activity, dynamic power losses significantly increased. These days, the VLSI industry's top concerns are both production costs and chip size. Every industry in the digital age needs a gadget that is affordable and small. Therefore, reducing the amount of silicon space occupied by CPUs is an important factor. To reduce the additional strain on the clock signal, flip-flop designs consuming less power are employed. In order to reduce clock burden, the flip-flop design using known as TSPC logic has been created [48][49][50]. The desire for innovative flip-flop designs that function at lower voltage rises as a result of technology scaling [42], [51].

Both the popular TGFF design and the traditional SR flip-flop include further circuitry for producing the clock signal [54]. Due to the fact that just one clock signal is used for conduction of twelve transistors, the TGFF experiences increased clock loading issues. A significant increase in power usage is the result of clock overloading. In contrast to the traditional SRFF design, the ACFF design uses a differential latch to obtain true single phase clock operation as shown in Figure 4.11. Two-level restoring (LR) circuits and pass transistor take the place of transmission gate. Each LR circuit consist of aNMOS and a PMOS transistors connected in parallel. Out of the used 22 transistors, the clock drives just 4 of them [49].

The ACFF slave latch has a problem with data conflicts. It declines even faster when switching activity rises. We can decrease the clock loading by using a multiplexer to replace the master of the SRFF, which results in a TSPC FF design [41]. Due to the employment of a single clock signal in a topologically compressed system, power dissipation is decreased [88-93]. In comparison to SRFF, the transistor count was also decreased by the TCFF.

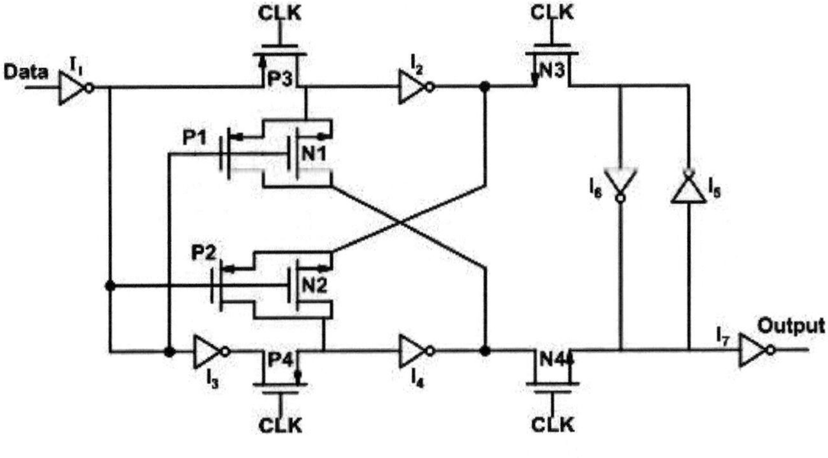

Figure 4.11 *ACFF CMOS design [49].*

The transistor having clock signal as input is further reduced to three by a topologically compressed flip-flop, bringing the total number of transistors down to 21 in Figure 4.12. It also succeeds in producing a slave who is free from conflict. In essence, the TCFF is developed from the SR- latch [94-98]. However, TCFF has a clear latch when the clock is 0, making it unsuitable for high-frequency operation. If the voltage level at the source terminal of transistor P2 (drawn up by stacked devices) is less than VDD in TCFF, transistor P2's pull-up effort will be reduced. Only 19 transistors are used in logic reduction flip-flops, but they have higher power losses. The complementary pass transistor logic (CPL) design of the LRFF, an improved version of the TCFF, is used. This way the slave's TCFF discharging path is divided in two different paths [52].

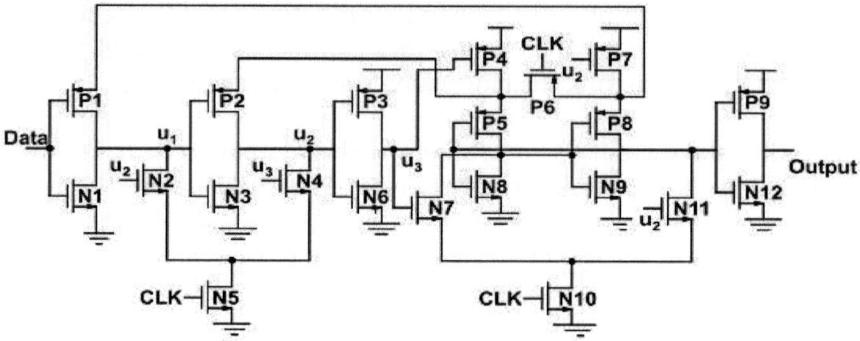

Figure 4.12 *TCFF circuit using CMOS [41].*

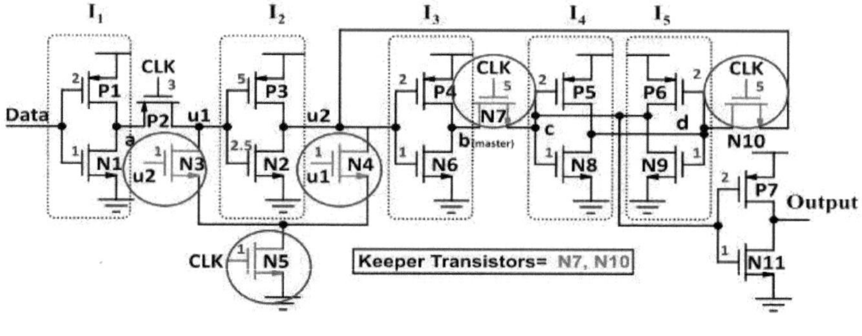

Figure 4.13 *ADTFF flip flop circuit [55].*

The CPL structure comprised of a pass transistor with node 2/node 3 acting like a sink and clock signal is acting as an input to the gate terminal. Two charging paths with a CPL structure are present on the slave side and comprise of two pass transistors. The nodes 4 and 5 are charged to logic high "1" with its assistance. A conditional clock approach (CC-LCFF) single edge implicit pulse-triggered level-converting flip flop design is presented. This is appropriate for dual-VDD low-power, non-critical paths [53]. Flip flop circuit consist of master and slave circuit [55]. These are having seven P-MOS and eleven N-MOS devices in total. Fewer PMOS transistors are used to attain the reduced complexity [99-103]. This results in a flip-flop that is quick and space-efficient. Pass transistor logic helps to decrease power usage, save space, and speed up operation. Another flip flop designed with eighteen transistors is presented in [56] and the leakage power consumption is greatly reduced.

4.7 Proposed Dynamic Flip Flop Circuit

Different latches are used in designing of Master and slave of the proposed flip flop circuit. A total of 23 MOS devices are used. Out of these 23 transistors 12 are the PMOS transistors and 11 are NMOS transistors. Use of single clock in this flip-flop removes the clock overloading issues. This flip flop aims to reduce power consumption and also to minimize clock-to-Q-delay and area. Proposed flip-flop circuit is depicted in the Figure 4.14. Working of the implemented flip-flop is as given below:

High-to-Low Transition:

When the clock is equal to zero, charging of node n1 is done through P1 and P2 to supply voltage, node X2 is charged through P7 and P5uptosupply voltage, the node X3 is discharged through N11 and N12 and the output will be high [104-107]. When the clock will be high, node X2 will be discharged through N4 and N5 to the ground, and then the transistor N1 will be turned off which isolates the flip-flop if there are any changes in the input data. Node X1 will be high through P3, while node X2 remains discharged through N4 and N5 during the rising edge of the clock.

Low-to-High Transition:

When the clock is equal to high, node X2 is charged through P6 and P5 up to the supply voltage, node X1 will be discharged through N1 and N2 to the ground, the node X3 will be charged through P11 and P12 up to supply voltage, and the output goes to discharged. At the positive edge of the clock cycle, the input data gets isolated from node n1 through transistor P2, and the node X3 keeps low through transistors N7 and N8, then the output (Vout/Q) will be equal to one. The voltage of node X1 is discharged to the ground through transistors N3 and N2, while node X2 goes high through transistor P4 during the rising edge of the clock cycle.

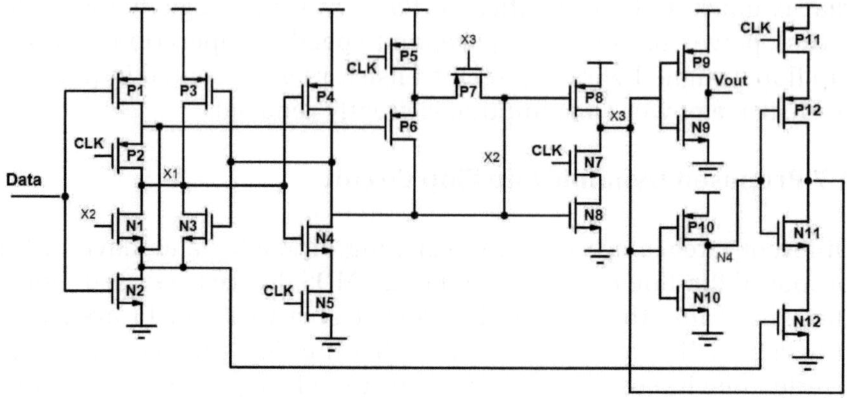

***Figure 4.14** Proposed Flip Flop Circuit.*

Transient response for both high to low and low to high transition of the proposed Flip flop is shown in Figure 4.15. Four data latching conditions are shown in following figures. Figure 4.16 is representing the state of transistors when clock signal is at low and the data input is zero. In this case node X1 and X2 are at logic high while X3 is at logic low [108-112]. Second condition is depicted in Figure 4.17 where clock is at logic high, and data is still at logic low. Here, node X2 goes at logic low while X3 is reaching to logic high. At the same time when the clock is at zero and the input is at logic one, the node X1 goes to logic low through transistors N1 and N2 as depicted in Figure 4.18. When clock is at logic one and data is also at logic one. This condition of Data and Clock inputs is in Figure 4.19.

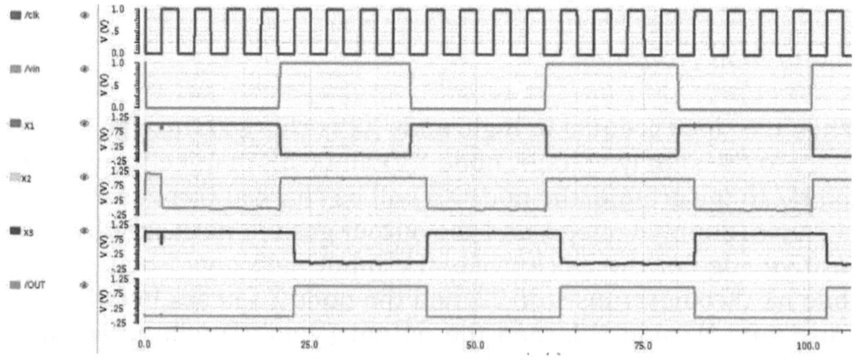

***Figure 4.15** Transient response of Proposed Flip flop design.*

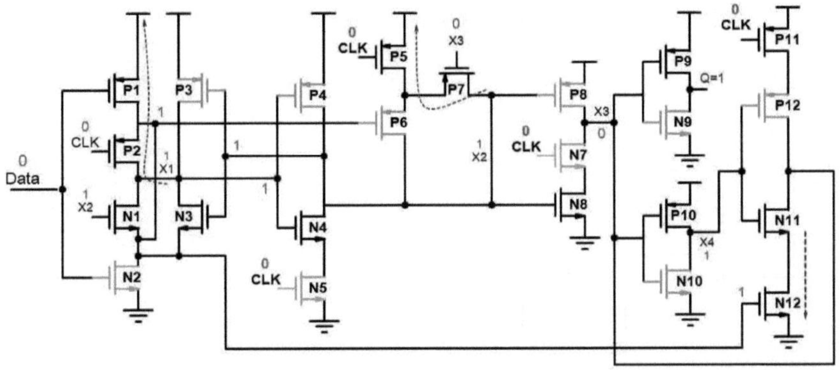

Figure 4.16 *State of circuit for CLK = 0 and DATA = 0.*

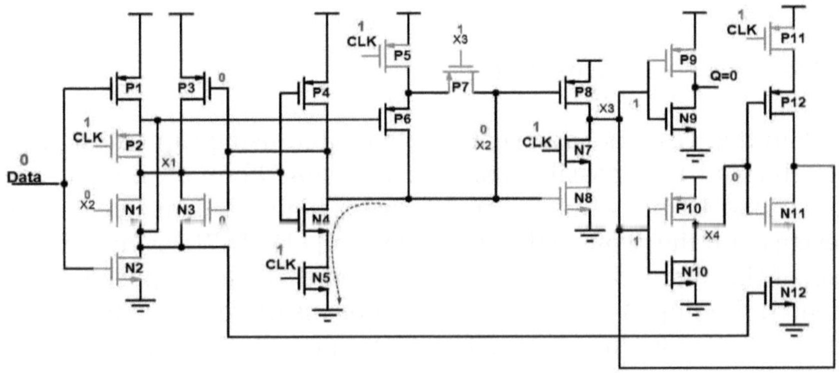

Figure 4.17 *State of circuit for CLK = 1 and DATA = 0.*

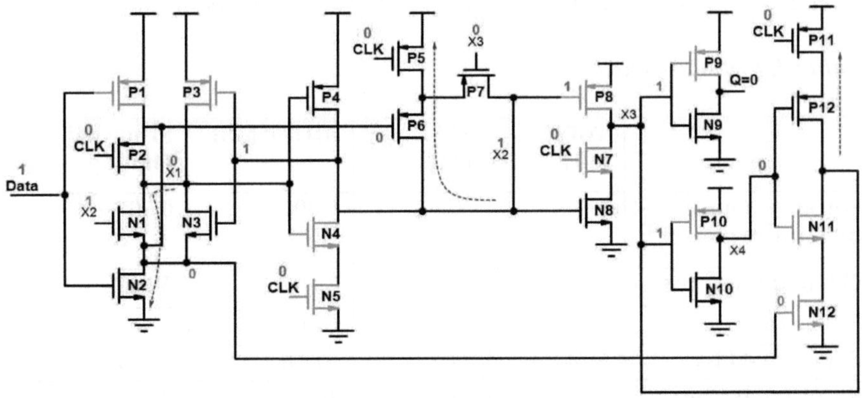

Figure 4.18 *State of circuit for CLK = 0 and DATA = 1.*

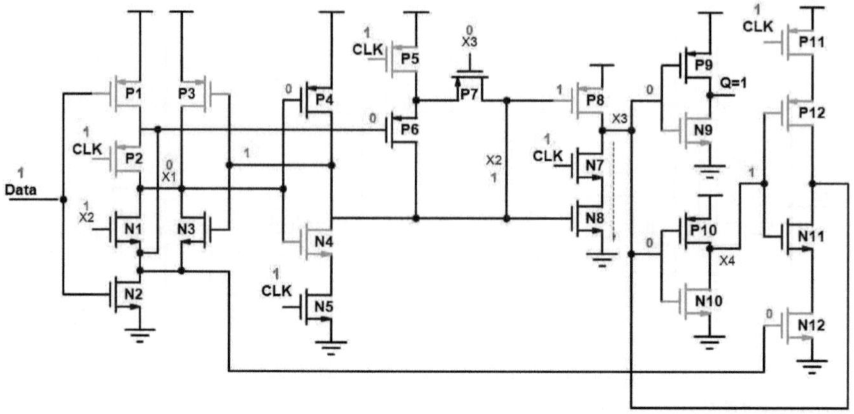

Figure 4.19 *State of circuit for CLK = 1 and DATA = 1.*

4.8 Results and Discussion

All Flip flop circuits have been implemented using same design parameters for best possible comparison using 90nm design parameters. Table 4.1 presents design metric for high performance flip flop designs and the proposed flip flop design [113-117]. From the analysis of the simulation data, it is observed that the proposed implemented flip flop is performing well in terms of power consumption as well as delay. Up to 56.58% reduction is obtained in power consumption along with up to 22.9% reduction in clock to Q delay. PDP and EDP are minimum with the proposed design. Comparative analysis is graphically represented in Figure 4.20 using bar chart.

Table 4.1 *Design metric analysis*

	TGFF [44]	**TCFF [41]**	**ADTFF [55]**	**Proposed Flipflop**
Power Consumption (µwatt)	1.398	0.663	0.895	0.607
CLK to Q Delay (ps)	64.64	63.99	75.62	58.61
PDP (x10^{-18}) J	90.37	42.4	67.68	35.57
EDP (x10^{-28}) J-s	58.41	27.13	51.18	20.84
No. of Transistors	24	21	18	23

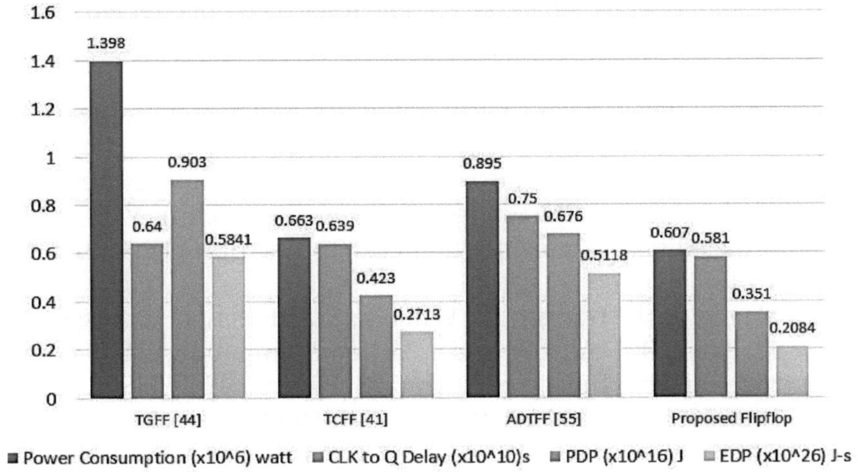

Figure 4.20 *Graphical representation of delay, power consumption and PDP for different design techniques.*

4.9 Conclusion

This Chapter discusses in detail about the low power circuits and systems. It primarily focuses on dynamic complementary metal oxide semiconductor logic design for low-power and high-speed electronic gadgets. Detailed study of combinational and sequential logic is presented. Further, a low power dynamic flip flop is proposed that requires small area. Through the experimental results, we noticed that the implemented design achieves the reduction in power consumption and Clock-to-Q delay is also less when comparing with TGFF and TCFF.

References

1. Weste HE, Weste Neil. CMOS VLSI design: a circuits and systems perspective, 3/E. Pearson Education India, 2006.
2. Chandrakasan, Anantha P., and Robert W. Brodersen "Minimizing power consumption in digital CMOS circuits." Proceedings of the IEEE 83, no. 4 (1995): 498-523.

3. Chandrakasan, Anantha P., and Robert W. Brodersen "Minimizing power consumption in digital CMOS circuits." Proceedings of the IEEE 83, no. 4 (1995): 498-523.
4. Kang, Sung-Mo, and Yusuf Leblebici. CMOS digital integrated circuits. Tata McGraw-Hill Education, 2003.
5. Athas, William C., and NestorasTzartzanis "Energy recovery for low-power CMOS." In Advanced Research in VLSI, 1995. Proceedings., Sixteenth Conference on, pp. 415-429. IEEE, 1995.
6. Slager, J., K. Gudger, T. Williams, L. Heller, C. Svensson, L. Madden, D. J. Allstot, and J. Yetter "Competing CMOS circuit techniques." Digest of Technical Papers. 40th ISSCC., 1993 IEEE International, pp. 80-81. IEEE, 1993.
7. Dobberpuhl, Daniel W., Richard T. Witek, Randy Allmon, Robert Anglin, David Bertucci, Sharon Britton, Linda Chao et al. "A 200-MHz 64-bit Dualissue CMOS Microprocessor." Digital Technical Journal 4, no. 4 (1992).
8. Silberman, Joel, Naoaki Aoki, David Boerstler, Jeffrey L. Burns, Sang Dhong, Axel Essbaum, Uttam Ghoshal et al. "A 1.0-GHz single-issue 64-bit PowerPC integer processor." IEEE Journal of Solid-State Circuits 33, no. 11 (1998): 1600-1608.
9. Fair, H., and D. Bailey "Clocking design and analysis for a 600 MHz alpha microprocessor." In Solid-State Circuits Conference, 1998. Digest of Technical Papers. 1998 IEEE International, pp. 398-399. IEEE, 1998.
10. Larsson, Patrik, and Christer Svensson "Noise in digital dynamic CMOS circuits." Pulse 20 (1994): 10.
11. Jiren, Yuan and Christer Svensson "High-speed CMOS circuit technique." IEEE Journal of Solid-State Circuits 24, no. 1 (1989): 62-70.
12. Gary, Y. E. A. P. "Practical low power digital VLSI design." Kluwer Academic Publishers 4 (1998): 103.
13. Y. Suzuki 1973 Yasoji, KaichiroOdagawa, and Toshio Abe "Clocked CMOS calculator circuitry." IEEE Journal of Solid-State Circuits 8, no. 6 (1973): 462-469.
14. Krambeck R. H., Charles M. Lee, and H-FS Law "High-speed compact circuits with CMOS." IEEE Journal of Solid-State Circuits 17, no. 3 (1982): 614-619
15. Goncalves Nelson F., and H. De Man "NORA: A racefree dynamic CMOS technique for pipelined logic structures." IEEE Journal of Solid-State Circuits 18, no. 3 (1983): 261-266.

16. R. Rogenmoser, N. Felber, Q. Huang, and W. Fichtner, "1.16 GHz dual-modulus 1.2 nm CMOS prescaler," in Proc. IEEE Custom Integrated Circuits Conf., 1993, pp. 27.6.1–27.6.4.
17. OklobdzijaVojin G., and Robert Kevin Montoye "Design-performance trade-offs in CMOS-domino logic." IEEE journal of solid-state circuits 21, no. 2 (1986): 304-306.
18. Ji-Ren, Yuan, Ingemar Karlsson, and Christer Svensson "A true single-phaseclock dynamic CMOS circuit technique." IEEE Journal of Solid-State Circuits 22, no. 5 (1987): 899-901.
19. Karlsson, Ingemar "True single-phase clock dynamic CMOS circuit technique." In Circuits and Systems, 1988., IEEE International Symposium on, pp. 475-478. IEEE, 1988.
20. Yuan, Jiren, C. Svensson, and K. Chen. "High speed pipelined CMOS encoder for flash A/D converters." In Circuits and Systems, 1989., IEEE International Symposium on, pp. 2088-2091. IEEE, 1989.
21. Afghahi, Morteza, and Christer Svensson "A unified single-phase clocking scheme for VLSI systems." IEEE Journal of Solid-State Circuits 25, no. 1 (1990): 225-233.
22. Ismail Enisungan and Murat Askar, "A gate array architecture for pipeline structure", Elektonik,Vol 1,No 2, 1993.
23. Schindler, Volker "A low-power true single phase clocked (TSPC) full-adder." In Proceedings of the 22nd ESSCIRC, pp. 72-75. 1996.
24. Seokjin Kim and Ramalingam Sridhar "Buffered single-phase clocked logic for high-speed CMOS pipelined circuits." In Circuits and Systems, 1997. ISCAS'97., Proceedings of 1997 IEEE International Symposium on, vol. 3, pp. 1900-1903. IEEE, 1997.
25. A Srivastava, "Back gate bias method of threshold voltage control for the design of low voltage CMOS ternary logic circuits." Microelectronics Reliability 40, no. 12 (2000): 2107-2110.
26. Cheng, Kuo-Hsing, and Yung-Chong Huang "The non-full voltage swing TSPC (NSTSPC) logic design." In ASICs, 2000. AP-ASIC 2000. Proceedings of the Second IEEE Asia Pacific Conference on, pp. 37-40. IEEE, 2000.
27. Cheng, Kuo-Hsing, Wen-Shiuan Lee, and Yung-Chong Huang "A 1.2 V 500 MHz 32-bit carry-lookahead adder." In Electronics, Circuits and Systems, 2001. ICECS 2001. The 8th IEEE International Conference on, vol. 2, pp. 765-768. IEEE, 2001.

28. Mendoza-Hernandez, F., M. Linarea, and V. H. Champac "Dynamic logic styles with improved noise-immunity." In Devices, Circuits and Systems, 2002. Proceedings of the Fourth IEEE International Caracas Conference on, pp. C031-C031. IEEE, 2002.
29. Srivastava, A., and D. Govindarajan. "A fast ALU design in CMOS for low voltage operation." VLSI Design 14, no. 4 (2002): 315-327.
30. Kursun, Volkan, and Eby G. Friedman "Domino logic with variable threshold voltage keeper." IEEE Transactions on Very Large-Scale Integration (VLSI) Systems 11, no. 6 (2003): 1080-1093.
31. Zhang, Mingyan, Jiangmin Gu, and Chip-Hong Chang "A novel hybrid pass logic with static CMOS output drive full-adder cell." In Circuits and Systems, 2003. ISCAS'03. Proceedings of the 2003 International Symposium on, vol. 5, pp. V-V. IEEE, 2003.
32. Kursun, Volkan, and Eby G. Friedman "Forward body biased keeper for enhanced noise immunity in domino logic circuits." In Circuits and Systems, 2004. ISCAS'04. Proceedings of the 2004 International Symposium on, vol. 2, pp. II-917. IEEE, 2004.
33. Mahmoodi-Meimand, Hamid, and Kaushik Roy "Diode-footed domino: A leakage-tolerant high fan-in dynamic circuit design style." IEEE Transactions on Circuits and Systems I: Regular Papers 51, no. 3 (2004): 495-503.
34. Liu, Zhiyu, and Volkan Kursun "Bidirectional dynamic node low voltage swing domino logic." In Circuits and Systems, 2005. 48th Midwest Symposium on, pp. 295-298. IEEE, 2005.
35. Wu, Ke, Song Jia, Zhongjian Chen, and Xuewen Gan "New dynamic threshold mos structures for low-energy true-single-phase-clocking circuits." In Electron Devices and Solid-State Circuits, 2005 IEEE Conference on, pp. 427-430. IEEE, 2005.
36. Mendoza-Hernandez, F., M. Linarea, and V. H. Champac "Dynamic logic styles with improved noise-immunity." In Devices, Circuits and Systems, 2002. Proceedings of the Fourth IEEE International Caracas Conference on, pp. C031-C031. IEEE, 2002.
37. Frustaci, Fabio, Marco Lanuzza, Paolo Zicari, Stefania Perri, and Pasquale Corsonello "Low-power split-path data-driven

dynamic logic." IET circuits, devices systems 3, no. 6 (2009): 303-312.

38. Fang Tang, Amine Bermak "Lower-power TSPC-based domino logic circuit design with 2/3 clock load." Energy Procedia 14 (2012): 1168-1174.
39. Ahn, Sang-Yun, and Kyoungrok Cho. "Small-swing domino logic based on twist-connected transistors." Electronics letters 50, no. 15 (2014): 1054-1056.
40. Xue, H., and S. Ren. "Low power-delay-product dynamic CMOS circuit design techniques." Electronics Letters 53, no. 5 (2017): 302-304.
41. N. Kawai et al., "A fully static topologically-compressed 21-transistor flip-flop with 75% power saving," IEEE J. Solid-State Circuits, vol. 49, no. 11, pp. 2526–2533, Nov. 2014.
42. N. Nedovic and V. G. Oklobdzija, "Hybrid latch flip-flop with improved power efficiency," in Proc. 13th Symp. Integr. Circuits Syst. Design, Manaus, Brazil, 2000, pp. 211–215.
43. J.-F. Lin, M.-H. Sheu, Y.-T. Hwang, C.-S. Wong, and M.-Y. Tsai, "Lowpower 19-transistor true single-phase clocking flip-flop design based on logic structure reduction schemes," IEEE Trans. Very Large Scale Integr. (VLSI) Syst., vol. 25, no. 11, pp. 3033–3044, Nov. 2017.
44. E. Consoli, G. Palumbo, and M. Pennisi, "Reconsidering high-speed design criteria for transmission-gate-based master–slave flip-flops," IEEE Trans. Very Large Scale Integr. (VLSI) Syst., vol. 20, no. 2, pp. 284–295, Feb. 2012.
45. H. Kawaguchi and T. Sakurai, "A reduced clock-swing flip-flop (RCSFF) for 63% power reduction," IEEE J. Solid-State Circuits, vol. 33, no. 5, pp. 807–811, May 1998.
46. C. K. Teh, T. Fujita, H. Hara, and M. Hamada, "A 77% energy-saving 22-transistor single-phase-clocking d-flip-flop with adaptive-coupling configuration in 40 nm CMOS," in Proc. IEEE Int. Solid-State Circuits Conf., San Francisco, CA, USA, 2011, pp. 338–340.
47. Y. Cai, A. Savanth, P. Prabhat, J. Myers, A. S. Weddell, and T. J. Kazmierski, "Ultra-low power 18-transistor fully static contentionfree single-phase clocked flip-flop in 65-nm CMOS," IEEE J. Solid-State Circuits, vol. 54, no. 2, pp. 550–559, Feb. 2019.
48. V. Stojanovic and V. G. Oklobdzija, "Comparative analysis of masterslave latches and flip-flops for high-performance and

low-power systems," IEEE J. Solid-State Circuits, vol. 34, no. 4, pp. 536–548, Apr. 1999.

49. J. Yuan and C. Svensson, "New single-clock CMOS latches and flipflops with improved speed and power savings," IEEE J. Solid-State Circuits, vol. 32, no. 1, pp. 62–69, Jan. 1997.

50. F. Stas and D. Bol, "A 0.4-v 0.66-fj/cycle retentive true-single-phaseclock 18t flip-flop in 28-nm fully-depleted SoI CMOS," IEEE Trans. Circuits Syst. I, Reg. Papers, vol. 65, no. 3, pp. 935–945, Mar. 2018.

51. U. Chopra, A. K. Mishra, and D. Vaithiyanathan, "Performance analysis of non-identical master slave flip flops at 65nm node," Int. J. Innovat. Technol. Exploring Eng., vol. 9, no. 1S, pp. 18–21, 2019.

52. K. Yamada, H. Maruoka, J. Furuta, and K. Kobayashi, "Radiationhardened flip-flops with low-delay overhead using pMOS passtransistors to suppress SET pulses in a 65-nm FDSOI process," IEEE Trans. Nucl. Sci., vol. 65, no. 8, pp. 1814–1822, Aug. 2018.

53. J. Tschanz, S. Narendra, Z. Chen, S. Borkar, M. Sachdev, and V. De, "Comparative delay and energy of single edge-triggered and dual edgetriggered pulsed flip-flops for high-performance microprocessors," in Proc. Int. Symp. Low Power Electron. Design, Huntington Beach, CA, USA, 2001, pp. 147–152.

54. N. H. Weste and D. Harris, CMOS VLSI Design: A Circuits and Systems Perspective, Pearson Educ. India, 2015.

55. Kumar Mishra, Alok, Dhandapani Vaithiyanathan, and Urvashi Chopra. "Design and analysis of ultra-low power 18T adaptive data track flip-flop for high-speed application." International Journal of Circuit Theory and Applications 49, no. 11 (2021): 3733-3747.

56. A. K. Mishra, U. Chopra and D. Vaithiyanathan, "A Partially Static High Frequency 18T Hybrid Topological Flip-Flop Design for Low Power Application," in IEEE Transactions on Circuits and Systems II: Express Briefs, vol. 69, no. 3, pp. 1592-1596, March 2022, doi: 10.1109/TCSII.2021.3107684.

57. Balwinder Raj, A. K. Saxena and S. Dasgupta, "Analytical Modeling for the Estimation of Leakage Current and Subthreshold Swing Factor of Nanoscale Double Gate FinFET Device" Microelectronics International, UK, vol. 26, pp. 53-63, 2009.

58. Soniya, Shailendra Singh, Girish Wadhwa and Balwinder Raj, "An Analytical Modeling for Dual Source Vertical Tunnel Field

Effect Transistor" submitted to 'International Journal of Recent Technology and Engineering (IJRTE)' at Volume-8 Issue-2, PP. 603-608, July 2019.

59. Shailendra Singh, B Raj, "Design and analysis of hetrojunction Vertical T-shaped Tunnel Field Effect Transistor", Journal of Electronics Material, Springer, Volume 48, Issue 10, pp 6253–6260, October 2019.
60. Candy Goyal, Jagpal Singh Ubhi and B Raj, "A Low Leakage CNTFET based Inexact Full Adder for Low Power Image Processing Applications", International Journal of Circuit Theory and Applications, Wiley, Volume47, Issue9, Pages 1446-1458, September 2019.
61. B Raj, A. K. Saxena and S. Dasgupta, "A Compact Drain Current and Threshold Voltage Quantum Mechanical Analytical Modeling for FinFETs" Journal of Nanoelectronics and Optoelectronics (JNO), USA, vol. 3, no. 2, pp. 163-170, 2008.
62. Girish Wadhwa, B Raj, "An Analytical Modeling of Charge Plasma based Tunnel Field Effect Transistor with Impacts of Gate underlap Region" Superlattices and Microstructures, Elsevier, Vol. 142, PP.106512, June 2020.
63. Shailendra Singh, B Raj,"Modeling and Simulation analysis of SiGe hetrojunction Double GateVertical t-shaped Tunnel FET", Superlattices and Microstructures, Elsevier Volume 142, PP. 106496, June 2020.
64. Amandeep Singh, Dinesh Kumar Saini, Dinesh Agarwal, Sajal Aggarwal, Mamta Khosla, BRaj, "Modeling and Simulation of Carbon Nanotube Field Effect Transistor and its Circuit Application ," Journal of Semiconductors (JoS), IOP Science, Vol.37, PP. 074001-6, July 2016.
65. Neeraj Jain, Balwinder Raj, "Device and Circuit Co-Design Perspective Comprehensive Approach on FinFET Technology - A Review, Journal of Electron Devices, Vol. 23, Num. 1, pp 1890-1901, 2016.
66. Sunil Kumar and BRaj, "Analysis of I_{ON} and Ambipolar Current for Dual-Material Gate-drain Overlapped DG-TFET," Journal of Nanoelectronics and Optoelectronics, American Scientific Publishers, USA, Vol.11, PP. 323-333, June 2016.
67. Naveed Anjum, Tarun Bali and B Raj, Design And Simulation Of Handwritten Multiscript Character Recognition, International Journal of Advanced Research in Computer and Communication Engineering, Vol.2 (7), PP. 2544-2549, July 2013.

68. Sanjeev Sharma, BRaj, Mamta Khosla, "A Gaussian Approach for Analytical Subthreshold Current Model of Cylindrical Nanowire FET with Quantum Mechanical Effects, Microelectronics Journal, Elsevier, Vol. 53, PP. 65-72, April 2016.
69. Karmjit Singh, BRaj, "Performance and analysis of temperature dependent Multi-walled Carbon Nanotubes as Global Interconnects at different technology nodes," Journal of Computational Electronics, Springer, Volume 14 Issue 2, pp. 469-476, June 2015.
70. Sunil Kumar and BRaj, "Compact channel potential analytical Modeling of DG-TFET based on Evanescent–mode Approach," Journal of Computational Electronics, Springer Vol. 14, No. 2, pp.820-827, July 2015.
71. Karmjit Singh, BRaj, "Temperature Dependent Modeling and Performance Evaluation of Multi-Walled CNT and Single-Walled CNT as Global Interconnects," Journal of Electronic Materials, Springer, Vol. 44, Issue 12, pp 4825-4835, Dec 2015.
72. V.K. Sharma, M. Pattanaik and B.Raj, "INDEP approach for leakage reduction in nanoscale CMOS circuits", International Journal of Electronics, Taylor & Francis, Vol.102, No.2, pages 200-215, 2014.
73. Karmjit Singh, BRaj, "Influence of Temperature on MWCNT bundle, SWCNT bundle and Copper interconnects for Nanoscaled Technology nodes," Journal of Materials Science: Materials in Electronics. Springer, Vol. 26, No. 8, pp. 6134-6142, 2015.
74. Naveed Anjum, Tarun Bali and B Raj, "Design and Simulation of Handwritten Gurumukhi and Devanagri Numerical Recognition, International Journal of Computer Applications, Published by Foundation of Computer Science, New York, USA, Vol. 73(12), pp. 16-21, 2013.
75. Khandelwal, S., Gupta, V., BRaj, and Gupta, R D., Process Variability Aware Low Leakage Reliable Nano Scale DG-FinFET SRAM Cell Design Technique, Journal of Nanoelectronics and Optoelectronics, Volume 10, Number 6, pp. 810-817, December 2015.
76. V.K. Sharma, M. Pattanaik and B Raj, "ONOFIC Approach: Low Power High Speed Nanoscale VLSI Circuits Design", International Journal of Electronics, Taylor & Francis, Vol. 101, issue 1, pages 61-73, 2014.

77. Khandelwal, S., Balwinder Raj, and Gupta, R D., FinFET based 6T SRAM Cell Design: Analysis of Performance Metric, Process Variation and Temperature Effect. Journal of Computational and Theoretical Nanoscience, ASP, USA, Vol. 12, PP. 2500-2506, 2015.
78. Sumit Singh; Yadav, Shekhar; Rahul, Jagdeep; Srivastava, Anurag; BRaj, Impact of HfO_2 in Graded Channel Dual Insulator Double Gate MOSFET, Journal of Computational and Theoretical Nanoscience, American Scientific Publishers, Volume 12, Number 6, pp. 950-953, April 2015.
79. Vijay Kumar Sharma, Manisha Pattanaik, BRaj, PVT variations aware low leakage INDEP approach for nanoscale CMOS Circuits, Microelectronics Reliability, Elsevier, Vol. 54, pp 90-99, 2014.
80. B Raj, A. K. Saxena and S. Dasgupta, "Quantum Mechanical Analytical Modeling of Nanoscale DG FinFET: Evaluation of Potential, Threshold Voltage and Source/Drain Resistance ", Elsevier's Journal of Material Science in Semiconductor Processing, Elsevier, Vol. 16, issue 4, pp. 1131- 1137, 2013.
81. Maisagalla Gopal, Siva Sankar D Prasad and Balwinder Raj, 8T SRAM Cell Design for Dynamic and Leakage Power Reduction, International Journal of Computer Applications, Published by Foundation of Computer Science, New York, USA, Vol. 71(9), pp. 43-48, June 2013.
82. Manisha Pattanaik, B Raj, Shashikant Sharma and Anjan Kumar, "Diode Based Trimode Multi-Threshold CMOS Technique for Ground Bounce Noise Reduction in Static CMOS Adders", Advanced Materials Research, Trans Tech Publications, Switzerland, vol. 548, pp 885-889, 2012.
83. Balwinder Raj, A. K. Saxena and S. Dasgupta, "Nanoscale FinFET Based SRAM Cell Design: Analysis of Performance metric, Process variation, Underlapped FinFET and Temperature effect" IEEE Circuits and System Magazine, vol. 11, issue 2, pp. 38- 50, 2011.
84. V.K. Sharma, M. Pattanaik & Balwinder Raj, "Leakage Current ONOFIC Approach for Deep Submicron VLSI Circuit Design", International Journal of Electrical, Computer, Electronics and Communication Engineering, World Academy of Sciences, Engineering and Technology, Vol. 7, issue 4, pages 239-244, 2013.

85. Tulika Chawla, Mamta Khosla, BRaj, Design and simulation of Triple metal Double-gate Germanium on Insulator Vertical Tunnel Field Effect Transistor, Microelectronics Journal, Elsevier; Vol. 114, PP. 105125, August 2021.
86. Parminder Kaur, Sandeep Singh Gill and B Raj, "Comparative Analysis of OFETs Materials and Devices for Sensor Applications", Journal of Silicon, Springer, Vol 14, PP. 4463–4471, 2022.
87. Sanjeev Kumar Sharma, Parveen Kumar, Balwant Raj and BRaj, "$In_{1-x}Ga_xAs$ Double Metal Gate-Stacking Cylindrical Nanowire MOSFET for highly sensitive Photo detector", *Journal of Silicon, Springer Vol.*14, PP. 3535–3541, 2022.
88. B Raj, A. K. Saxena and S. Dasgupta, "Analytical Modeling of Quasi Planar Nanoscale Double Gate FinFET with Source/Drain Resistance and field dependent carrier mobility: A quantum mechanical study" Journal of Computer (JCP),Academy Publisher, FINLAND, vol. 4, issue 9, PP.1-8, 2009.
89. Bhushan, S., Khandelwal, S. and Raj, B., (2012). Analyzing Different Mode FinFET based Memory Cell at Different Power Supply for Leakage Reduction, Seventh International Conference on Bio-Inspired Computing: Theories and Application, (BIC-TA 2012) Advances in Intelligent Systems and Computing, Volume 202, p89-100 (2013).
90. Jeetendra Singh, BRaj, "Temperature Dependent Analytical Modeling and Simulations of Nanoscale Memristor, Journal: Engineering Science and Technology, an International Journal, Elsevier's, Vol. 21, PP- 862-868, Oct 2018.
91. Shradhya Singh, Shashi Bala, Balwant Raj, B Raj, "Improved Sensitivity of Dielectric Modulated Junctionless Transistor for Nanoscale Biosensor Design", Sensor Letter, ASP, Vol.18, PP.328–333, Apl 2020.
92. Vivek Kumar, Santosh Kumar Vishvakarma, and B Raj, "Design and Performance Analysis of ASIC for IoT Applications" Sensor Letter ASP, Vol. 18, PP. 31–38, Jan 2020.
93. Akanksha Jaiswal, R. K. Sarin, B Raj and Shikha Sukhija, "A Novel Circular Slotted Microstrip-fed Patch Antenna with Three Triangle Shape Defected Ground Structure for Multiband Applications", Advanced Electromagnetic (AEM), Vol. 7, Issue 3, PP. 56-63, Aug 2018.

94. Girish Wadhwa, B Raj, "Label Free Detection of Biomolecules using Charge-Plasma-Based Gate Underlap Dielectric Modulated Junctionless TFET" Journal of Electronic Materials (JEMS), Springer, Volume 47, Issue 8, pp 4683–4693, August 2018.
95. Gurmohan Singh, R. K. Sarin and B Raj, "**Design and Performance Analysis of a New Efficient Coplanar Quantum-Dot Cellular Automata Adder**", Indian Journal of Pure & Applied Physics (IJPAP), Vol. 55, pp. 97-103, February 2017.
96. Amandeep Singh, Mamta Khosla, Balwinder Raj, "Design and Analysis of Electrostatic Doped Schottky Barrier CNTFET Based Low Power SRAM," International Journal of Electronics and Communications, (AEÜ), Elsevier, Vol. 80, PP.67–72, 2017.
97. Parminder Kaur, Vikas Pandey and B Raj, "Comparative Study of Efficient Design, Control and Monitoring of Solar Power using IoT", Sensor Letter, ASP Vol. 18, PP. 419–426, May 2020.
98. Anil Kumar Bhardwaj, Sumeet Gupta and B Raj, Development & Analysis of compact model for Double Gate Schottky Barrier CNTFET, Journal of Nanoelectronics and Optoelectronics, ASP, Vol. 15, pp. 1199–1208, Aug 2020.
99. Girish Wadhwa, Priyanka Kamboj; Jeetendra Singh, B Raj, "Design and Investigation of Junctionless DGTFET for Biological Molecule Recognition" Transactions on Electrical and Electronic Materials, Springer, Vol. , PP. 8 Aug, 2020.
100. Tulika Chawla, Mamta Khosla, B Raj, "Optimization of Double-gate Dual material GeOI-Vertical TFET for VLSI Circuit Design, IEEE VLSI Circuits and Systems Letter, Vol. 6, issue-2, PP.13-25, Aug 2020.
101. Sachin Kumar Verma, Shailendra Singh, Girish Wadhwa and B Raj, "Detection of Biomolecules using Charge-Plasma based Gate Underlap Dielectric Modulated Dopingless TFET" Transactions on Electrical and Electronic Materials (TEEM), Springer, Vol. 21, PP. 528-535, June 2020.
102. Neeraj Jain, BRaj, "Impact of underlap spacer region variation on Electrostatic and analog/RF performance of symmetrical high-k SOI FinFET at 20 nm channel length", Journal of Semiconductors (JoS), IOP Science, Vol.38, Issue 12, PP- 122002, Dec 2017.
103. Shailendra Singh, B Raj," Analytical Modeling and Simulation analysis of T-shaped III-V heterojunction Vertical T-FET", Su-

perlattices and Microstructures, Elsevier, Vol. 147, PP. 106717, Nov 2020.
104. Gurmohan Singh, R. K. Sarin and B Raj, "Design and Analysis of Area Efficient QCA Based Reversible Logic Gates", Journal of Microprocessors and Microsystems, Elsevier, Vol. 52, PP.59-68, May 2017.
105. Amandeep Singh, Mamta Khosla, BRaj, "Compact Model for Ballistic Single Wall CNTFET under Quantum Capacitance Limit," Journal of Semiconductors (JoS), IOP Science, Vol.37, PP. 104001-8, Oct 2016.
106. Sonal Singh; Mamta Khosla, Girish Wadhwa; B Raj, "Design and analysis of double-gate junctionless vertical TFET for gas sensing applications", Applied Physics A, Springer, Vol.127, No.16, 2 Jan 2021.
107. Inderjit Singh, BRaj, Mamta Khosla, and Brajesh Kumar Kaushik, Potential MRAM Technologies for Low Power SoCs, SPIN World Scientific Publisher, SCIE; Vol. 10, No. 04, PP.2050027, Dec 2020.
108. Shailendra Singh, B Raj,"Parametric variation analysis on hetero-junction Vertical t-shape TFET for Supressing ambipolar conduction", Indian Journal of Pure and Applied Physics, Vol. 58,pp. 478-485 June 2020.
109. Soniya, Shailendra Singh, Girish Wadhwa and Balwinder Raj, Design and Analysis of Dual Source Vertical Tunnel Field Effect Transistor for high performance, Transactions on Electrical and Electronics Materials, Springer, Vol. 21, Pages 74–82, Oct 2019.
110. Manjit Kaur; Neena Gupta; Sanjeev Kumar; BRaj; Arun Kumar Singh, "RF Performance Analysis of Intercalated Graphene Nanoribbon Based Global Level Interconnects" Journal of Computational Electronics, Springer, Vol. 19, PP.1002–1013, June 2020
111. Girish Wadhwa, BRaj, "Design and Performance Analysis of Junctionless TFET Biosensor for high sensitivity" IEEE Nanotechnology, Vol.18, PP. 567 – 574, 2019.
112. Jeetendra Singh, BRaj, "Enhanced Nonlinear Memristor Model Encapsulating Stochastic Dopant Drift, JNO, ASP 14, 958–963 (2019)
113. Hamendra Singh, Pankaj Kumar, Balwinder Raj, "Performance Analysis of Majority Gate SET Based 1-bit Full Adder", *International Journal of Computer and Communication Engineer-*

ing (IJCCE), IACSIT Press Singapore, ISSN: 2010-3743, Volume 2, No. 4, 2013.

114. Anil Kumar Bhardwaj, Sumeet Gupta and B Raj, "Investigation of Parameters for Schottky Barrier (SB) Height for Schottky Barrier Based Carbon Nanotube Field Effect Transistor Device", Journal of Nanoelectronics and Optoelectronics, ASP, Vol. 15, PP. 783–791, July 2020.
115. Priya Bansal, BRaj, "MEMRISTOR: A VERSATILE NONLINEAR MODEL FOR DOPANT DRIFT AND BOUNDARY ISSUES," JCTN, American Scientific Publishers, Vol. 14 (5), PP. 2319-2325, May 2017.
116. Neeraj Jain, BRaj, "An Analog and Digital Design Perspective Comprehensive Approach on Fin-FET (Fin-Field Effect transistor) Technology - A Review", Reviews in Advanced Sciences and Engineering (RASE), ASP, Vol. 5, PP. 1-14, 2016.
117. Sanjeev Sharma, B Raj, Mamta Khosla, "Subthreshold Performance of In1-xGaxAs based Dual Metal with Gate Stack Cylindrical/Surrounding Gate Nanowire MOSFET for Low Power Analog Applications, Journal of Nanoelectronics and Optoelectronics, American Scientific Publishers, USA, Vol.12, PP. 171–176, 2017.

5

Tunnel FET and its Applications

Kamal Kishor Jha
Indian Institute of Information Technology Vadodara, 382028, India

5.1 Introduction

A Tunnel Field-Effect Transistor (TFET) is a type of field-effect transistor that uses the quantummechanical phenomenon of tunneling to control the flow of current through the device. Unlike traditional metal-oxide-semiconductor field-effect transistors (MOSFETs), which use thermionicemission to inject electrons from the source to the channel, TFETs rely on tunneling, which allows for a more efficient and faster transfer of charge carriers. The unique properties of TFETs, such as their high transconductance, low subthreshold swing, and high current on/off ratio, make them attractive for a variety of applications. These include low-power and high-frequency digitallogic, low-power analog circuits, low-power memory devices, and low-power sensor interfaces. Additionally, TFETs are also being investigated as a potential replacement for MOSFETs (see Figure 5.1) in complementary metal-oxide-semiconductor (CMOS) technology, which is the backboneof modern integrated circuits.

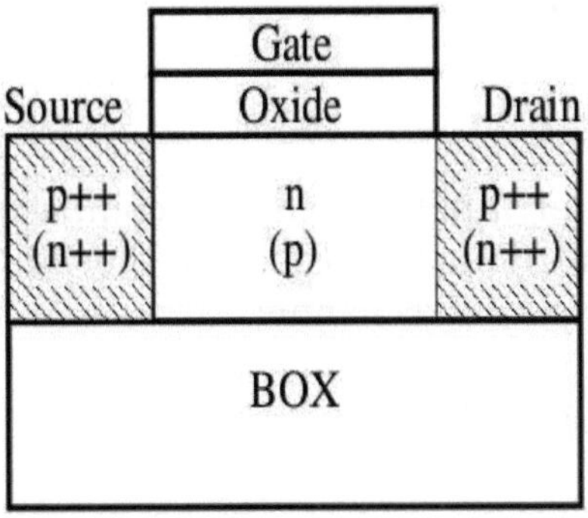

Figure 5.1 *Device structure of MOSFET.*

In order to fully utilize the advantages of TFETs, it requires finding appropriate materials, understanding the underlying physics, and advancing the fabrication techniques. The most recentresearch focus on TFETs mainly focus on finding new materials for the channel, developing new fabrication techniques, and optimizing the device design for improved performance. The new materials research including emerging 2D materials, such as transition metal dichalcogenides, have been proposed as an alternative to conventional semiconductors for the channel of TFETs. These materials have unique electronic and optical properties that are not found in conventional semiconductors, which makes them ideal for use in TFETs [1-5].

TFETs have certain properties that make them attractive for use in a wide range of electronic devices, such as low power consumption, high switching speed, and high input impedance [6-10].

Low power consumption:

TFETs have a low subthreshold swing (the amount of voltage required to turn the device on) which can be less than 60mV/decade compared to the typical subthreshold swing of 600mV/decade for MOSFETs. This results in TFETs consuming less power and making them suitable for use in low-power electronic devices [11-16].

High switching speed:

TFETs have a higher transconductance (the amount of current that can flow through the device) than MOSFETs, which results in faster switching speed. This makes TFETs attractive for use in high-frequency applications [17-22].

High input impedance:

TFETs have a higher input impedance compared to MOSFETs because they rely on tunnelinginstead of thermionic emission to inject charge carriers into the channel.

Applications of Tunnel FETs:

Low-Power Logic Circuits: TFETs can be used to build low-power logic circuits such as digital gates, flip-flops, and multiplexers. They can also be used to build low-power memory devices such as SRAM andflash memory [23-28].

Low-Power Analog Circuits: TFETs can be used to build low-power analog circuits such as amplifiers, filters, and oscillators. This makes TFETs suitable for use in low-power wireless communication systems andlow-power sensor interfaces.

Low-Power Sensors Interface: TFETs can be used to build low-power interfaces for sensors, such as temperature sensors, pressure sensors, and chemical sensors. The low power consumption of TFETs makes themsuitable for use in battery-powered sensors [29-35].

High-Frequency Circuits: TFETs have a higher transconductance than MOSFETs, which results in faster switching speed.This makes TFETs attractive for use in high-frequency circuits such as oscillators, mixers, and modulators.

It is worth noting that TFETs are still in the research stage and commercial applications are limited, current research is focused on finding new materials for the channel, developing new fabrication techniques, and optimizing the device design for improved performance.

There are several ways that the performance of TFETs can be improved [7].

- Improving the quality of the tunneling region: The quality of the tunneling region is critical for the performance of TFETs. Research is ongoing to improve the material quality of the tunneling region, such as reducing defects and impurities in the material.
- Increasing the band-to-band tunneling current: The current flowing through a TFET is directly proportional to the band-to-band tunneling current. Research is ongoing to increase this current by using materials with a higher bandgap or by optimizing the band alignment at the source-tunneling region and drain-tunneling region interfaces.
- Enhancing the gate control: Improving the gate control can increase the on-off ratio and transconductance of the TFET. The gate control can be enhanced by using advanced gate dielectrics, such as high-k materials, and by optimizing the gate geometry.

- Improving the contact resistance: The contact resistance at the source and drain terminals can be a major limiting factor in the performance of TFETs. Research is ongoing to reduce the contact resistance by using advanced metal contacts, such as Schottky contacts, and by optimizing the contact geometry.
- Optimizing the device structure: The device structure can be optimized to improve the performance of TFETs. For example, research is ongoing to develop new device structures, such as double-gate TFETs and FinFETs, that can improve the on-off ratio andtransconductance of the TFET [36-43].

Types of TFETS

Single gate TFET

A single-gate transistor with ferroelectric tunnel junctions (TFET) is a type of TFET that uses a single gate electrode to control the flow of current through the device. In contrast, traditional FETs such as metal-oxide-semiconductor field-effect transistors (MOSFETs) use a gate electrode separated from the channel by an insulating material to control the flow of current. Single-gate TFETs have some advantages over their multi-gate counterparts, such as a simpler fabrication process, fewer materials and less processing steps which lower the cost of fabrication. Figure 5.2 shows the single gate TFET [1-2].

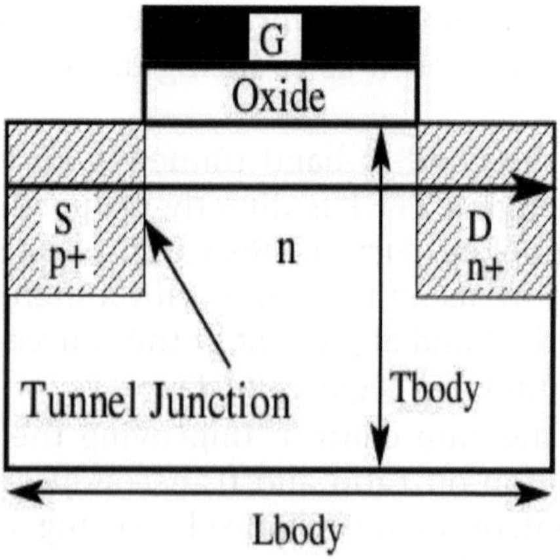

***Figure 5.2** Single Gate TFET*

Double Gate TFET

A double-gate transistor (DGT) is a type of field-effect transistor (FET) that has two gates, rather than the typical one gate found in a single-gate FET. The two gates allow for more precise control of the current flowing through the channel of the transistor, which can result in improved performance in certain applications, such as digital logic circuits and low-power electronic devices [44-49]. One of the most popular variants of DGT is Double Gate Tunnel FET (DG-TFET) which uses Tunneling effect for charge conduction. It has very low leakage current compared to regular FETs and also has lower subthreshold swing [7-8].

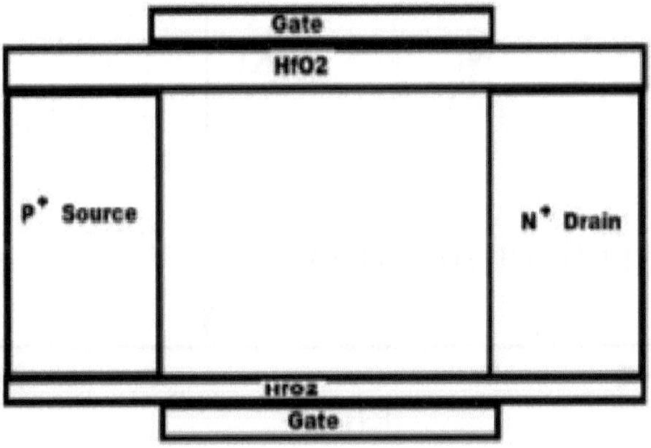

Figure 5.3 *Double Gate TFET.*

Pocket doped TFET

A pocket doped Tunnel FET (TFET) is a type of transistor that uses quantum tunneling tocontrol the flow of current through the device. In a pocket doped TFET as shown in Figure 5.4., a pocket of dopant atoms is added to the channel region of the device, rather than at the source or drain terminals [50-56]. This pocket doping allows for the formation of a very narrow, high-mobility channel that can improve the device's performance, such as a lowersubthreshold swing and a higher on-current. Compared to traditional MOSFETs, which rely on thermionic field emission to control the flow of current, TFETs have a subthreshold slope that is independent of temperature, and are expected to consume significantly less power at low voltage operation, making them a promising candidate for low-power electronics [57-62]. However, TFETs are still under research and development, and there are still challenges that need to be overcome before they can be widely used in commercial applications, such as developing reliable fabrication methods for the devices,and improving their off-state current [9-10].

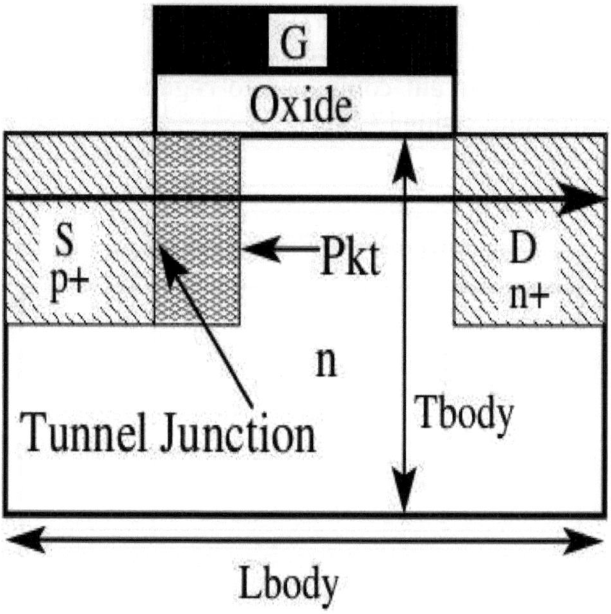

Figure 5.4 *Pocket doped TFET.*

Junctionless TFET

VerticalA junctionless (JL) TFET is a type of TFET that does not rely on junctions formed by diffusing dopants into the channel region of the device. Instead, the channel ofa JL TFET is formed by a thin undoped semiconductor layer that is sandwiched between source and drain terminals that are heavily doped. By using an undoped channel, the device's performance can be improved, such as lower subthreshold swing and higher on-current, compared to traditional TFETs which use junction doped channels [63-70]. JL TFETs are also expected to have other advantages over traditional TFETs, such as improved scalability, better immunity to variations in the manufacturing process, and reduced source-to-drain leakage current. However, this technology is still under research and development, and the practical implementation of JL TFETs is still facing some technical challenges such as precise control over the channel layer thickness and controlling the band-bending near the source and drain [71-76].

Vertical Tunnel FET

A Vertical Tunnel FET (VTFET) is a type of TFET (See Figure 5.5) that uses a vertical channel architecture rather than a planar one. This architecture allows for the source and drain terminals to be located on opposite sides of the channel region, resulting in a much shorter channel length, which can improve the device's performance, such as a lower subthreshold swing and a higher on-current. VTFETs can be fabricated using a variety of techniques such as grown-on-substrate, epitaxial lift-off, and bonded wafer technologies. These fabrication methods can also allow multiple VTFETs to be stacked on top of each other to form 3D structures, which can further improve the device's performance and power efficiency [77-84].

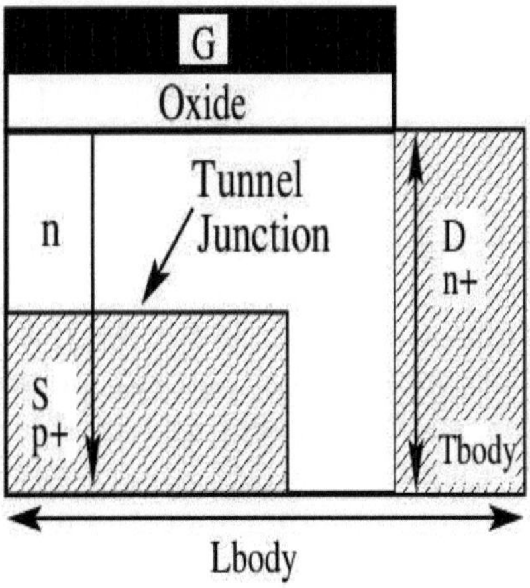

Figure 5.5 *Vertical tunnel FET.*

Sandwich Tunnel Barrier FET (STB-FET)

A Sandwich Tunnel Barrier FET (STB-FET) as shown in Figure 5.6 is a type of TFET which includes a sandwich layer of insulator between the channel and the source/drain terminals [85-92]. This insulator layer, called the Tunnel Barrier (TB), is used to enhance the band-to-band tunneling that occurs in the channel of a TFET, which results in improved performance. The main benefit of this design is that it can reduce the leakage current of the device, thereby reducing power consumption and increasing the potential for low-power devices. In addition, the thickness of the TB can be easily adjusted to optimize the device's performance, and the use of a wide-band-gap material as the TB can further increase the device's breakdown voltage, resulting in more robust and reliable operation

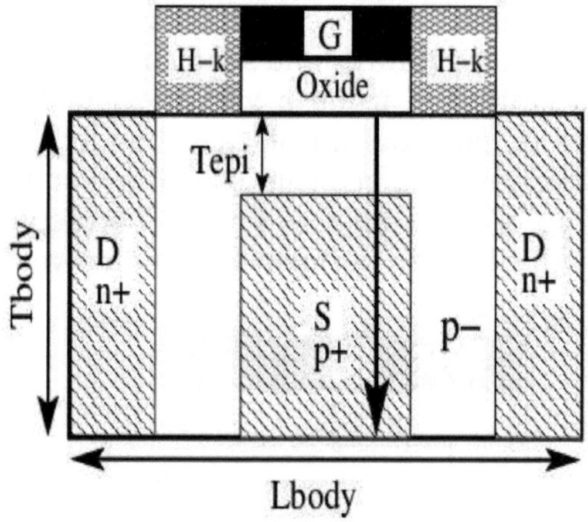

Figure 5.6 *Sandwich tunnel barrier FET.*

STB-FETs are still under research and development, and challenges that need to be overcome before they can be widely used in commercial applications, such as developingreliable methods for fabricating high-quality Tunnel Barrier layers and achieve efficient band-to-band tunneling, and the operational temperature of the devices [93-97].

Electron-Hole Bilayer Tunnel FET (EHBT-TFET)

An Electron-Hole Bilayer Tunnel FET (EHBT-TFET) is a type of TFET that uses a bilayer channel structure, consisting of both an electron channel and a hole channel, to control the flow of current through the device. The electron and hole channels are typically made from different semiconductor materials and are separated by a thininsulating layer called a tunnel barrier [98-103].

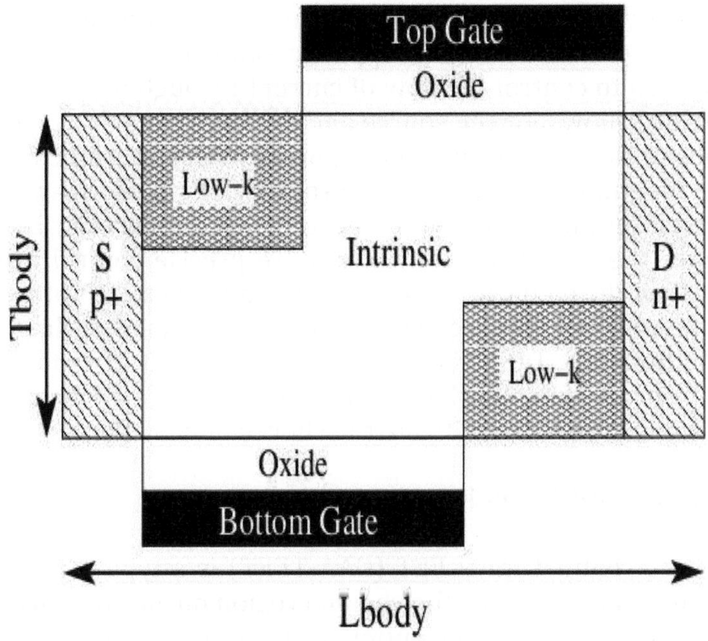

Figure 5.7 *Electron-Hole Bilayer Tunnel FET.*

In this device, the electron channel is used to turn on the device, while the hole channel isused to turn it off. When the device is on, electrons tunnel from the electron channel to the hole channel through the tunnel barrier, and when the device is off, the current is blocked by the insulating barrier. EHBT-TFETs are expected to have improved device performance compared to traditional TFETs, such as lower subthreshold swing, higher on-current and off-current, and better short-channel behavior, due to the presence of both an electron and hole Sandwich Tunnel Barrier FET Sandwich Tunnel Barrier FET Sandwich Tunnel Barrier FET channel [104-108]. However, this technology is still under research and development, and practical implementation of EHBT-TFETs is still facing some technical challenges such as, achieving efficient electron-hole tunneling across the Tunnel barrier and integration of two different semiconductors material and achieving the electrical compatibility between them. Figure 5.7 shows the Electron-Hole Bilayer Tunnel FET device structure [14].

Double-Gate Vertical Tunnel FET (DG-VTFET)

A Double-Gate Vertical Tunnel FET (DG-VTFET) is a type of TFET that uses a vertical channel architecture and two gates instead of one. The gates are located on opposite sidesof the channel region and are used to control the flow of current through the device. This architecture allows for the source and drain terminals to be located on opposite sides of the channel region, resulting in a much shorter channel length, which can improve the device's performance, such as a lower subthreshold swing and a higher on-current. By using two gates, the DG-VTFET can offer more precise control over the current flow through the channel, allowing for better subthreshold characteristics, and improved drain-source voltage scaling [109-103]. Additionally, DG-VTFETs also have a benefit of reducing the source-to-drain tunneling by creating a potential barrier near the drain terminal, so it can mitigate the short channel effects [15].

Gate-All-Around Tunnel FET (GAA-TFET)

A Gate-All-Around Tunnel FET (GAA-TFET) is a type of TFET that uses a gate that surrounds the channel region on all sides, instead of just on one side as in conventional planar FETs. This allows for more precise control over the flow of current through the channel, which can improve the device's performance, such as a lower sub-threshold swing and a higher on-current[GAA-TFETs are typically fabricated using a combination of planar and vertical fabrication techniques, such as growing the channel layer on a substrate and then selectively etching away the substrate material to leave the channel standing up, so that it can be surrounded by the gate on all sides [104-108]. The use of a GAA structure can offer several benefits over traditional planar TFETs, such as improved subthreshold characteristics, reduced short-channel effects, and increased gate-to-channelcoupling which can result in higher on-current, and lower subthreshold swing [109-114].

References

1. Q. Zhang, W. Zhao, and A. Seabaugh, "Low-subthreshold-swing tunnel transistors," IEEE Electron Device Letters, vol. 27, no. 4, pp. 297–300, 2006.

2. A. C. Seabaugh and Q. Zhang, "Low-voltage tunnel transistors for beyond CMOSlogic," Proceedings of the IEEE, vol. 98, no. 12, pp. 2095–2110, 2010.
3. A. M. Ionescu and H. Riel, "Tunnel field-effect transistors as energy-efficient electronic switches," Nature, vol. 479, no. 7373, pp. 329–337, 2011.
4. S. Rajput and S. S. Jamuar, "Low voltage analog circuit design techniques," IEEECircuits and Systems Magazine, vol. 2, no. 1, pp. 24–42, 2002.
5. R. Narang, M. Saxena, R. Gupta, and M. Gupta, "Assessment of ambipolarbehavior of a tunnel FET and influence of structural modifications," Journal of Semiconductor Technology and Science, vol. 12, no. 4, pp. 482–491, 2012.
6. L. McGarry, "Standby power challenge," in Proceedings of the IEEE Asian GreenElectronics Conference (AGEC); China, 2004, pp. 56–62.
7. K. Boucart and A. M. Ionescu, "Double gate tunnel FET with ultrathin silicon body and high-k gate dielectric," in Proceedings of the IEEE European Solid-State Device Research Conference (ESSDERC); Switzerland, 2006, pp.383–386.
8. K. Boucart and A. M. Ionescu, "Double-gate tunnel FET with high-κ gate dielectric," IEEE Transaction on Electron Devices, vol. 54, no. 7, pp. 1725–1733, 2007.
9. R. Jhaveri, V. Nagavarapu, and J. C. Woo, "Effect of pocket doping and annealingschemes on the source-pocket tunnel field-effect transistor," IEEE Transaction on Electron Devices, vol. 58, no. 1, pp. 80–86, 2011.
10. K.-H. Kao, A. S. Verhulst, W. G. Vandenberghe, B. Soree, W. Magnus, D. Leonelli, G. Groeseneken, and K. De Meyer, "Optimization of gate-on- source-only tunnel FETs with counter-doped pockets," IEEE Transactions on Electron Devices, vol. 59, no. 8, pp. 2070–2077, 2012.
11. M. J. Kumar and S. Janardhanan, "Doping-less tunnel field effect transistor: Design and investigation," IEEE Transactions on Electron Devices, vol. 60, no. 10, pp. 3285–3290, 2013.
12. L. De Michielis, L. Lattanzio, P. Palestri, L. Selmi, and A. Ionescu, "Tunnel FETarchitecture with improved performance due to enhanced gate modulation of the tunneling barrier," in Proceedings of the IEEE Annual Device Research Conference (DRC); USA, 2011, pp. 111–112.
13. R. Asra, M. Shrivastava, K. V. Murali, R. K. Pandey, H. Gossner,

and V. Ramgopal Rao, "A tunnel FET for scaling below 0.6 V with a CMOS comparableperformance," IEEE Transactions on Electron Devices, vol. 58, no. 7, pp. 1855–1863, 2011.

14. L. Lattanzio, L. De Michielis, and A. M. Ionescu, "Complementary germanium electron–hole bilayer tunnel FET for sub-0.5 V operation," IEEE Electron Device Letters, vol. 33, no. 2, pp. 167–169, 2012.
15. G. B. Beneventi, E. Gnani, A. Gnudi, S. Reggiani, and G. Baccarani, "Dual metal-gate InAs tunnel FET with enhanced turn-on steepness and high on current," IEEE Transactions on Electron Devices, vol. 61, no. 3, pp. 776–784, 2014.
16. J. H. Seo, Y. J. Yoon, S. Lee, J.-H. Lee, S. Cho, and I. M. Kang, "Design and analysis of Si-based arch-shaped gate-all-around (GAA) tunneling field effect transistors (TFET)," Current Applied Physics, vol. 15, no. 3, pp. 208-212, 2015.
17. Anjana Bhardwaj, Pradeep Kumar, B. Raj, Sunny Anand, "Design and Performance Optimization of doping-less Vertical Nanowire TFET using Gate Stack Technique", Journal of Electronic Materials (JEMS) Springer, Vol. 41, Issuse 7, PP. 4005-4013, 2022.
18. Jeetendra Singh, BRaj, "Tunnel Current Model of Asymmetric MIM Structure Levying Various Image Forces to Analyze the Characteristics of Filamentary Memristor", Applied Physics A, Springer, Vol. 125, No.3, pp. 203.1 to 203.11 Feb 2019.
19. Candy Goyal, Jagpal Singh Ubhi and B Raj, Low Leakage Zero Ground Noise Nanoscale Full Adder using Source Biasing Technique, "Journal of Nanoelectronics and Optoelectronics", American Scientific Publishers, Vol. 14, PP. 360–370, March 2019.
20. Gurmohan Singh, R. K. Sarin and B Raj, "A Novel Robust Exclusive-OR Function Implementation in QCA Nanotechnology with Energy Dissipation Analysis", Journal of Computational Electronics, Springer, Volume 15, Issue 2, PP. 455-465, June 2016.
21. Tanu Wadhera, Deepti Kakkar, Girish wadhwa, B Raj, " Recent Advances and Progress in Development of the Field Effect Transistor Biosensor: A Review" Journal of ELECTRONIC MATERIALS, Springer, Volume 48, Issue 12, pp 7635–7646, December 2019.
22. Girish Wadhwa, Priyanka Kamboj, Balwinder Raj" Design optimisation of junctionless TFET biosensor for high sensitivity"

Advances in Natural Sciences: Nanoscience and Nanotechnology, Vol. 10, PP. 045001 (7), (2019).

23. Priya Bansal, BRaj, "MEMRISTOR MODELING AND ANALYSIS FOR LINEAR DOPANT DRIFT KINETICS," Journal of Nanoengineering and Nanomanufacturing, American Scientific Publishers, Vol. 6, pp. 1–7, 2016.
24. Amandeep Singh, Mamta Khosla, BRaj, "Circuit Compatible Model for Electrostatic Doped Schottky Barrier CNT-FET,"Journal of Electronic Materials, Springer, Vol. 45, Issue 12, pp 4825-4835, 2016.
25. Ashima, D Vaithiyanathan, B Raj," Performance Analysis of Charge Plasma induced Graded Channel Si nanotube", Journal of Engineering Research (JER), EMSME Special Issue pp. 146-154 Aug 2021.
26. Abhishek Singh Tomar, Vijay Kumar Magraiya, and B Raj, Scaling of Access and Data Transistor for High Performance DRAM Cell Design, Quantum Matter, Vol. 2, PP. 412-416 Oct. 2013.
27. Neeraj Jain, BRaj, "Parasitic Capacitance and Resistance Model Development and Optimization of Raised Source/Drain SOI FinFET Structure for Analog Circuit Applications, Journal of Nanoelectronics and Optoelectronins, ASP, USA, Vol. 13, PP. 531–539, Apl 2018.
28. Shradhya Singh, S. K. Vishvakarma and B Raj, "Analytical Modeling of Split-Gate Junction-Less Transistor for a Biosensor Application", Sensing and Bio-sensing, Elsevier, Vol. 18, Pages 31-36, April 2018.
29. Maisagalla Gopal, and Balwinder Raj, Low Power 8T SRAM Cell Design for High Stability Video Applications, ITSI Transaction on Electrical and Electronics Engineering, Vol. 1(5), pp. 91-97, 2013.
30. Balwinder Raj, Jatin Mitra, Deepak Kumar Bihani, Rangharajan V, A.K. Saxena and S. Dasgupta, "Analysis of Noise Margin, Power and Process Variation for 32 nm FinFET Based 6T SRAM Cell" Journal of Computer (JCP), Academy Publisher, FINLAND, vol. 5, issue 6, PP.1-8, 2010.
31. Divya Sharma, Rajesh Mehra and B Raj, "Comparative Analysis of Photovoltaic Technologies for High Efficiency Solar Cell Design", Superlattices and Microstructures, Elsevier, Volume 153, PP. 106861, May 2021.
32. Pawandeep Kaur, Avtar Singh Buttar, B Raj, "A comprehen-

sive Analysis of Nanoscale Transistor based Biosensor: A Review, Indian Journal of Pure and Applied Physics, Vol. 59, pp. 304-318, April 2021.

33. Divya Yadav, Balwant Raj and B Raj, "Design and Simulation of Low Power Microcontroller for IoT Applications", Journal of Sensor Letters, ASP, Vol. 18, PP. 401–409, May 2020.
34. Shailendra Singh, B Raj,"A 2-D Analytical Surface Potential and Drain current Modeling of Double-Gate Vertical t-shaped Tunnel FET", Journal of Computational Electronics, Springer, Vol. 19, PP.1154–1163, Apr 2020.
35. Jeetendra Singh, B Raj, "An Accurate and Generic Window function for Non-linear Memristor Model" Journal of Computational Electronics, Springer, Volume 18 Issue 2, PP. 640647, June 2019.
36. Manjit Kaur; Neena Gupta; Sanjeev Kumar; BRaj; Arun Kumar Singh, "Comparative RF and Crosstalk Analysis of Carbon Based Nano Interconnects" IET Circuits, Devices & Systems, Vol.15(6), PP. 493-503, Feb 2021.
37. Nehru Kandasamy, Firdous Ahmad, D. Ajitha, B Raj, and Nagarjuna Telagam "Quantum Dot Cellular Automata based Scan Flip Flop and Boundary Scan Register, IETE Journal of Research, Volume 66, PP. 535-548, 2020.
38. Sharma, S. K.., Raj, B., Khosla, M., "Enhanced Photosensivity of Highly Spectrum Selective Cylindrical Gate In1-xGaxAs Nanowire MOSFET Photodetector, Modern Physics letter-B, Vol. 33, No. 12, PP. 1950144 (2019).
39. Jeetendra Singh, BRaj, "Design and Investigation of 7T2M NVSARM with Enhanced Stability and Temperature Impact on Store/Restore Energy", IEEE Transactions on Very Large Scale Integration Systems, Vol. 27, Issue 6, PP. 1322 - 1328 June 2019.
40. Anil Kumar Bhardwaj, Sumeet Gupta, BRaj, Amandeep Singh, "Impact of Double Gate Geometry on the Performance of Carbon Nanotube Field Effect Transistor Structures for Low Power Digital Design", Computational and Theoretical Nanoscience, ASP, Vol. 16, PP. 1813–1820, 2019.
41. Neeraj Jain, BRaj, "Thermal Stability Analysis and Performance Exploration of Asymmetrical Dual-k underlap Spacer (ADKUS) SOI FinFET for Security and Privacy Applications", Indian Journal of Pure & Applied Physics (IJPAP), Vol. 57, pp. 352-360, May 2019.

42. Amandeep Singh, Mamta Khosla, BRaj, "Design and Analysis of Dynamically Configurable Electrostatic Doped Carbon Nanotube Tunnel FET"," Microelectronics Journal, Elesvier, Volume 85, Pages 17-24, March 2019.
43. Neeraj Jain and Balwinder Raj, "Dual-k Spacer Region Variation at the Drain Side of Asymmetric SOI FinFET Structure: Performance Analysis towards the Analog/RF Design Applications", communicated in "Journal of Nanoelectronics and Optoelectronics", American Scientific Publishers, Vol. 14, PP. 349–359, March 2019.
44. Jeetendra Singh, Sanjeev Sharma, B Raj, Mamta Khosla Analysis of barrier layer thickness on performance of In1-xGaxAs based Gate Stack Cylindrical Gate Nanowire MOSFE," JNO, ASP, Vol. 13, PP 1473-1477, Oct 2018.
45. Neeraj Jain, B Raj, "Analysis and Performance Exploration of High-k SOI FinFETs Over the Conventional Low-k SOI FinFET toward Analog/RF Design", Journal of Semiconductors (JoS), IOP Science, Vol.39, Issue 12, PP- 124002-1-7, Dec 2018.
46. Candy Goyal, Jagpal Singh Ubhi and BRaj, A reliable leakage reduction technique for approximate full adder with reduced ground bounce noise, Journal of Mathematical Problems in Engineering, Hindawi, Volume 2018, Article ID 3501041, 16 pages, 15 Oct 2018.
47. Anuradha, Jeetendra Singh, B Raj and Mamta Khosla, "Design and Performance Analysis of Nano-scale Memristor-based Nonvolatile SRAM" Journal of Sensor Letter", American Scientific Publishers, Vol. 16, PP 798–805, Oct 2018.
48. Girish Wadhwa, BRaj, "Parametric Variation Analysis of Charge-Plasma-based Dielectric Modulated JLTFET for Biosensor Application" IEEE Sensor Journal, VOL. 18, NO. 15, AUGUST 1, PP. 6070 – 6077, 2018 (SCI)
49. Jeetendra Singh, BRaj, "Comparative Analysis of Memristor Models for Memories Design", JoS, IoP, Vol. 39, No. 7, PP. 074006-1-12, July 2018.
50. Divya Yadav, Shailesh Singh Chouhan, Santosh Kumar Vishvakarma and BRaj, "Application Specific Microcontroller Design for IoT based WSN", Sensor Letter, ASP, Vol. 16, PP. 374–385, May 2018.
51. Gurmohan Singh, R. K. Sarin and BRaj, "Fault-Tolerant Design and Analysis of Quantum-Dot Cellular Automata Based Circuits", IEEE/IET Circuits, Devices & Systems, Vol. 12, PP. 638

- 64, 2018.

52. Jeetendra Singh, BRaj, "Modeling of Mean Barrier Height Levying Various Image Forces of Metal Insulator Metal Structure to Enhance the Performance of Conductive Filament Based Memristor Model", IEEE Nanotechnology, Vol. 17, No. 2, PP. 268-267, March 2018.
53. Aakash Jain, Sanjeev Sharma, BRaj, "Analysis of Triple Metal Surrounding Gate (TM-SG) III-V Nanowire MOSFET for Photosensing Application", Opto-electronics Journal, Elsevier, Volume 26, Issue 2, Pages 141-148, May 2018.
54. Aakash Jain, Sanjeev Sharma, B Raj, "Design and Analysis of High Sensitivity Photosensor Using Cylindrical Surrounding Gate MOSFET For Low Power Sensor Applications", Engineering Science and Technology, an International Journal, Volume 19, Issue 4, Pages 1864–1870, December 2016.
55. Amandeep Singh, Mamta Khosla, B Raj, "Analysis of Electrostatic Doped Schottky Barrier Carbon Nanotube FET for Low Power Applications," Journal of Materials Science: Materials in Electronics, Springer, Vol.28, PP. 1762–1768, 2017.
56. G. Saiphani Kumar, Amandeep Singh, BRaj, "Design and Analysis of Gate All Around CNTFET based SRAM cell Design", Journal of Computational Electronics, Springer, Volume 17, Issue 1, pp 138–145, March 2018.
57. Gurinder pal Singh, B. S. Sohi, Balwinder Raj, "Material Properties Analysis of Graphene Base Transistor (GBT) for VLSI Analog Circuits, Indian Journal of Pure & Applied Physics (IJPAP), Vol. 55, pp. 896-902, Dec-2017.
58. Amandeep Singh, Mamta Khosla, B Raj, "Comparative Analysis of Carbon Nanotube Field Effect Transistor and Nanowire Transistor for Low Power Circuit Design," Journal of Nanoelectronics and Optoelectronics, American Scientific Publishers, USA, Vol. 11, PP. 388-393, June 2016.
59. Sunil Kumar and B Raj, "Estimation of Stability and Performance metric for Inward Access Transistor based 6T SRAM Cell Design using n-type/p-type DMDG-GDOV TFET" IEEE VLSI Circuits and Systems Letter, Volume3, Issue2, PP 25-39, June2017
60. Shashikant Sharma, Anjan Kumar, Manisha Pattanaik and B Raj, "Forward Body Biased Multimode Multi-Threshold CMOS Technique for Ground Bounce Noise Reduction in Static CMOS Adders" International Journal of Information and Electronics

Engineering, pp. 567-572, Vol. 3, issue 3, 2013.

61. Hamendra Singh, Pankaj Kumar, Balwinder Raj, "Performance Analysis of Majority Gate SET Based 1-bit Full Adder", *International Journal of Computer and Communication Engineering* (IJCCE), IACSIT Press Singapore, ISSN: 2010-3743, Volume 2, No. 4, 2013.
62. Anil Kumar Bhardwaj, Sumeet Gupta and B Raj, "Investigation of Parameters for Schottky Barrier (SB) Height for Schottky Barrier Based Carbon Nanotube Field Effect Transistor Device", Journal of Nanoelectronics and Optoelectronics, ASP, Vol. 15, PP. 783–791, July 2020.
63. Priya Bansal, BRaj, "MEMRISTOR: A VERSATILE NONLINEAR MODEL FOR DOPANT DRIFT AND BOUNDARY ISSUES," JCTN, American Scientific Publishers, Vol. 14 (5), PP. 2319-2325, May 2017.
64. Neeraj Jain, BRaj, "An Analog and Digital Design Perspective Comprehensive Approach on Fin-FET (Fin-Field Effect transistor) Technology - A Review", Reviews in Advanced Sciences and Engineering (RASE), ASP, Vol. 5, PP. 1-14, 2016.
65. Sanjeev Sharma, B Raj, Mamta Khosla, "Subthreshold Performance of In1-xGaxAs based Dual Metal with Gate Stack Cylindrical/Surrounding Gate Nanowire MOSFET for Low Power Analog Applications, Journal of Nanoelectronics and Optoelectronics, American Scientific Publishers, USA, Vol.12, PP. 171–176, 2017.
66. Balwinder Raj, A. K. Saxena and S. Dasgupta, "Analytical Modeling for the Estimation of Leakage Current and Subthreshold Swing Factor of Nanoscale Double Gate FinFET Device" Microelectronics International, UK, vol. 26, pp. 53-63, 2009.
67. Soniya, Shailendra Singh, Girish Wadhwa and Balwinder Raj, "An Analytical Modeling for Dual Source Vertical Tunnel Field Effect Transistor" submitted to 'International Journal of Recent Technology and Engineering (IJRTE)' at Volume-8 Issue-2, PP. 603-608, July 2019.
68. Shailendra Singh, B Raj, "Design and analysis of hetrojunction Vertical T-shaped Tunnel Field Effect Transistor", Journal of Electronics Material, Springer, Volume 48, Issue 10, pp 6253–6260, October 2019.
69. Candy Goyal, Jagpal Singh Ubhi and B Raj, "A Low Leakage CNTFET based Inexact Full Adder for Low Power Image Processing Applications", International Journal of Circuit Theory

and Applications, Wiley, Volume47, Issue9, Pages 1446-1458, September 2019.

70. B Raj, A. K. Saxena and S. Dasgupta, "A Compact Drain Current and Threshold Voltage Quantum Mechanical Analytical Modeling for FinFETs" Journal of Nanoelectronics and Optoelectronics (JNO), USA, vol. 3, no. 2, pp. 163-170, 2008.
71. Girish Wadhwa, B Raj, "An Analytical Modeling of Charge Plasma based Tunnel Field Effect Transistor with Impacts of Gate underlap Region" Superlattices and Microstructures, Elsevier, Vol. 142, PP.106512, June 2020.
72. Shailendra Singh, B Raj, "Modeling and Simulation analysis of SiGe hetrojunction Double Gate Vertical t-shaped Tunnel FET", Superlattices and Microstructures, Elsevier Volume 142, PP. 106496, June 2020.
73. Amandeep Singh, Dinesh Kumar Saini, Dinesh Agarwal, Sajal Aggarwal, Mamta Khosla, BRaj, "Modeling and Simulation of Carbon Nanotube Field Effect Transistor and its Circuit Application," Journal of Semiconductors (JoS), IOP Science, Vol.37, PP. 074001-6, July 2016.
74. Neeraj Jain, Balwinder Raj, "Device and Circuit Co-Design Perspective Comprehensive Approach on FinFET Technology - A Review, Journal of Electron Devices, Vol. 23, Num. 1, pp 1890-1901, 2016.
75. Sunil Kumar and BRaj, "Analysis of I_{ON} and Ambipolar Current for Dual-Material Gate-drain Overlapped DG-TFET," Journal of Nanoelectronics and Optoelectronics, American Scientific Publishers, USA, Vol.11, PP. 323-333, June 2016.
76. Naveed Anjum, Tarun Bali and B Raj, Design and Simulation Of Handwritten Multiscript Character Recognition, International Journal of Advanced Research in Computer and Communication Engineering, Vol.2 (7), PP. 2544-2549, July 2013.
77. Sanjeev Sharma, BRaj, Mamta Khosla, "A Gaussian Approach for Analytical Subthreshold Current Model of Cylindrical Nanowire FET with Quantum Mechanical Effects, Microelectronics Journal, Elsevier, Vol. 53, PP. 65-72, April 2016.
78. Karmjit Singh, BRaj, "Performance and analysis of temperature dependent Multi-walled Carbon Nanotubes as Global Interconnects at different technology nodes," Journal of Computational Electronics, Springer, Volume 14 Issue 2, pp. 469-476, June 2015.
79. Sunil Kumar and BRaj, "Compact channel potential analytical

Modeling of DG-TFET based on Evanescent–mode Approach," Journal of Computational Electronics, Springer Vol. 14, No. 2, pp.820-827, July 2015.

80. Karmjit Singh, BRaj, "Temperature Dependent Modeling and Performance Evaluation of Multi-Walled CNT and Single-Walled CNT as Global Interconnects," Journal of Electronic Materials, Springer, Vol. 44, Issue 12, pp 4825-4835, Dec 2015.
81. V.K. Sharma, M. Pattanaik and B.Raj, "INDEP approach for leakage reduction in nanoscale CMOS circuits", International Journal of Electronics, Taylor & Francis, Vol.102, No.2, pages 200-215, 2014.
82. Karmjit Singh, BRaj, "Influence of Temperature on MWCNT bundle, SWCNT bundle and Copper interconnects for Nanoscaled Technology nodes," Journal of Materials Science: Materials in Electronics. Springer, Vol. 26, No. 8, pp. 6134-6142, 2015.
83. Naveed Anjum, Tarun Bali and B Raj, "Design and Simulation of Handwritten Gurumukhi and Devanagri Numerical Recognition, International Journal of Computer Applications, Published by Foundation of Computer Science, New York, USA, Vol. 73(12), pp. 16-21, 2013.
84. Khandelwal, S., Gupta, V., BRaj, and Gupta, R D., Process Variability Aware Low Leakage Reliable Nano Scale DG-FinFET SRAM Cell Design Technique, Journal of Nanoelectronics and Optoelectronics, Volume 10, Number 6, pp. 810-817, December 2015.
85. V.K. Sharma, M. Pattanaik and B Raj, "ONOFIC Approach: Low Power High Speed Nanoscale VLSI Circuits Design", International Journal of Electronics, Taylor & Francis, Vol. 101, issue 1, pages 61-73, 2014.
86. Khandelwal, S., Balwinder Raj, and Gupta, R D., FinFET based 6T SRAM Cell Design: Analysis of Performance Metric, Process Variation and Temperature Effect. Journal of Computational and Theoretical Nanoscience, ASP, USA, Vol. 12, PP. 2500-2506, 2015.
87. Sumit Singh; Yadav, Shekhar; Rahul, Jagdeep; Srivastava, Anurag; BRaj, Impact of HfO_2 in Graded Channel Dual Insulator Double Gate MOSFET, Journal of Computational and Theoretical Nanoscience, American Scientific Publishers, Volume 12, Number 6, pp. 950-953, April 2015.

88. Vijay Kumar Sharma, Manisha Pattanaik, BRaj, PVT variations aware low leakage INDEP approach for nanoscale CMOS Circuits, Microelectronics Reliability, Elsevier, Vol. 54, pp 90-99, 2014.
89. B Raj, A. K. Saxena and S. Dasgupta, "Quantum Mechanical Analytical Modeling of Nanoscale DG FinFET: Evaluation of Potential, Threshold Voltage and Source/Drain Resistance", Journal of Material Science in Semiconductor Processing, Elsevier, Vol. 16, issue 4, pp. 1131- 1137, 2013.
90. Maisagalla Gopal, Siva Sankar D Prasad and Balwinder Raj, 8T SRAM Cell Design for Dynamic and Leakage Power Reduction, International Journal of Computer Applications, Published by Foundation of Computer Science, New York, USA, Vol. 71(9), pp. 43-48, June 2013.
91. Manisha Pattanaik, B Raj, Shashikant Sharma and Anjan Kumar, "Diode Based Trimode Multi-Threshold CMOS Technique for Ground Bounce Noise Reduction in Static CMOS Adders", Advanced Materials Research, Trans Tech Publications, Switzerland, vol. 548, pp 885-889, 2012.
92. Balwinder Raj, A. K. Saxena and S. Dasgupta, "Nanoscale FinFET Based SRAM Cell Design: Analysis of Performance metric, Process variation, Underlapped FinFET and Temperature effect" IEEE Circuits and System Magazine, vol. 11, issue 2, pp. 38- 50, 2011.
93. V.K. Sharma, M. Pattanaik & Balwinder Raj, "Leakage Current ONOFIC Approach for Deep Submicron VLSI Circuit Design", International Journal of Electrical, Computer, Electronics and Communication Engineering, World Academy of Sciences, Engineering and Technology, Vol. 7, issue 4, pages 239-244, 2013.
94. Tulika Chawla, Mamta Khosla, BRaj, Design and simulation of Triple metal Double-gate Germanium on Insulator Vertical Tunnel Field Effect Transistor, Microelectronics Journal, Elsevier; Vol. 114, PP. 105125, August 2021.
95. Parminder Kaur, Sandeep Singh Gill and B Raj, "Comparative Analysis of OFETs Materials and Devices for Sensor Applications", Journal of Silicon, Springer, Vol 14, PP. 4463–4471, 2022.
96. Sanjeev Kumar Sharma, Parveen Kumar, Balwant Raj and BRaj, "$In_{1-x}Ga_xAs$ Double Metal Gate-Stacking Cylindrical Nanowire MOSFET for highly sensitive Photo detector", *Journal of*

*Silicon, Springer Vol.*14, PP. 3535–3541, 2022.

97. B Raj, A. K. Saxena and S. Dasgupta, "Analytical Modeling of Quasi Planar Nanoscale Double Gate FinFET with Source/Drain Resistance and field dependent carrier mobility: A quantum mechanical study" Journal of Computer (JCP), Academy Publisher, FINLAND, vol. 4, issue 9, PP.1-8, 2009.
98. Bhushan, S., Khandelwal, S. and Raj, B., (2012). Analyzing Different Mode FinFET based Memory Cell at Different Power Supply for Leakage Reduction, Seventh International Conference on Bio-Inspired Computing: Theories and Application, (BIC-TA 2012) Advances in Intelligent Systems and Computing, Volume 202, p89-100 (2013).
99. Jeetendra Singh, BRaj, "Temperature Dependent Analytical Modeling and Simulations of Nanoscale Memristor, Journal: Engineering Science and Technology, an International Journal, Elsevier's, Vol. 21, PP- 862-868, Oct 2018.
100. Shradhya Singh, Shashi Bala, Balwant Raj, B Raj, "Improved Sensitivity of Dielectric Modulated Junctionless Transistor for Nanoscale Biosensor Design", Sensor Letter, ASP, Vol. 18, PP.328–333, Apl 2020.
101. Vivek Kumar, Santosh Kumar Vishvakarma, and B Raj, "Design and Performance Analysis of ASIC for IoT Applications" Sensor Letter ASP, Vol. 18, PP. 31–38, Jan 2020.
102. Akanksha Jaiswal, R. K. Sarin, B Raj and Shikha Sukhija, "A Novel Circular Slotted Microstrip-fed Patch Antenna with Three Triangle Shape Defected Ground Structure for Multiband Applications", Advanced Electromagnetic (AEM), Vol. 7, Issue 3, PP. 56-63, Aug 2018.
103. Girish Wadhwa, B Raj, "Label Free Detection of Biomolecules using Charge-Plasma-Based Gate Underlap Dielectric Modulated Junctionless TFET" Journal of Electronic Materials (JEMS), Springer, Volume 47, Issue 8, pp 4683–4693, August 2018.
104. Gurmohan Singh, R. K. Sarin and B Raj, "Design and Performance Analysis of a New Efficient Coplanar Quantum-Dot Cellular Automata Adder", Indian Journal of Pure & Applied Physics (IJPAP), Vol. 55, pp. 97-103, February 2017.
105. Amandeep Singh, Mamta Khosla, Balwinder Raj, "Design and Analysis of Electrostatic Doped Schottky Barrier CNTFET Based Low Power SRAM," International Journal of Electronics and Communications, (AEÜ), Elsevier, Vol. 80, PP.67–72,

2017.
106. Parminder Kaur, Vikas Pandey and B Raj, "Comparative Study of Efficient Design, Control and Monitoring of Solar Power using IoT", Sensor Letter, ASP Vol. 18, PP. 419–426, May 2020.
107. Anil Kumar Bhardwaj, Sumeet Gupta and B Raj, Development & Analysis of compact model for Double Gate Schottky Barrier CNTFET, Journal of Nanoelectronics and Optoelectronics, ASP, Vol. 15, pp. 1199–1208, Aug 2020.
108. Girish Wadhwa, Priyanka Kamboj; Jeetendra Singh, B Raj, "Design and Investigation of Junctionless DGTFET for Biological Molecule Recognition" Transactions on Electrical and Electronic Materials, Springer, 8 Aug, 2020.
109. Tulika Chawla, Mamta Khosla, B Raj, "Optimization of Double-gate Dual material GeOI-Vertical TFET for VLSI Circuit Design, IEEE VLSI Circuits and Systems Letter, Vol. 6, issue-2, PP.13-25, Aug 2020.
110. Sachin Kumar Verma, Shailendra Singh, Girish Wadhwa and B Raj, "Detection of Biomolecules using Charge-Plasma based Gate Underlap Dielectric Modulated Dopingless TFET" Transactions on Electrical and Electronic Materials (TEEM), Springer, Vol. 21, PP. 528-535, June 2020.
111. Neeraj Jain, BRaj, "Impact of underlap spacer region variation on Electrostatic and analog/RF performance of symmetrical high-k SOI FinFET at 20 nm channel length", Journal of Semiconductors (JoS), IOP Science, Vol.38, Issue 12, PP- 122002, Dec 2017.
112. Shailendra Singh, B Raj," Analytical Modeling and Simulation analysis of T-shaped III-V heterojunction Vertical T-FET", Superlattices and Microstructures, Elsevier, Vol. 147, PP. 106717, Nov 2020.
113. Gurmohan Singh, R. K. Sarin and B Raj, "Design and Analysis of Area Efficient QCA Based Reversible Logic Gates", Journal of Microprocessors and Microsystems, Elsevier, Vol. 52, PP.59-68, May 2017.
114. Amandeep Singh, Mamta Khosla, BRaj, "Compact Model for Ballistic Single Wall CNTFET under Quantum Capacitance Limit," Journal of Semiconductors (JoS), IOP Science, Vol.37, PP. 104001-8, Oct 2016.
115. Sonal Singh; Mamta Khosla, Girish Wadhwa; B Raj, "Design and analysis of double-gate junctionless vertical TFET for gas sensing applications", Applied Physics A, Springer,

Vol.127, No.16, 2 Jan 2021.

116. Inderjit Singh, BRaj, Mamta Khosla, and Brajesh Kumar Kaushik, Potential MRAM Technologies for Low Power SoCs, SPIN World Scientific Publisher, SCIE; Vol. 10, No. 04, PP.2050027, Dec 2020.
117. Shailendra Singh, B Raj,"Parametric variation analysis on hetero-junction Vertical t-shape TFET for Supressing ambipolar conduction", Indian Journal of Pure and Applied Physics, Vol. 58, pp. 478-485, June 2020.
118. Soniya, Shailendra Singh, Girish Wadhwa and Balwinder Raj, Design and Analysis of Dual Source Vertical Tunnel Field Effect Transistor for high performance, Transactions on Electrical and Electronics Materials, Springer, Vol. 21, Pages 74–82, Oct 2019.
119. Manjit Kaur; Neena Gupta; Sanjeev Kumar; BRaj; Arun Kumar Singh, "RF Performance Analysis of Intercalated Graphene Nanoribbon Based Global Level Interconnects" Journal of Computational Electronics, Springer, Vol. 19, PP.1002–1013, June 2020.
120. Girish Wadhwa, BRaj, "Design and Performance Analysis of Junctionless TFET Biosensor for high sensitivity" IEEE Nanotechnology, Vol.18, PP. 567 – 574, 2019.
121. Jeetendra Singh, BRaj, "Enhanced Nonlinear Memristor Model Encapsulating Stochastic Dopant Drift, JNO, ASP 14, 958–963 (2019)

6

Study of Charge Plasma based JLTFET for Biosensor Application

Girish Wadhwa and Balwinder Raj*
[1]Department of Electronics and Communication Engineering, NIT Jalandhar, India
**girishwadhwa12@gmail.com*

Abstract: As CMOS technology is facing many severe problems as a sensor application like- power scaling, short channel effects, sensitivity limitation and many fabrication issues etc. So, junction less transistor replaces CMOS technology because of its immunity to short channel effects and outstanding characteristics. But sensitivity is limited for junction less transistor devices either for charged biomolecule or for the neutral biomolecule. So, there is a need to enhance the sensitivity of the device for the detection of biomolecules that is why in this chapter work the focus is to study the sensitivity of the TFET devices for the detection of the biomolecules. In the proposed device, a cavity is introduced on side of the gate metal to attain high sensitivity for biomedical applications. The immobilization of biomolecules within the cavity induces variation in surface potential.

6.1 Introduction

In the Complementary Metal-Oxide - Semiconductor (CMOS) technology, over the past four decades, scaling has been a gateway to continuous development in the silicone-based semiconductor industry. However, as the nanometre-sized technology development framework for the goal of building ultra-high-density integrated electronic computers and expanding performance, CMOS devices face fundamental problems such as increased leakage currents, large variations in process parameters, short channel effects, higher production costs, etc. The new technology will need to be energy-efficient, dense and allow more features of the tool, per unit area and time. The present chapter discusses and summarizes advances in semiconductor devices production of low power vertical design Tunnel Field-Effect Transistors (TFETs). Owing to its steeper subthreshold slope (SS), TFETs are useful device for ultra-low power applications having very low OFF current. Their operations are based on the controlled switching of quantum-mechanical Band-to-Band Tunneling (BTBT), instead of the thermionic emissions.

6.2 Challenges and Development of Semiconductor device

With the intent to continue the growth of semiconductor technology and to keep Moore's law [1-5] alive, the recent research trends focus on miniaturization, low power operations, and cost-effective technologies with enhanced performance. However, the miniaturization of semiconductor devices especially those that rely on Metal-Oxide-Semiconductor (MOS)below nanometer regime causes several Short Channel Effects (SCEs) such as threshold variability, mobility degradation, Gate Induced Drain Leakage (GIDL), Drain Induced Barrier Lowering (DIBL),high leakage etc., which demean the performance of the devices [6], [7]. Because conventional FETs have failed to deliver the needed results in the nanoscale area, various nanoscale architectures have previously been proposed for low-power devices. We chose a nanowire charge plasma based junctionless vertical Tunnel FET biosensor for technology innovation because it has a higher Ion/Ioff current ratio, lower leakage current, and a steeper subthreshold slope than MOSFETs [8], [9]. The doped nanowire TFET has so many problems concerning heavily doped devices as the performance degradation of the device due to random dopant fluctuations (RDFs). Higher RDFs, variation in threshold voltage, large leakage current and high thermal budget of the device reduce the reliability of the device. However, for band to band tunneling to occur it requires abrupt junctions and creating abrupt junctions is not an easy task. So, to overcome the aforementioned problems, dopingless (charge plasma based) techniques have been investigated [10], [11].

6.3 Conventional Devices and Their Limitations

In 1959 Kahang and Attala invented the first basic working metal oxide semiconductor field effect transistor (MOSFET) [12-15].But after three decades, the MOSFETs became very popular electronic devices, which were being used in manufacturing of integrated circuits(IC's) for different types of applications with Si-SiO_2 by Kahang, in which silicon dioxide (SiO_2) provides good isolation and source to drain conduction current is controlled by gate terminal. But we know that the number of transistors is doubled on a chip in almost every two years. This exponential behavior of VLSI technology is known as Moore's law, as shown in Figure 6.1.

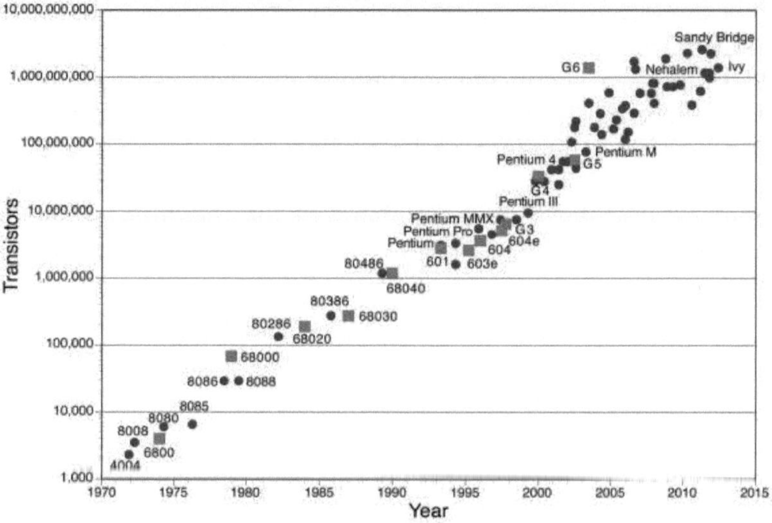

Figure 6.1 *Progress of number of transistors counts in CPU commercial chip versus year of introduction [Source: Intel Corporation]*

Key points of 2019/2020 International Roadmap for Device and System (IRDS)

This is a completely new part that puts the relationship between IRDS expectations and the state of the electronics sector in context.

I. In 2019, the electronics manufacturing industries stayed in good shape. Consumer electronics gained 10% year over year in 2019 and is expected to rise another 13% in 2020 to $365 billion.

II. R&D spending on semiconductors increased to 14.6 percent of income in 2019, with a slightly higher level of investment predicted in 2020.

III. For at least another ten years, new device designs using both vertical monolithic and heterogeneous integration will sustain expanded capability at historical rate but expect additional innovation beyond that.

IV. Logic vendors are still experimenting with the idea of stacked transistors. The drain current of nano-stacked ribbons is higher than that of FinFETs [16-19], according to Intel at the VLSI Symposia 2020.

6.4 Simulation Tool

For structuring transistors, technology computer aided design (TCAD) shows a significant job. Block diagram representation is shown in Figure 6.2. These are 2 dimensional or 3 dimensional recreation devices dependent on finite element technique, where the structure creation and its electrical attributes can be assessed dependent on the physical mechanisms of the handling of electrical attributes in the device. To predict the actual behavior of the transistors which involves virtual fabrication and characterization without actually fabricating it helps to save time and cost of the device. For the electrical characterization, the simulator provides both static and dynamic solutions.

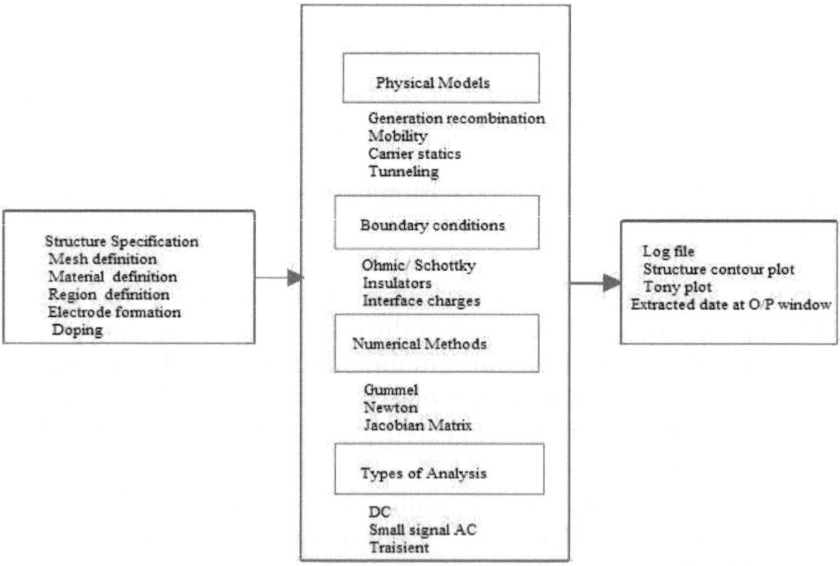

Figure 6.2 *Block diagram description for designing of structure and its characterization.*

6.4.1 Physical Models

Mobility

The mobility in bulk semiconductor region involves, (a) characterizing electrons and holes mobility as a function of lattice temperature and doping concentration, (b) illustrating saturation velocity as a function of lattice temperature (c) transition between low-field mobility. When an electric field is applied across the structure, the charge carriers are accelerated from valance band to conduction band and the momentum of these carrier losses. This is related to a variety of scattering phenomena, including impurity scattering, lattice fluctuations, surface roughness, and other flaws inside a material. The mobility mechanism is divided into four categories: (a) low-field movement, (b) high-field movement, (c) mobility in the bulk semiconductor area, and (d) mobility in the inversion region.

a) Several models are available for low field mobility such as concentration-dependent mobility (CONMOB) model which describes the dependency of mobility on the charge carriers (electron and hole) concentrations.
b) When an electric field is increased, mobility of carriers also increases but when electric field is reached critically high at the junction, there is sluggish rise in drift velocity and mobility is degraded after critical electric field which is occurred in short channel device. Thus, high field mobility model in the device is defined by Field dependent mobility (FLDMOB) model [11].
c) The mobility in the bulk semiconductor region involves two steps the first one is characterizing electrons and holes mobility is a function of lattice temperature and doping concentration, and the second is velocity saturation characterization which is a function of temperature.
d) Lombardi Constant Voltage and Temperature (CVT) [12] mobility model considers the dependency which is occurred due to transverse field, longitudinal field, doping concentration and lattice temperature in inversion mode of devices.

Generation Recombination

(i) Shockley-Read Hall (SRH) recombination

This model considers confined state while evolution of electrons inside the band gap. This is useful for indirect band-gap materials such as silicon and germanium. The model additionally defines the leakage current produced because of minority carriers. The trap charges at the intersection of Si-SiO_2additionally should be accounted for to precisely model the structure parameters. SRH model additionally called trap assisted recombination measure takes these trap charges into account [15], [16].

(ii) Auger recombination
In this model, there is a third transition which is energized excited at an advanced energy level where the mobile carrier can either be trapped or transmitted. The model is considered for a heavily doped structure. However, the basis of third transition is still needed to be explored.

Band Gap Narrowing

The experimental result shows that a heavily doped semiconductor material becomes doping dependent (for doping >10^{18}cm^{-3}) [17-19]. By increasing the level of doping the conduction band starts lowering and valance band rose by the same amount. This results in the narrowing of the band gap. To resemble this effect band gap narrowing (BGN) model has to be considered to simulate the heavily doped device.

The spatial variation in the intrinsic carrier concentration is defined in equation 6.1 as

$$n_{ie}^2 = n_i^2 \exp\left({\Delta E_G}/{kT} \right) \tag{6.1}$$

Here $\Delta E_G$is the variation in the band gap.

Direct Quantum Tunneling

The aggressive scaling of gate oxide thickness to have better control over channel leads to an increase in the leakage current. The leakage current in modern MOS devices is the result of direct tunneling process. Even if the energy of an electron is less than the barrier height, there is a finite probability that the particle penetrates the barrier. The tunneling of electrons through the triangular barrier is called as Fowler-Nordheim (FN) tunneling is a special case of direct tunneling process. However, for a very thin gate oxide FN tunneling is invalid rather more accurately the tunneling takes place through the trapezoidal barrier. Here the effective mass approximation is used to formulate the Schrödinger equation for the calculation of transmission probability.

6.5 Dual Material Double Gate Transistor

Although a double gate metal oxide semiconductor field effect transistor provides improved short channel characteristics as compared to single gate MOSFET because of better electrostatic control over the channel. However, for significantly small gate length the potential expansion along the channel results in lowering of source channel potential barrier causing threshold voltage roll-off and DIBL. Moreover, due to non-uniform electric field in the channel and the occurrence of electric field near drain end the velocity of charge carrier accelerating from source towards drain side is low. A dual material gate (DMG) MOSFET was proposed [8], [20-24]. Based upon the concept of dual gate and split gate architecture. The DMG MOSFET helps to enhance the electron transport efficiency and suppress DIBL [8]. In a DMG device metal gate (top/bottom) splits into gates M1 and M2 of two different gate workfunctions (i.e. two different metal gates). Metal gate M1 is close to source side having higher gate workfunction compared to M2 which is close to drain side [23]. Due to the different gate workfunctions of M1 and M2 the potential abruptness occurs at the junction between the two gates.

6.6 Junctionless Transistor

A multiple gate MOSFET such as double gate MOSFET [13], FinFET [16] and gate all around devices [21] provide better electrostatic control because more than one gate is controlling the channel. Despite having better SCE, the creation of such a sharp junction for an ultra-scaled structure is reasonably challenging. A junctionless transistor (JLT) [25] recently emerged as a promising device having characteristics similar to conventional MOSFET and is uniformly doped device with either n+ for n-JLT or p+ for p-JLT [27]. Therefore, the fabrication steps are simpler as compared to conventional MOSFET. It can be observed that in a JLT the entire silicon film is doped with n+ impurity for n-type JLT (or p+ impurity for p-type JLT). Due to the work function mismatch between p+ gate and n+ thin silicon film, the n-JLT device is entirely depleted and there is a barrier between source and channel off-state [25-27]. Conduction occurs when the gate bias is increased, forming a channel within the bulk. [9]. Ina conventional double gate MOSFET since it is based on the formation of an inversion region or the channel is inverted in on-state condition, it is often called an inversion mode transistor. J. Lelienfield first-ever proposed and patented the concept of junctionless transistor in 1930. Due to unavailability of advanced fabrication technology for the deposition of thin film, its fabrication could not be made possible at that time. Recently, Colingeet. al. [1] have successfully fabricated junctionless transistors and demonstrated characteristics similar to that of inversion mode transistors.

6.7 Emerging Semiconductor Biosensor

Research in investigating the different biomedical biomarkers such as nutritional imbalances of brain disorders is expanding in addition to research on their etiology. A biosensor [28-31] focusing on the biomarkers, for example, Amino Acids (AAs) and DNA is a significant requirement for early analysis and mediation of disorder. The biosensors can detect the presence of a specific biomolecule in a given target sample such as human sample (urine, blood and saliva) or food and environmental sample (water, soil). For instance, a blood glucose detector non-invasively senses the presence of glucose in blood sample. In the same vein, to trace the neurological and psychiatric problems the enzymatic microbiosensor, using D-amino acid oxidase as an enzyme, has been designed that detects the concentration of d-serine Amino Acid (AA) in human brain [30]. Biosensors are mainly constructed (i) to identify a component within tissue, which has been observed in several studies screening for biomarkers that are indicators of a particular disease, discovering drugs, as well as monitoring the disease. (ii) to characterize novel molecular interactions between different biological molecules [31]. The biosensors have a number of advantages over the time-consuming laboratory-based medical diagnostics such as high precision and accuracy, ultra-specificity and sensitivity, fast, simple to use, very low reagent consumption, automatic optimization of measurable parameters and offers reusability of sensing elements.

Amino Acids (AAs) are building blocks of the human body that can play a vital role in normal functioning of humans, for example, Glycine performs several physiological functionalities such as a neurotransmitter in the central nervous system [32-34], arginine is considered as Miracle Molecule and most potent nutraceutical with powerful healing properties [32], tryptophan modulates the neurological and immunological activities through multiple metabolites which includes serotonin and melatonin [33]. Depending upon synthesis of protein and nitrogen balance, the amino acids have been categorized into essential (nine in number) and non-essential (eleven in number). The proper balance of these acids is essential for emotional and cognitive development, healthy functioning of nervous system and vital organs. The nano regime device offers a new avenue for bio-sensing applications [29], however, their emergence brings a whole new set of challenges which include Random Dopant Fluctuation (RDF) and thermal budget. FET-based biosensors have also shown their capability in clinical diagnostic field by detecting the targets (such as nucleotides or cells) which are important biomarkers of diseases such as cardiac, diabetes and infectious diseases. Biosensors based on Field Effect Transistor (FETs) [35] have shown enormous potential in a number of fields that includes medical, nanotechnological, biotechnological, pharmaceutical, agricultural and various other industry sectors [36].

6.7.1 Label-Based Detection

The word 'label' represents any foreign biological analyte linked chemically or temporarily to recognize the desired analyte [35]. During the bio-recognition of the objective analyte on sensor surface, the tagging is a mandatory step to be used in conjunction with bio-recognition unit for the amplifying the signal as well as sensitivity of the biosensor. The labeling will lead to more efficient detection of analyte as it develops the number of binding sites. Moreover, electrochemical sensors can easily identify the labels.

Based on the labels/reagents such as enzymes, nanomaterials, quantum dots, and dyes, the label-Based detection can be categorized as:

(i) Fluorescence Labeling
(ii) Chemiluminescence
(iii) Radioactive/ Isotope Labeling

(iv) Electrochemically Active Probe Labeling
(v) Nanoparticle Labeling

6.7.2 Label-free Detection

To simplify the review process and lower cost while providing more qualitative and quantitative information, label-free methods are proposed. The label-free recognition, with no need for linking biological analyte, examines the target biological analyte based on physical or chemical characteristics of target biological analyte by noticing and depending exclusively on binding sensitivity, selectivity and refractive index. The existence of the target biological analyte is determined by intrinsic physical qualities such as dielectric constants, mass, volume, and conductivity. The elimination of confounding molecular labels has reduced the sample preparation time, procedure complexity, need for trained personnel as well as analysis cost [36], [37]. The label-free biosensing has great potential to be fabricated to micro-scale systems and deliver give superior quality sensors that are highly flexible to remote diagnostics and can test any complex analyte with minimal assay development. The label-free methods can be categorized as:
(i) Surface Plasmon Resonance (SPR)
(ii) Quartz Crystal Microbalance
(iii) Mass Spectrometry
(iv) Microcantilevers

6.8 Tunnel Field Effect Transistor (TFET)

To improve device efficiency, researchers began looking for solutions that could eventually replace MOSFETs. Tunnel Field Effect Transistor (TFET) [38] is a type of PIN diode wherein the conduction is controlled by the electric field of the gate metal. TFET relies on band-to-band tunneling (BTBT) mechanism instead of thermionic diffusion (conduction mechanism of conventional MOSFET) for current conduction. Short-channel performance is improved with TFETs, and leakage current is reduced [39] and at the same time, the BTBT conduction mechanism used by TFETs permits the subthreshold slope (SS) of the device to be below 60mv/dec.

The above-mentioned merits of TFET indicate its usefulness as an ultra-low power structure in the near future. Yet some issues remain unresolved which reduce the performance and usefulness of TFETs. Some major ones among them include the low ON drive current (I_{ON}) as compared to traditional MOSFETs and high ambipolarity (conduction for both positive gate voltages and negative gate voltages) [38-41].

Several techniques are introduced in past few decades to improve the ON current of structure like use of low bandgap materials, high k-dielectric, hetero-structures and nanowire Tunnel FET [38-42]. Another undesirable characteristic of Tunnel FETs is the requirement of sharp junctions for efficient band to band tunneling mechanism [39], [42]. This mandates a high thermal budget in case of doped source/drain due to thermal annealing and ion-implantation technique. For physically doped structures, diffusion of charge carriers to the channel region from drain (or source) region makes an abrupt junction profile very difficult to achieve. Also continuous reduction in size of doped Tunnel FET having heavy doping profile results in higher random dopant fluctuations (RDF) affecting variable structure performance [40], [41].

6.8.1 Charge plasma based TFET

The conventional TFETs are suffered from random dopant fluctuations (RDF) and abrupt junction problems. Therefore we move to another type of electronic device which is used another type of technique for source and drain formation in the intrinsic silicon semiconductor. This type of TFET is called the doping less TFET [41]. The charge plasma technique for creating n-type or p-type regions in TFET without doping (Doping-Less TFET) is introduced to conquer this problem [43]. The metal electrode having a suitable work function has been used to obtain N^+type or P^+type regions on intrinsic material body in dopingless technique [44]. DL-TFETs are fabricated at low temperatures because they do not require high temperature annealing. The thickness of the silicon film is given by the Debye length and given by eq. (1.2).

$$L_d = (\frac{\epsilon_{si} V_T}{Q} + 60n)^{1/2} \quad (1.2)$$

Where V_T, q, N and $\varepsilon_{si}$are thermal voltage, electronic charge, carrier concentration and permittivity of the silicon. A thicker high-k dielectric material between source metal electrode and silicon substrate for avoiding the silicide formation is used.

6.8.2 Junction less tunnel FET

The JLTFET is made of silicon with a heavily doped N type substrate. The JLTFET [26], [45] in this study employs the dielectric modulation approach, which facilitates the bio-transistor in distinguishing between neutral and charged analytes (biological molecules). The P^{+}source region is introduced by applying metal work function on highly doped N type silicon substrate. To limit the impact of short channels on device properties, a dual metal gate structure is built by using two different metal gate electrodes [27], [39]. By selecting an appropriate metal electrode work function for the source (S), one can construct the necessary p+ source region in an n-type silicon body that has been heavily doped. ($n_i = 10^{19} cm^{-3}$)

6.9 Charge-plasma-based dielectric modulated gate underlap JLTFET utilized as biosensor

A gate underlap dielectric modulated JLTFET has been intended for the label free electrical recognition of biological molecules like myoglobin, Protein, apomyoglobin, staphylococcal nuclease so forth including the Dielectric Modulation (DM) method [46].

The competence of DM-JLTFET [46-48] structure to distinguish the occurrence of biomolecules in nanocavity is corresponding to electrical attributes sensitivity. When immobilized biological molecules are available in the gate underlap channel area, a CPB DM JLTFET device shows a superior stage for recognition of biological molecules. A few significant highlights viz. label-free recognition, selectivity, high sensitivity (for neutral and charged analysts) features the device as a better option in comparison with other available-devices accessible right now [49]. Consequently, for reduction in cost of biomedical equipment, a CPB GU DM-JLTFET shows high sensitivity and superior selectivity by arrangement of cavity length and thickness close to the source-channel intersection at a particular input voltage. By considering the effect of biological molecules on the electrical quantities of the CPB GU DM-JLTFET structure, it is discovered that it shows less short channel effect and low leakage current.

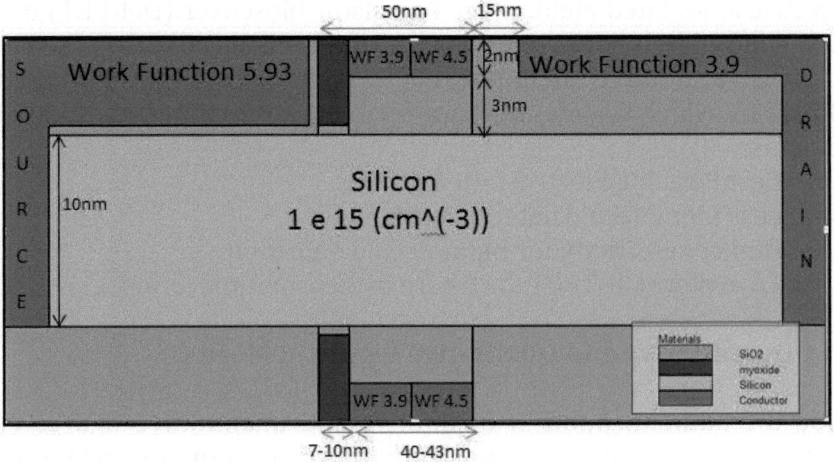

Figure 6.3 *Cross-sectional device structure of proposed charge-plasma-based dielectric modulated Junctionless Tunnel FET.*

6.10 Research Contribution

As CMOS technology is facing many severe problems as a sensor application like- power scaling, short channel effects, Sensitivity limitation and many fabrication issues etc. So junction less transistor replaces CMOS technology [50-53] because of its immunity to short channel effects and outstanding characteristics. But sensitivity is limited for junction less transistor devices either for charged biomolecule or for the neutral biomolecule. So there is a need to enhance the sensitivity of the device for the detection of biomolecules that is why in this chapter work the main focus is to enhance the sensitivity of the proposed device for the detection of the biomolecules. From the literature survey, it has been found that conventional JLTFET [1] faces many issues. It has been observed from the detailed literature survey that the research work related to the charge plasma based dielectric modulated Tunnel-FET [12], [16], [27] and its biosensor application is not yet limited. Following research gaps are observed after a survey of the literature. The main challenge in TFET is to realize high ON currents compared with CMOS counterparts. The tunneling distance, the band gap, and the oxide thickness of the source material affect the drive current and the impact of the drain field on the tunneling distance affects the output conductance and the saturation voltage of the devices.

Dielectric Modulated Field Effect Transistor biosensor (DM-FET) has a few limitations, which limits the utilization of sensors that use capacitance variation as the input [89] so that a charge-plasma-based DM-JLTFET which is gate underlap, overcomes the above-mentioned problems.

a) Best Possible Electrostatic Control
b) Less biomolecule detection time
c) Higher sensitivity for biomolecule detection
d) To explore CP-TFET for future biosensor applications.

6.11 Quantitative and Qualitative Figure of Merits

First, the authentication of qualitative and quantitative figures of merit is a basic level to demonstrate the capacity of biological sensors [50-52]. To standard the device, guarantee superior performance, good selectivity and sensitivity of the structure, it is obligatory to put some more effort into the parameters as talked about beneath:

(i) Short-Channel Impacts (SCIs): The SCIs profoundly affect the presentation of the detecting deice, i.e., affectability and are progressively serious in thicker oxide layers [53]. It minimizes the threshold voltage and output current of FET biological sensors because of its foremost wonder thermionic emission which brings about low affectability of the device. To defeat the SCEs:

a) Numerous device designing methodologies have been utilized, for example, gate dielectric study or different channel doping.
b) A different elective structure has been proposed, for example, the fabricating channel profounds into the substrate or creates silicon on insulator (SOI) FET.

(ii) Immobilization methods: The immobilization methods should be picked such that the chosen technique ought to create a powerful and solid sensor area that has low absorption, high explicitness/selectivity, and high binding productivity of target biomolecule [54]. The properties of the sensor surface choose the technique for immobilization that can be utilized for superior operational stabilities. The chosen technique should upgrade the consistency, thickness, stability, reusability and dispersion of the detecting surface. The utilization of polymers and nanomaterials can accomplish immobilization of analytes permitting high affectability and recognition range of sensing surface.

(iii) Specificity: The specificity or selectivity is a trademark that is similarly significant as affectability to portray the biological sensor structure. The absorption of the biological molecules on the recognition surface changes the structureelectrical attributes [55]. Also, it has accounted that the exposed channel prompts absorption of the objective biological molecules, and along with this prompting interference of the non-required biological molecules i.e., non-specificity.

(iv) Signal-to-Noise Ratio (SNR): Fundamentally, SNR mirrors the loyalty and relies on the impedance coordination between various sub-frameworks of FET-based biological sensors [56]. If the noise signal is produced from the testing condition, at that point it is called signal to background proportion (Sbr= Change in drain current / I_D, where change in drain current is the variation in drain current when including the objective biomolecules and I_D is the drain current before including biomolecule).

(v) Nanocavity Filling Area (NFA): The sensitivity of the FET biological molecules sensor based on the nanocavitylength and its filling basis [11]. Regardless of the nature and size of target biological molecules, the device investigation is commonly done with hypothesis that the nanocavity is filled by the biological molecules or air [57]. Much after this presumption, a significant number of empty spaces stay exist in the nanocavity of the device, thus, influencing sensitivity. The N_{FA} can give the nanocavity zone filled by the biological molecules in equation 1.3 and solved by utilizing following condition:

$$N_{FA} = \frac{P_{FA}}{F_{FA}} \times 100 \tag{1.3}$$

Here N_{FA} denotes to nanocavity filling zone, P_{FA} is a partially filled zone, F_{FA} is a filled area. As shown in Figure 6.4, there can be various manners by which a filled or incompletely filled nanogap can be portrayed i.e., uniform and non-uniform ways.

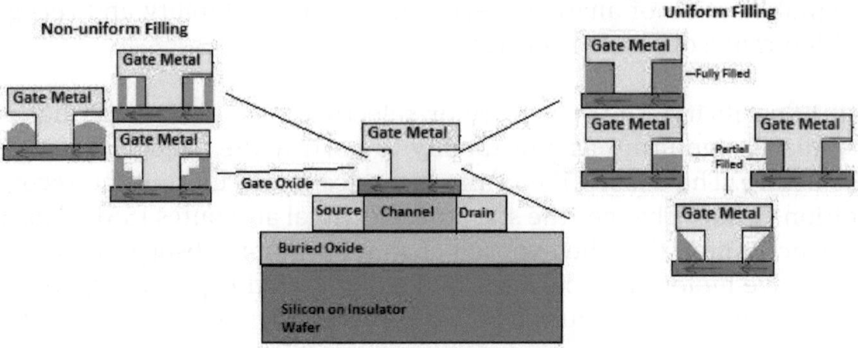

Figure 6.4 *Typical structure of DM-FET biosensor reflecting various methods for non-uniform and uniform filling of the nanogap region.*

With filled nanocavity, the structure is more delicate to surface charge just as dielectric steady, since sensitivity is closely related to these two impacts.

6.12 FET based dielectric modulate Biosensor

FETs are a good choice for biosensors, but its sensitivity is limited by short-channel effect (SCE) that restricts the downscaling of the FETs in nanoscale and Sub-threshold Swing (SS) non-scalability (60mv/dec) [17-19]. A better alternative to FET seems to be Tunnel FET (TFET) [17] which can improve the sensor performance by virtue of its different current generation mechanism called as Band to Band Tunneling (BTBT). Aside from the cost factor, the pragmatic issue, for example, poor reliability (i.e., quality affirmation during manufacture) additionally blocks the commercialization of FET biosensors and should be settled. The qualities of Si-SiO2 interface straightforwardly add to the dependability of the biosensor. A portion of the issues happening inside the oxide, for example, traps, fixed charge likewise lead to poor dependability and streamlining of the device. The above talked about parameters can improve the sensitivity and reliability of the biosensor. Changes in the electrical attributes of the device before and after immobilization of the biomolecule in TFET biosensor are used to determine the TFET's sensitivity. For TFET sensitivity, measures such as the ratio of on current to off current ratio (Ion/Ioff), surface potential, and subthreshold swing, among others, are used.

Tunnel FET is extensively used for biosensor label free detection [35], [43] because scaling of FET introduces various shortcomings such as short channel effects, High Power consumption, minimum I_{ON}/I_{OFF} and scaling of threshold with respect to power supply, the biomolecules detection time is large but due to high doping concentration of TFET constitute the fabrication process not easy to analyse.

a) Another limitation of junctionless transistors is the latch phenomena [1]. Due to this problem, transistor is latched to on-state and fails to turn off. To encounter such problem of poor switch-off capability, keeping the device in junctionless regime needs a distinctive approach.
b) Dielectric Modulated FET has a few shortcomings, which limits the use of sensors that utilize capacitance variation as the signal, sensitivity, linearity, high cost, and low binding probability in the cavity region.

In future, alongside these figures of merit, the thought of structure viewpoint is extremely basic as an appropriately planned biosensor would not influence the device sensitivity. For instance, the sensitivity can be upgraded by:

(i) selecting TFET-based biosensor structures with subthreshold slope less than 60 mV/dec
(ii) Proficient surface immobilization

6.13 Motivation behind this work

As TFET is a heavily doped device so problem concerning heavily doped devices is the performance degradation of the device due to random dopant fluctuations (RDFs) [2], [6]. Due to RDFs threshold voltage of the device varies which increases the leakage currents of the device. However, for band to band tunneling to occur it requires abrupt junctions and creating abrupt junctions is not an easy task. So, in order to overcome the aforementioned problems, doping-less (charge plasma based) techniques have been investigated.

Dielectric Modulated Field Effect Transistor biosensor (DM-FET) has a few limitations, which limits the utilization of sensors that use capacitance variation as the input [89]so that a charge-plasma-based DM-JLTFET which is gate underlap, overcomes the above-mentioned problems.

a) Best Possible Electrostatic Control
b) Less biomolecule detection time
c) Resolve the problem of higher doping control during fabrication
d) Higher RDF in JL-TFET
e) High thermal budget,
f) Problem of threshold voltage variability
g) Higher sensitivity for biomolecule detection
h) To explore CP-TFET for future biosensor applications.

Commercial Viability: The following greatest test is the cost factor since the appropriation and wide acknowledgment of FET-biosensors [89] for most applications in each area (particularly in pharmaceuticals) relies on the expense of the device. The difficulty level between the imaginative research thoughts at scholastic level and commercialization is because of the cost factor. The label free strategy offers a good favorable position of decreased expense over label based detection. The cost parameters rely on: (i) transducer, (ii) bundling, (iii) testing, (iv) producing, (v) extra reagents, (vi) temperature control requirement, (vii)Industrial handling methods. Hence, the essential worry for commercialization should be the conventional primary concern that is growing ease biosensors, while improving a portion of the exhibition measurements. The potential approaches to decrease the expense and improve the sensitivity of the biosensors.

(i) Use of natural materials: The execution of natural materials in framing transducer by supplanting the channel dielectric layer with polymer can yield minimal effort because of diminished creation costs, power utilization, straightforwardness and high adaptability. The natural slender film FETs have given another option and a potential answer for an approach for financially savvy biosensors.

(ii) Regeneration and large-scale manufacturing: The recovery and large scale manufacturing of institutionalized sensors by devouring cheap techniques and materials likewise offer a delivery making biosensors savvy. Consequently, a multidisciplinary group involving physicists, specialists, scholars and experts is expected to address the sensor advancement from the earliest starting point to plan and assemble a biosensor that is modest, convenient and equipped for being utilized by semi- skilled operators.

6.14 Key Parameters of Biosensor Sensitivity

The sensitivity and recognition speed are the important factors of apprehension for the designing of a biological sensor. The sensitivity of biomolecule detection increases as the as the size of the cavity enhances. The Tunnel FET (TFET) centered dielectric modulated biological sensors proved itself as the most potential candidate for the design of the next-generation biosensor with minimum short-channel effect (SCE) and low subthreshold swing (SS) (less than 60 mV/decade). The low subthreshold slop help to improve the bio-molecule detection time of the device.

The sensitivity is limited for junction less transistor devices either for charged biomolecule or for the neutral biomolecule. So, there is a need to enhance the sensitivity of the device for the detection of biomolecules that is why in this chapter work the main focus is to enhance the sensitivity of the charge plasma based device for the detection of the biomolecules.

Scaling down of device: The sensitivity of biomolecule detection increases as the device dimensions are scaled down. The utilization of nanostructured materials and configuration approaches can scale down the biosensors and decrease the expense because of less pre-requisites of much vitality, materials and endeavors. For instance, the utilization of silicon nanowire, graphene and other two-dimensional materials in FET biosensors has diminished expense as well as gave appealing highlights, for example, scaling down, light-weight, and multiplexed recognition.

6.15 Conclusions

In this book chapter we have studied that when immobilised biological molecules are existing in the gate underlap channel region then dopingless junctionless Tunnel FET device provides an improved stage for biomolecule detection. Numerous key advantages, such as label-free recognition and maximal sensitivity, distinguish the device as a superior choice above other sensing devices currently available. A dopingless gate underlap DM-JLTFET demonstrates superior enactment for cost-effective improvement of biomedical analysis devices, as it achieves increased sensitivity through modest adjustment of structural parameters of cavity region near the source-channel junction for optimal biasing circumstances. The impact of biomolecules on the electrical attributes of the proposed device structure, such as drain current, surface potential, Ion/Ioff ratio, and subthreshold slope, is discovered to have low leakage current and a high sensitivity factor for charged biomolecules when compared to other FET devices. The CPB DM-JLTFET device has a low impact on short channel effect and leakage currents. An analytical potential model of CPB DM-JLTFET for all the ten sections has been evaluated and validated by comparing it with TCAD simulation results.

6.16 Future Scopes

The point-of-care-diagnostics play a crucial role in providing an instant response with their rapid and real-time applications. These days, to detect neurological and psychiatric problems, there is a need for efficient and accurate diagnostic tools. The FET-based biosensors have also shown their capability in the clinical diagnostic field by detecting the targets (such as nucleotides or cells) which are important biomarkers of diseases such as cardiac, diabetes, and infectious diseases. The proposal aims to propose a Junctionless Nanowire Vertical TFET Biosensor with Bioreceptor with a final goal to provide a diagnostic tool. To benchmark the device structure, high sensitivity, specificity, stability, accuracy, reusability, and insensitivity to physical parameters such as pH and environmental factors, it is necessary to focus further on the characteristics discussed below:

i. Short Channel Effects (SCEs)
ii. Immobilization Technique
iii. Specificity
iv. Linearity
v. Signal-to-Noise Ratio (SNR)
vi. Regeneration or Reusability

There is a Risk of increased expression of ACE2 membrane protein in patients with hypertension. Research investigating the different biomedical biomarkers such as ACE2 membrane protein of hypertensive retinopathy is expanding and researching their etiology. A biosensor focuses on the biomarkers; for example, Amino Acids (AAs) and DNA is significant requirements for early analysis and mediation of blood pressure and heart rate. The biosensors can detect a specific biomolecule in a given target sample, such as a human sample (saliva). In the same vein, to trace the neurological and hypertensive retinopathy problems, the enzymatic micro biosensor, using D-amino acid oxidase as an enzyme, has been designed that detects the concentration of d-serine Amino Acid (AA), Glutamic acid and Leptin in the human body. Glutathione peroxidase is partially responsible for determining the susceptibility of cells to oxidative stress. Point-of-care (POC) biosensors such as chip-based sensors are portable, cost-effective, and user-friendly, and can use for rapid detection of high blood pressure. These days, to detect neurological and hypertensive retinopathy problems, there is a need for efficient and accurate diagnostic tools. The proposal proposes a Non-Invasive Screening for hypertension by detecting ACE2 membrane protein in saliva using bio-receptor immobilized on Junctionless Nanowire Vertical TFET Biosensor Model.

The nano regime device offers a new avenue for bio-sensing applications, however, their emergence brings a whole new set of challenges which include Random Dopant Fluctuation (RDF) and thermal budget. FET-based biosensors have also shown their capability in clinical diagnostic field by detecting the targets (such as nucleotides or cells) which are important biomarkers of diseases such as cardiac, diabetes and infectious diseases. Biosensors based on Field Effect Transistor (FETs) have shown enormous potential in several fields that include medical, nanotechnological, biotechnological, pharmaceutical, agricultural and various other industry sectors.

References

[1] J. P. Colinge, C.-W.Lee, A. Afzalian, ND Akhavanl, Ran Yan, Isabelle Ferain, Pedram Rajavi, Brendan O'Neill, Alan Blake, Mary White, Anne-Marie Kellehar, Brendan McCarthy and Richard Marphy, "Nanowire transistors without junctions," Nature Nanotechnol., vol. 5, no. 3, pp. 225–229, Mar. 2010.

[2] G. Leung and C. O. Chui, "Variability impact of random dopant fluctuation on nanoscale junctionless FinFETs," IEEE Electron Device Letters, vol. 33, no. 6, pp. 767–769, Jun. 2012.

[3] Dhiman, R., & Chandel, R. (Eds.). (2020). *Nanoscale VLSI: Devices, Circuits and Applications*. Springer Nature.

[4] Kumar, S., Jantsch, A., Soininen, J. P., Forsell, M., Millberg, M., Oberg, J., ... & Hemani, A. (2002, April). A network on chip architecture and design methodology. In *Proceedings IEEE Computer Society Annual Symposium on VLSI. New Paradigms for VLSI Systems Design. ISVLSI 2002* (pp. 117-124). IEEE.

[5] Srivastava, S., & Rathod, S. S. (2016, March). Silicon neuron-analog CMOS VLSI implementation and analysis at 180nm. In *2016 3rd International Conference on Devices, Circuits and Systems (ICDCS)* (pp. 28-32). IEEE.

[6] Gnudi, S. Reggiani, E. Gnani, G.Baccarani, "Analysis of threshold voltage variability due to random dopant fluctuations in junctionless FETs," IEEE Electron Device Letters., vol. 33, no. 3, pp. 336–338, Mar. 2012.

[7] S. Verhulst, D. Leonelli, R. Rooyackers, and G. Groeseneken, "Drain voltage dependent analytical model of tunnel field-effect transistors," Journal of Applied Physics, vol. 110, no. 2, p. 024510, 2011.

[8] Wei Long, HaijiangOu, Jen-Min Kuo, and Ken K. Chin, "Dual-material gate (DMG) field effect transistor", IEEE Transactions on Electron Devices, vol. 46, no. 5, pp. 865-870, May 1999.

[9] Sahu, and J. Singh, "Charge-plasma based process variation immune junctionless transistor," IEEE Electron Device Letters, vol. 35, no. 3, pp. 411-413, Mar. 2014.

[10] Dutta, U., Soni, M. K., & Pattanaik, M. (2018). Design and Analysis of Tunnel FET for Low Power High Performance Applications. International Journal of Modern Education & Computer Science, 10(1).

[11] Anjana Bhardwaj, Pradeep Kumar, B. Raj, Sunny Anand, "Design and Performance Optimization of doping-less Vertical Nanowire TFET using Gate Stack Technique", Journal of Electronic Materials (JEMS) Springer, Vol. 41, Issuse 7, PP. 4005-4013, 2022.

[12] Jeetendra Singh, BRaj, "Tunnel Current Model of Asymmetric MIM Structure Levying Various Image Forces to Analyze the Characteristics of Filamentary Memristor", Applied Physics A, Springer, Vol. 125, No.3, pp. 203.1 to 203.11 Feb 2019.

[13] Candy Goyal, Jagpal Singh Ubhi and B Raj, Low Leakage Zero Ground Noise Nanoscale Full Adder using Source Biasing Technique, "Journal of Nanoelectronics and Optoelectronics", American Scientific Publishers, Vol. 14, PP. 360–370, March 2019.

[14] Gurmohan Singh, R. K. Sarin and B Raj, "A Novel Robust Exclusive-OR Function Implementation in QCA Nanotechnology with Energy Dissipation Analysis", Journal of Computational Electronics, Springer, Volume 15, Issue 2, PP. 455-465, June 2016.

[15] Tanu Wadhera, Deepti Kakkar, Girish wadhwa, B Raj, " Recent Advances and Progress in Development of the Field Effect Transistor Biosensor: A Review" Journal of ELECTRONIC MATERIALS, Springer, Volume 48, Issue 12, pp 7635–7646, December 2019.

[16] Girish Wadhwa, Priyanka Kamboj, Balwinder Raj" Design optimisation of junctionless TFET biosensor for high sensitivity" Advances in Natural Sciences: Nanoscience and Nanotechnology, Vol. 10, PP. 045001 (7), (2019).

[17] Priya Bansal, BRaj, "MEMRISTOR MODELING AND ANALYSIS FOR LINEAR DOPANT DRIFT KINETICS," Journal of Nanoengineering and Nanomanufacturing, American Scientific Publishers, Vol. 6, pp. 1–7, 2016.

[18] Amandeep Singh, Mamta Khosla, BRaj, "Circuit Compatible Model for Electrostatic Doped Schottky Barrier CNTFET,"Journal of Electronic Materials, Springer, Vol. 45, Issue 12, pp 4825-4835, 2016.

[19] Ashima, D Vaithiyanathan, B Raj," Performance Analysis of Charge Plasma induced Graded Channel Si nanotube", Journal of Engineering Research (JER), EMSME Special Issue pp. 146-154 Aug 2021.

[20] Abhishek Singh Tomar, Vijay Kumar Magraiya, and B Raj, Scaling of Access and Data Transistor for High Performance DRAM Cell Design, Quantum Matter, Vol. 2, PP. 412-416 Oct. 2013.

[21] Neeraj Jain, BRaj, "Parasitic Capacitance and Resistance Model Development and Optimization of Raised Source/Drain SOI FinFET Structure for Analog Circuit Applications, Journal of Nanoelectronics and Optoelectronins, ASP, USA, Vol. 13, PP. 531–539, Apl 2018.

[22] Shradhya Singh, S. K. Vishvakarma and B Raj, "Analytical Modeling of Split-Gate Junction-Less Transistor for a Biosensor Application", Sensing and Bio-sensing, Elsevier, Vol. 18, Pages 31-36, April 2018.

[23] Maisagalla Gopal, and Balwinder Raj, Low Power 8T SRAM Cell Design for High Stability Video Applications, ITSI Transaction on

Electrical and Electronics Engineering, Vol. 1(5), pp. 91-97, 2013.

[24] Balwinder Raj, Jatin Mitra, Deepak Kumar Bihani, Rangharajan V, A.K. Saxena and S. Dasgupta, "Analysis of Noise Margin, Power and Process Variation for 32 nm FinFET Based 6T SRAM Cell" Journal of Computer (JCP), Academy Publisher, FINLAND, vol. 5, issue 6, PP.1-8, 2010.

[25] Divya Sharma, Rajesh Mehra and B Raj, "Comparative Analysis of Photovoltaic Technologies for High Efficiency Solar Cell Design", Superlattices and Microstructures, Elsevier, Volume 153, PP. 106861, May 2021.

[26] Pawandeep Kaur, Avtar Singh Buttar, B Raj, "A comprehensive Analysis of Nanoscale Transistor based Biosensor: A Review, Indian Journal of Pure and Applied Physics, Vol. 59, pp. 304-318, April 2021.

[27] Divya Yadav, Balwant Raj and B Raj, "Design and Simulation of Low Power Microcontroller for IoT Applications", Journal of Sensor Letters, ASP, Vol. 18, PP. 401–409, May 2020.

[28] Shailendra Singh, B Raj,"A 2-D Analytical Surface Potential and Drain current Modeling of Double-Gate Vertical t-shaped Tunnel FET", Journal of Computational Electronics, Springer, Vol. 19, PP.1154–1163, Apl 2020.

[29] Jeetendra Singh, B Raj, "An Accurate and Generic Window function for Non-linear Memristor Model" Journal of Computational Electronics, Springer, Volume 18 Issue 2, PP. 640-647 June 2019.

[30] Manjit Kaur; Neena Gupta; Sanjeev Kumar; BRaj; Arun Kumar Singh, "Comparative RF and Crosstalk Analysis of Carbon Based Nano Interconnects" IET Circuits, Devices & Systems, Vol.15(6), PP. 493-503, Feb 2021.

[31] Nehru Kandasamy, Firdous Ahmad, D. Ajitha, B Raj, and Nagarjuna Telagam "Quantum Dot Cellular Automata based Scan Flip Flop and Boundary Scan Register, IETE Journal of Research, Volume 66, PP. 535-548, 2020.

[32] Sharma, S. K.., Raj, B., Khosla, M., "Enhanced Photosensivity of Highly Spectrum Selective Cylindrical Gate In1-xGaxAs Nanowire MOSFET Photodetector, Modern Physics letter-B, Vol. 33, No. 12, PP. 1950144 (2019).

[33] Jeetendra Singh, BRaj, "Design and Investigation of 7T2M NVSARM with Enhanced Stability and Temperature Impact on Store/Restore Energy", IEEE Transactions on Very Large

Scale Integration Systems, Vol. 27, Issure 6, PP. 1322 - 1328 June 2019.
[34] Anil Kumar Bhardwaj, Sumeet Gupta, BRaj, Amandeep Singh, "Impact of Double Gate Geometry on the Performance of Carbon Nanotube Field Effect Transistor Structures for Low Power Digital Design", Computational and Theoretical Nanoscience, ASP, Vol. 16, PP. 1813–1820, 2019.
[35] Neeraj Jain, BRaj, "Thermal Stability Analysis and Performance Exploration of Asymmetrical Dual-k underlap Spacer (ADKUS) SOI FinFET for Security and Privacy Applications", Indian Journal of Pure & Applied Physics (IJPAP), Vol. 57, pp. 352-360, May 2019.
[36] Amandeep Singh, Mamta Khosla, BRaj, "Design and Analysis of Dynamically Configurable Electrostatic Doped Carbon Nanotube Tunnel FET"," Microelectronics Journal, Elesvier, Volume 85, Pages 17-24, March 2019.
[37] Neeraj Jain and Balwinder Raj, "Dual-k Spacer Region Variation at the Drain Side of Asymmetric SOI FinFET Structure: Performance Analysis towards the Analog/RF Design Applications", communicated in "Journal of Nanoelectronics and Optoelectronics", American Scientific Publishers, Vol. 14, PP. 349–359, March 2019.
[38] Jeetendra Singh, Sanjeev Sharma, B Raj, Mamta Khosla Analysis of barrier layer thickness on performance of In1-xGaxAs based Gate Stack Cylindrical Gate Nanowire MOSFET, " JNO, ASP, Vol. 13, PP 1473-1477, Oct 2018.
[39] Neeraj Jain, B Raj, "Analysis and Performance Exploration of High-k SOI FinFETs Over the Conventional Low-k SOI FinFET toward Analog/RF Design", Journal of Semiconductors (JoS), IOP Science, Vol.39, Issue 12, PP- 124002-1-7, Dec 2018.
[40] Candy Goyal, Jagpal Singh Ubhi and BRaj, A reliable leakage reduction technique for approximate full adder with reduced ground bounce noise, Journal of Mathematical Problems in Engineering, Hindawi, Volume 2018, Article ID 3501041, 16 pages, 15 Oct 2018.
[41] Anuradha, Jeetendra Singh, B Raj and Mamta Khosla, "Design and Performance Analysis of Nano-scale Memristor-based Nonvolatile SRAM" Journal of Sensor Letter", American Scientific Publishers, Vol. 16, PP 798–805, Oct 2018.
[42] Girish Wadhwa, BRaj, "Parametric Variation Analysis of Charge-Plasma-based Dielectric Modulated JLTFET for Bio-

sensor Application" IEEE Sensor Journal, VOL. 18, NO. 15, AUGUST 1, PP. 6070 – 6077, 2018 (SCI)

[43] Jeetendra Singh, BRaj, "Comparative Analysis of Memristor Models for Memories Design", JoS, IoP, Vol. 39, No. 7, PP. 074006-1-12, July 2018.

[44] Divya Yadav, Shailesh Singh Chouhan, Santosh Kumar Vishvakarma and BRaj, "Application Specific Microcontroller Design for IoT based WSN", Sensor Letter, ASP, Vol. 16, PP. 374–385, May 2018.

[45] Gurmohan Singh, R. K. Sarin and BRaj, "Fault-Tolerant Design and Analysis of Quantum-Dot Cellular Automata Based Circuits", IEEE/IET Circuits, Devices & Systems, Vol. 12, PP. 638 - 64, 2018.

[46] Jeetendra Singh, BRaj, "Modeling of Mean Barrier Height Levying Various Image Forces of Metal Insulator Metal Structure to Enhance the Performance of Conductive Filament Based Memristor Model", IEEE Nanotechnology, Vol. 17, No. 2, PP. 268-267, March 2018.

[47] Aakash Jain, Sanjeev Sharma, BRaj, "Analysis of Triple Metal Surrounding Gate (TM-SG) III-V Nanowire MOSFET for Photosensing Application", Opto-electronics Journal, Elsevier, Volume 26, Issue 2, Pages 141-148, May 2018.

[48] Aakash Jain, Sanjeev Sharma, B Raj, "Design and Analysis of High Sensitivity Photosensor Using Cylindrical Surrounding Gate MOSFET For Low Power Sensor Applications", Engineering Science and Technology, an International Journal, Elsevier's, Volume 19, Issue 4, Pages 1864–1870, December 2016.

[49] Amandeep Singh, Mamta Khosla, B Raj, "Analysis of Electrostatic Doped Schottky Barrier Carbon Nanotube FET for Low Power Applications," Journal of Materials Science: Materials in Electronics, Springer, Vol.28, PP. 1762–1768, 2017.

[50] G. Saiphani Kumar, Amandeep Singh, BRaj, "Design and Analysis of Gate All Around CNTFET based SRAM cell Design", Journal of Computational Electronics, Springer, Volume 17, Issue 1, pp 138–145, March 2018.

[51] Gurinder pal Singh, B. S. Sohi, Balwinder Raj, "Material Properties Analysis of Graphene Base Transistor (GBT) for VLSI Analog Circuits, Indian Journal of Pure & Applied Physics (IJPAP), Vol. 55, pp. 896-902, Dec-2017.

[52] Amandeep Singh, Mamta Khosla, B Raj, "Comparative Analysis of Carbon Nanotube Field Effect Transistor and Nanowire

Transistor for Low Power Circuit Design," Journal of Nanoelectronics and Optoelectronics, American Scientific Publishers, USA, Vol. 11, PP. 388-393, June 2016.

[53] Sunil Kumar and B Raj, "Estimation of Stability and Performance metric for Inward Access Transistor based 6T SRAM Cell Design using n-type/p-type DMDG-GDOV TFET" IEEE VLSI Circuits and Systems Letter, Volume3, Issue2, PP 25-39, June2017

[54] Shashikant Sharma, Anjan Kumar, Manisha Pattanaik and B Raj, "Forward Body Biased Multimode Multi-Threshold CMOS Technique for Ground Bounce Noise Reduction in Static CMOS Adders" International Journal of Information and Electronics Engineering, pp. 567-572, Vol. 3, issue 3, 2013.

[55] Hamendra Singh, Pankaj Kumar, Balwinder Raj, "Performance Analysis of Majority Gate SET Based 1-bit Full Adder", *International Journal of Computer and Communication Engineering* (IJCCE), IACSIT Press Singapore, ISSN: 2010-3743, Volume 2, No. 4, 2013.

[56] Anil Kumar Bhardwaj, Sumeet Gupta and B Raj, "Investigation of Parameters for Schottky Barrier (SB) Height for Schottky Barrier Based Carbon Nanotube Field Effect Transistor Device", Journal of Nanoelectronics and Optoelectronics, ASP, Vol. 15, PP. 783–791, July 2020.

[57] Priya Bansal, BRaj, "MEMRISTOR: A VERSATILE NONLINEAR MODEL FOR DOPANT DRIFT AND BOUNDARY ISSUES," JCTN, American Scientific Publishers, Vol. 14 (5), PP. 2319-2325, May 2017.

[58] Neeraj Jain, BRaj, "An Analog and Digital Design Perspective Comprehensive Approach on Fin-FET (Fin-Field Effect transistor) Technology - A Review", Reviews in Advanced Sciences and Engineering (RASE), ASP, Vol. 5, PP. 1-14, 2016.

[59] Sanjeev Sharma, B Raj, Mamta Khosla, "Subthreshold Performance of In1-xGaxAs based Dual Metal with Gate Stack Cylindrical/Surrounding Gate Nanowire MOSFET for Low Power Analog Applications, Journal of Nanoelectronics and Optoelectronics, American Scientific Publishers, USA, Vol.12, PP. 171–176, 2017.

[60] Balwinder Raj, A. K. Saxena and S. Dasgupta, "Analytical Modeling for the Estimation of Leakage Current and Subthreshold Swing Factor of Nanoscale Double Gate FinFET Device" Microelectronics International, UK, vol. 26, pp. 53-63, 2009.

[61] Soniya, Shailendra Singh, Girish Wadhwa and Balwinder Raj, "An Analytical Modeling for Dual Source Vertical Tunnel Field Effect Transistor" submitted to 'International Journal of Recent Technology and Engineering (IJRTE)' at Volume-8 Issue-2, PP. 603-608, July 2019.

[62] Shailendra Singh, B Raj, "Design and analysis of hetrojunction Vertical T-shaped Tunnel Field Effect Transistor", Journal of Electronics Material, Springer, Volume 48, Issue 10, pp 6253–6260, October 2019.

[63] Candy Goyal, Jagpal Singh Ubhi and B Raj, "A Low Leakage CNTFET based Inexact Full Adder for Low Power Image Processing Applications", International Journal of Circuit Theory and Applications, Wiley, Volume47, Issue9, Pages 1446-1458, September 2019.

[64] B Raj, A. K. Saxena and S. Dasgupta, "A Compact Drain Current and Threshold Voltage Quantum Mechanical Analytical Modeling for FinFETs" Journal of Nanoelectronics and Optoelectronics (JNO), USA, vol. 3, no. 2, pp. 163-170, 2008.

[65] Girish Wadhwa, B Raj, "An Analytical Modeling of Charge Plasma based Tunnel Field Effect Transistor with Impacts of Gate underlap Region" Superlattices and Microstructures, Elsevier, Vol. 142, PP.106512, June 2020.

[66] Shailendra Singh, B Raj,"Modeling and Simulation analysis of SiGe hetrojunction Double GateVertical t-shaped Tunnel FET", Superlattices and Microstructures, Elsevier Volume 142, PP. 106496, June 2020.

[67] Amandeep Singh, Dinesh Kumar Saini, Dinesh Agarwal, Sajal Aggarwal, Mamta Khosla, BRaj, "Modeling and Simulation of Carbon Nanotube Field Effect Transistor and its Circuit Application," Journal of Semiconductors (JoS), IOP Science, Vol.37, PP. 074001-6, July 2016.

[68] Neeraj Jain, Balwinder Raj, "Device and Circuit Co-Design Perspective Comprehensive Approach on FinFET Technology - A Review, Journal of Electron Devices, Vol. 23, Num. 1, pp 1890-1901, 2016.

[69] Sunil Kumar and BRaj, "Analysis of I_{ON} and Ambipolar Current for Dual-Material Gate-drain Overlapped DG-TFET," Journal of Nanoelectronics and Optoelectronics, American Scientific Publishers, USA, Vol.11, PP. 323-333, June 2016.

[70] Naveed Anjum, Tarun Bali and B Raj, Design And Simulation Of Handwritten Multiscript Character Recognition, Interna-

tional Journal of Advanced Research in Computer and Communication Engineering, Vol.2 (7), PP. 2544-2549, July 2013.

[71] Sanjeev Sharma, BRaj, Mamta Khosla, "A Gaussian Approach for Analytical Subthreshold Current Model of Cylindrical Nanowire FET with Quantum Mechanical Effects, Microelectronics Journal, Elsevier, Vol. 53, PP. 65-72, April 2016.

[72] Karmjit Singh, BRaj, "Performance and analysis of temperature dependent Multi-walled Carbon Nanotubes as Global Interconnects at different technology nodes," Journal of Computational Electronics, Springer, Volume 14 Issue 2, pp. 469-476, June 2015.

[73] Sunil Kumar and BRaj, "Compact channel potential analytical Modeling of DG-TFET based on Evanescent–mode Approach," Journal of Computational Electronics, Springer Vol. 14, No. 2, pp.820-827, July 2015.

[74] Karmjit Singh, BRaj, "Temperature Dependent Modeling and Performance Evaluation of Multi-Walled CNT and Single-Walled CNT as Global Interconnects," Journal of Electronic Materials, Springer, Vol. 44, Issue 12, pp 4825-4835, Dec 2015.

[75] V.K. Sharma, M. Pattanaik and B.Raj, "INDEP approach for leakage reduction in nanoscale CMOS circuits", International Journal of Electronics, Taylor & Francis, Vol.102, No.2, pages 200-215, 2014.

[76] Karmjit Singh, BRaj, "Influence of Temperature on MWCNT bundle, SWCNT bundle and Copper interconnects for Nanoscaled Technology nodes," Journal of Materials Science: Materials in Electronics. Springer, Vol. 26, No. 8, pp. 6134-6142, 2015.

[77] Naveed Anjum, Tarun Bali and B Raj, "Design and Simulation of Handwritten Gurumukhi and Devanagri Numerical Recognition, International Journal of Computer Applications, Published by Foundation of Computer Science, New York, USA, Vol. 73(12), pp. 16-21, 2013.

[78] Khandelwal, S., Gupta, V., BRaj, and Gupta, R D., Process Variability Aware Low Leakage Reliable Nano Scale DG-FinFET SRAM Cell Design Technique, Journal of Nanoelectronics and Optoelectronics, Volume 10, Number 6, pp. 810-817, December 2015.

[79] V.K. Sharma, M. Pattanaik and B Raj, "ONOFIC Approach: Low Power High Speed Nanoscale VLSI Circuits Design", Interna-

tional Journal of Electronics, Taylor & Francis, Vol. 101, issue 1, pages 61-73, 2014.

[80] Khandelwal, S., Balwinder Raj, and Gupta, R D., FinFET based 6T SRAM Cell Design: Analysis of Performance Metric, Process Variation and Temperature Effect. Journal of Computational and Theoretical Nanoscience, ASP, USA, Vol. 12, PP. 2500-2506, 2015.

[81] Sumit Singh; Yadav, Shekhar; Rahul, Jagdeep; Srivastava, Anurag; BRaj, Impact of HfO_2 in Graded Channel Dual Insulator Double Gate MOSFET, Journal of Computational and Theoretical Nanoscience, American Scientific Publishers, Volume 12, Number 6, pp. 950-953, April 2015.

[82] Vijay Kumar Sharma, Manisha Pattanaik, BRaj, PVT variations aware low leakage INDEP approach for nanoscale CMOS Circuits, Microelectronics Reliability, Elsevier, Vol. 54, pp 90-99, 2014.

[83] B Raj, A. K. Saxena and S. Dasgupta, "Quantum Mechanical Analytical Modeling of Nanoscale DG FinFET: Evaluation of Potential, Threshold Voltage and Source/Drain Resistance", Elsevier's Journal of Material Science in Semiconductor Processing, Elsevier, Vol. 16, issue 4, pp. 1131- 1137, 2013.

[84] Maisagalla Gopal, Siva Sankar D Prasad and Balwinder Raj, 8T SRAM Cell Design for Dynamic and Leakage Power Reduction, International Journal of Computer Applications, Published by Foundation of Computer Science, New York, USA, Vol. 71(9), pp. 43-48, June 2013.

[85] Manisha Pattanaik, B Raj, Shashikant Sharma and Anjan Kumar, "Diode Based Trimode Multi-Threshold CMOS Technique for Ground Bounce Noise Reduction in Static CMOS Adders", Advanced Materials Research, Trans Tech Publications, Switzerland, vol. 548, pp 885-889, 2012.

[86] Balwinder Raj, A. K. Saxena and S. Dasgupta, "Nanoscale FinFET Based SRAM Cell Design: Analysis of Performance metric, Process variation, Underlapped FinFET and Temperature effect" IEEE Circuits and System Magazine, vol. 11, issue 2, pp. 38- 50, 2011.

[87] V.K. Sharma, M. Pattanaik & Balwinder Raj, "Leakage Current ONOFIC Approach for Deep Submicron VLSI Circuit Design", International Journal of Electrical, Computer, Electronics and Communication Engineering, World Academy of Sciences, En-

gineering and Technology, Vol. 7, issue 4, pages 239-244, 2013.

[88] Tulika Chawla, Mamta Khosla, BRaj, Design and simulation of Triple metal Double-gate Germanium on Insulator Vertical Tunnel Field Effect Transistor, Microelectronics Journal, Elsevier; Vol. 114, PP. 105125, August 2021.

[89] Parminder Kaur, Sandeep Singh Gill and B Raj, "Comparative Analysis of OFETs Materials and Devices for Sensor Applications", Journal of Silicon, Springer, Vol 14, PP. 4463–4471, 2022.

[90] Sanjeev Kumar Sharma, Parveen Kumar, Balwant Raj and BRaj, "$In_{1-x}Ga_xAs$ Double Metal Gate-Stacking Cylindrical Nanowire MOSFET for highly sensitive Photo detector", *Journal of Silicon, Springer Vol.*14, PP. 3535–3541, 2022.

[91] B Raj, A. K. Saxena and S. Dasgupta, "Analytical Modeling of Quasi Planar Nanoscale Double Gate FinFET with Source/Drain Resistance and field dependent carrier mobility: A quantum mechanical study" Journal of Computer (JCP), Academy Publisher, FINLAND, vol. 4, issue 9, PP.1-8, 2009.

[92] Bhushan, S., Khandelwal, S. and Raj, B., (2012). Analyzing Different Mode FinFET based Memory Cell at Different Power Supply for Leakage Reduction, Seventh International Conference on Bio-Inspired Computing: Theories and Application, (BIC-TA 2012) Advances in Intelligent Systems and Computing, Volume 202, p89-100 (2013).

[93] Jeetendra Singh, BRaj, "Temperature Dependent Analytical Modeling and Simulations of Nanoscale Memristor, Journal: Engineering Science and Technology, an International Journal, Elsevier's, Vol. 21, PP- 862-868, Oct 2018.

[94] Shradhya Singh, Shashi Bala, Balwant Raj, B Raj, "Improved Sensitivity of Dielectric Modulated Junctionless Transistor for Nanoscale Biosensor Design", Sensor Letter, ASP, Vol. 18, PP.328–333, Apl 2020.

[95] Vivek Kumar, Santosh Kumar Vishvakarma, and B Raj, "Design and Performance Analysis of ASIC for IoT Applications" Sensor Letter ASP, Vol. 18, PP. 31–38, Jan 2020.

[96] Akanksha Jaiswal, R. K. Sarin, B Raj and Shikha Sukhija, "A Novel Circular Slotted Microstrip-fed Patch Antenna with Three Triangle Shape Defected Ground Structure for Multiband Applications", Advanced Electromagnetic (AEM), Vol. 7, Issue 3, PP. 56-63, Aug 2018.

[97] Girish Wadhwa, B Raj, "Label Free Detection of Biomolecules using Charge-Plasma-Based Gate Underlap Dielectric Modulated Junctionless TFET" Journal of Electronic Materials (JEMS), Springer, Volume 47, Issue 8, pp 4683–4693, August 2018.

[98] Gurmohan Singh, R. K. Sarin and B Raj, "Design and Performance Analysis of a New Efficient Coplanar Quantum-Dot Cellular Automata Adder", Indian Journal of Pure & Applied Physics (IJPAP), Vol. 55, pp. 97-103, February 2017.

[99] Amandeep Singh, Mamta Khosla, Balwinder Raj, "Design and Analysis of Electrostatic Doped Schottky Barrier CNTFET Based Low Power SRAM," International Journal of Electronics and Communications, (AEÜ), Elsevier, Vol. 80, PP.67–72, 2017.

[100] Parminder Kaur, Vikas Pandey and B Raj, "Comparative Study of Efficient Design, Control and Monitoring of Solar Power using IoT", Sensor Letter, ASP Vol. 18, PP. 419–426, May 2020.

[101] Anil Kumar Bhardwaj, Sumeet Gupta and B Raj, Development & Analysis of compact model for Double Gate Schottky Barrier CNTFET, Journal of Nanoelectronics and Optoelectronics, ASP, Vol. 15, pp. 1199–1208, Aug 2020.

[102] Girish Wadhwa, Priyanka Kamboj; Jeetendra Singh, B Raj, "Design and Investigation of Junctionless DGTFET for Biological Molecule Recognition" Transactions on Electrical and Electronic Materials, Springer, 8 Aug, 2020.

[103] Tulika Chawla, Mamta Khosla, B Raj, "Optimization of Double-gate Dual material GeOI-Vertical TFET for VLSI Circuit Design, IEEE VLSI Circuits and Systems Letter, Vol. 6, issue-2, PP.13-25, Aug 2020.

[104] Sachin Kumar Verma, Shailendra Singh, Girish Wadhwa and B Raj, "Detection of Biomolecules using Charge-Plasma based Gate Underlap Dielectric Modulated Dopingless TFET" Transactions on Electrical and Electronic Materials (TEEM), Springer, Vol. 21, PP. 528-535, June 2020.

[105] Neeraj Jain, BRaj, "Impact of underlap spacer region variation on Electrostatic and analog/RF performance of symmetrical high-k SOI FinFET at 20 nm channel length", Journal of Semiconductors (JoS), IOP Science, Vol.38, Issue 12, PP- 122002, Dec 2017.

[106] Shailendra Singh, B Raj," Analytical Modeling and Simulation analysis of T-shaped III-V heterojunction Vertical T-FET", Su-

perlattices and Microstructures, Elsevier, Vol. 147, PP. 106717, Nov 2020.

[107] Gurmohan Singh, R. K. Sarin and B Raj, "Design and Analysis of Area Efficient QCA Based Reversible Logic Gates", Journal of Microprocessors and Microsystems, Elsevier, Vol. 52, PP.59-68, May 2017.

[108] Amandeep Singh, Mamta Khosla, BRaj, "Compact Model for Ballistic Single Wall CNTFET under Quantum Capacitance Limit," Journal of Semiconductors (JoS), IOP Science, Vol.37, PP. 104001-8, Oct 2016.

[109] Sonal Singh; Mamta Khosla, Girish Wadhwa; B Raj, "Design and analysis of double-gate junctionless vertical TFET for gas sensing applications", Applied Physics A, Springer, Vol.127, No.16, 2 Jan 2021.

[110] Inderjit Singh, BRaj, Mamta Khosla, and Brajesh Kumar Kaushik, Potential MRAM Technologies for Low Power SoCs, SPIN World Scientific Publisher, SCIE; Vol. 10, No. 04, PP.2050027, Dec 2020.

[111] Shailendra Singh, B Raj,"Parametric variation analysis on hetero-junction Vertical t-shape TFET for Supressing ambipolar conduction", Indian Journal of Pure and Applied Physics, Vol. 58, pp. 478-485, June 2020.

[112] Soniya, Shailendra Singh, Girish Wadhwa and Balwinder Raj, Design and Analysis of Dual Source Vertical Tunnel Field Effect Transistor for high performance, Transactions on Electrical and Electronics Materials, Springer, Vol. 21, Pages 74–82, Oct 2019.

[113] Manjit Kaur; Neena Gupta; Sanjeev Kumar; BRaj; Arun Kumar Singh, "RF Performance Analysis of Intercalated Graphene Nanoribbon Based Global Level Interconnects" Journal of Computational Electronics, Springer, Vol. 19, PP.1002–1013, June 2020

[114] Girish Wadhwa, BRaj, "Design and Performance Analysis of Junctionless TFET Biosensor for high sensitivity" IEEE Nanotechnology, Vol.18, PP. 567 – 574, 2019.

[115] Jeetendra Singh, BRaj, "Enhanced Nonlinear Memristor Model Encapsulating Stochastic Dopant Drift, JNO, ASP 14, 958–963 (2019).

7

Analysis of Non Volatile Memory Using Artificial Intelligence

Mandeep Singh, Tarun Chaudhary and Balwinder Raj
NIT Jalandhar

Abstract: In this chapter the various types of non volatile memory using Artificial intelligence has been discussed. In this chapter various non volatile memory like FeFET, PCRAM, RRAM, MRAM has been discussed. Conventional memories like DRAM, SRAM, and NAND Flash have been progressively being overwhelmed by the enormous demand for information storage and transmissions, and the process miniaturization of conventional memories is getting more challenging. This is forcing the semiconducting company to create new memories with increased storage efficiency, cheaper prices, and system miniaturization. The MRAM, RRAM or PCRAM memories types stood out.

7.1 Introduction

Information technology has been enhanced by artificial intelligence (AI). The information age's modern economy had also influenced how we live today. Previously, researchers' focus has been focused on AI algorithms that have also been quickly and effectively used to solve technical challenges. However, due to the unpredictable nature of the search strategies, AI algorithms require a significant amount of computation time for massive and complicated problems. In order to find solutions in real-world applications with low resources, time, and money, it may be necessary to develop efficient algorithms. This special edition aims to provide the most recent improvements in artificial intelligence technology across the board, including face recognition, data mining, and machine learning human evaluation, agent based systems, and fuzzy logic [1].

A widely accepted concept and formatting technology for so many completely fictitious research programs is artificial intelligence, or AI. It's indeed, nevertheless, a very advanced field of study and advancement. Poole, Mackworth, and Goebel [2] provide the actual definition of artificial intelligence, stating that it is the "investigating or developing intelligent systems." Inside this context, the agent is merely a thing which gathers information from its surroundings and acts obeying logical processing and input analysis. With our everyday appliances, this form of system communication is available to the community. For cold storage to maintain the required temperatures, refrigeration units must adjust the temperature. Someone using of framework items often react to player actions is common in computer games [3-9].

It is asserted that artificial intelligence is becoming more prevalent in management research and operations strategy fields of study. Frequently, people define intelligence as the capacity to acquire knowledge and use knowledge for specific purpose regarding problems. Intelligent machines will soon take over many skill sets in the coming years. The research and creation of intelligent machine intelligence that really can sense, gain knowledge, collect information, interact, exploit, and interpret objects is known as artificial intelligence. The focus on computation sets artificial intelligence apart from psychoanalytic theory, and its primary focus on perspective, argument, and initiative sets it apart from computer science. Machines become wiser and much more helpful as a result [10-16].

AI is related to the development and application of computer systems that can solve issues that typically call for human intelligence. These issues relate to functions that are naturally complex or both (e.g., vision or natural language understanding). Typically, they cannot be resolved using traditional algorithmic techniques. AI apps manipulate respondents reported rather than just numerical information, as is customary in computer science, to solve them. AI uses many different types of knowledge about such an application field. Therefore, central to AI research and development are the issues of representation of knowledge, implementation, and use. One of the really active subfields is the execution of knowledge-based systems. The multidisciplinary nature of AI is a significant feature [17-23]. Although AI and computer science have close ties, the development of Autonomous projects also requires ideas from logic, language teaching, psycholinguistics, etc. In addition, a wide range of application areas, including finance and banking, related industries, and medicine, have been using AI [24-30]. Unlike some other emerging technologies, AI techniques present entities with both novel, unique possibilities as well as major development challenges. Initially, once implemented in an organization, Technology differs in their ability to restrain, supplement, and/or replace workers at work. It is important to develop an understanding of how people and AI technologies communicate in new and different ways that provide a stabilizing influence, an important attributes of work, or the emergence of novel forms of work since this shifts the locus of intervention, preference, regulate, and control away from humans' unique domain. Second, the primary distinction between human and machine capabilities is fundamentally challenged by AI technologies. A variety of human skills, including impression, detecting and realizing feelings, discussion, and even creative thinking, can now be accomplished by modern AI technologies [31-35]. With the help of these new abilities, AI is now able to interact with areas that have previously been reserved for humans alone, such as algorithmic management, the creation of new products, and face recognition. Although There's still contention over how computers should act or think, recent trends in AI abilities have given rise to a number of conflicts beyond just novel interactions between people. Third, the sophistication of AI technologies is rising, which frequently produces a lot of unusual dual results. Although AI technologies have many advantages for businesses, they frequently have numerous negative (or designed) consequences on both individuals and organizations. Often, AI technologies lead to the networks' inevitable stakeholders from the process of designing, implementing, and using Often, AI technologies lead to the networks' inevitable failure because the impact of AI implementation differs greatly among stakeholders [36-41]. A multi-stakeholder viewpoint is necessary to account for this complication given the wide variety of concerned parties. Information systems (IS) scientific literature has a plethora of chances given the unique consequences of AI technologies in organizations. The four main business functionality of digitization, involvement, understanding, and invention are used to investigate these possibilities [42-47]. As shown in Figure 7.1 the AI can be classified into following two categories:

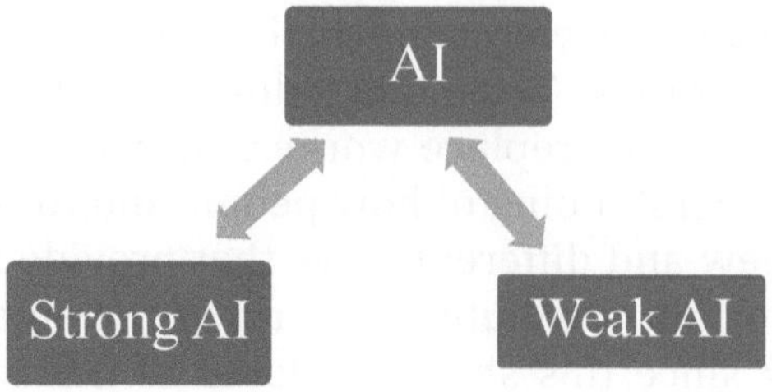

Figure 7.1 *Classification of AI on the basis of philosophy.*

7.1.1 Strong AI

Strong AI is based on the idea that, in the coming years, machines may be made to think, or, to put it another way, may substitute for human brains. According to Strong AI, humans will soon be enclosed by machines that can function exactly like humans and may even possess intelligence comparable to that of humans [48-53]. If so, then all those machines would be capable of thinking, and perform all of the tasks that humans can. Strong AI is still far from being possible with current studies, and there is a heated debate about whether this is sometimes feasible.

7.1.2 Weak AI

Minimal AI is predicated upon that straightforward notion that intelligence behavior within computers may be programmed. Weak AI merely claims that's it is simple to add cognitive capabilities to computers in order to make them extra helpful applications, and that this has already begun to happen. A human player may believe that the computer is simply making brilliant moments when playing chess against a computer, for instance. However, management and reasoning are in no way involved in the game of chess [54-60]. A individual inputs every move it tends to make into the computer beforehand, ensuring that the system will make the appropriate moves at the appropriate times. Witness knowledge - based systems, drive-by-wire vehicles, and speech processing systems are additional examples of Weak AI. Artificial intelligence, also known as AI is the capacity of a computer to do things that would typically only be performed by the human brain. The ability for learning and the ability to obtain it are among such operations. Along with judgment and connection comprehension, it also includes the capacity for independent idea [61-66].

7.1.3 Advantages and Disadvantages

(i) The major advantage of artificial intelligence is that it bases choices on facts rather than emotions. Despite our best attempts, our instincts will always exert a negative influence on the decisions we make as humans.

(ii) Extremely easy knowledge dissemination when an artificial brain has been trained towards something, it could be very easily replicated to some others, saving time that would otherwise be wasted teaching new information towards other beings.

(iii) Artificially intelligent machines, apart from humans, don't really require sleep, eliminating the adverse influence of human fatigue [67-73].

(iv) Blind dependence on AI can lead to problems since it cannot justify why it came up with a certain reactions and can offer wrong responses as a consequence of any error.

(v) Inadequately inventive reactions

(vi) In the beginning of growth, AI cannot acknowledge whenever a particular issue seems to have no answer.

(vii) Major issues could also arise from thinking process that lacks intelligence.

7.2 Sub-fields of Artificial Intelligence

The Artificial intelligence can be divided into various sub fields which is shown in Figure 7.2. The various sub field of AI are like Machine learning, neural network, deep learning, computer network, cognitive computing [74-80].

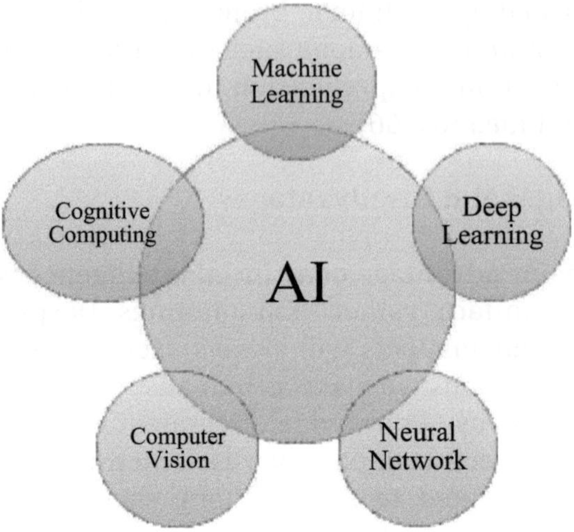

Figure 7.2 *Sub-fields of Artificial Intelligence.*

7.2.1 Machine Learning

In the context of information analyzation and computational science, artificial intelligence (AI), and now in specific machine learning (ML), had also risen quickly in recent years. These techniques usually increased the ease to function intelligently. The most widely used latest techniques in the 4th industrial revolution, machine learning (ML) typically gives systems the ability to learn and improve from experience immediately without even being specifically programmed [81-87]. The ongoing technology of traditional industrial and manufacturing strategies, such as explorative data acquisition, utilizing modern digital innovations like machine learning is known as "Industry 4.0" mechanization. Consequently, to thoughtfully assess these data and to construct the correlating practical applications and the step is to learn methodologies [88-96]. The machine learning can be classified into four categories as shown in Figure 7.3.

(i) **Supervised:** A feature which maps an input to an outcome is typically employing calibration and validation pairs to be trained through learning algorithm. It infers a function using learning algorithm and a set of training dataset.

(ii) **Unsupervised learning:** Unsupervised learning, or a data-driven methodology, analyses unsupervised datasets without any need for human involvement. This is frequently used as generative feature extraction, useful pattern and framework identification, result clusters, and explorative reasons.

(iii) **Semi-supervised:** Semi-supervised learning combines the supervised and unsupervised techniques mentioned above since it employs both supervised and unsupervised.. This places it in the middle of learning "with mentorship" and "with no mentorship."

(iv) **Reinforcement:** A category of machine learning algorithm called reinforcement learning enables software agents and machines to instantly assess the most appropriate behavior in a given circumstance in order to increase the efficiency of the system.

7.2.2 Deep Learning

A group of machine learning (ML) called deep learning (DL), also known as neural networks and deep learning is defined as a set of schemes that aim to design high-level layers in information. These algorithms build an elaborate, layered architecture for having to learn and identifying information. The deep, layered process of learning used by the human brain's main multisensory regions of the neocortical, which extracts characteristics and generalizations from original data, is reproduced by artificial intelligence in this traditional teaching and learning architecture. According to DL algorithms naturally learn representations of data using a greedy layer-wise approach and are useful for gathering a multitude of unlabeled data [97-104]. Additionally, there is the issue of learning, in which the computer examines and evaluates the user input using various methods till its locate the sole expected results. This is also referred to as machine self-learning. The ability of cell phones to identify various voice tones and translate them into interesting observations is just an illustration of deep learning in action [105-108].

7.2.3 Cognitive Computing

A new approach of advanced computational research methods and systems known as "cognitive computing" uses independent conclusions and impressions that resemble the activity of the brain to achieve cognition technology. Upcoming cognitive computers and intelligent independent devices with thought or emotion could be created and expressed in the form of cognitive computing [109-111].

Challenges of Cognitive Computing

(i) **Relation lapse between understanding, pondering, and emotion:** For acquiring and modeling domain knowledge, many methods are used. Advanced software paradigms and computing gadgets both contribute to the capacity for thought. Different psychological and numerical methods have been equipped with the ability to express emotion. Even still, attention must be paid to the integration of these cognitive abilities.

(ii) **Integration of applications and brain-like equipment:** Software frameworks' "brain-like machinery" concentrates on modeling the capacity for emotion and thought as well as trying to represent the knowledge and understanding. The following components require a lot of attention: - (a) To incorporate domain expertise into the faculty of thoughts. (b) Relationship between feeling and reasoning in the brain.

(iii) **Social cognitive computing:** In addition to brain activity, other factors such as social activity are also important in human society. This is referred to as social cognitive technologies. Therefore, it is important to concentrate on social cognitive computing, which improves cognitive agents' cognitive abilities [112-115].

7.2.4 Computer Network

A data service or computer network is an interconnected network that enables the data transfer of computers. Networked computing devices transmit data through data connections in computer systems. Cable or cellular technologies are used to make links (network links) among base stations. The most well-known computer network is the Internet. Data is created, routed, and terminated by network elements. Personal computers, smart phones, servers, and networking equipment are examples of nodes. Not whether two nodes are directly connected to one another, they have been shown to be networked together when someone is able to share information with another. Computer networks allow shared use of servers for implementation and backup, scanners, and fax, as well as application areas connect directly to the Web [116-120]. The various component of computer network are shown in Figure 7.3. Like modem, hub, switch, national interference card etc.

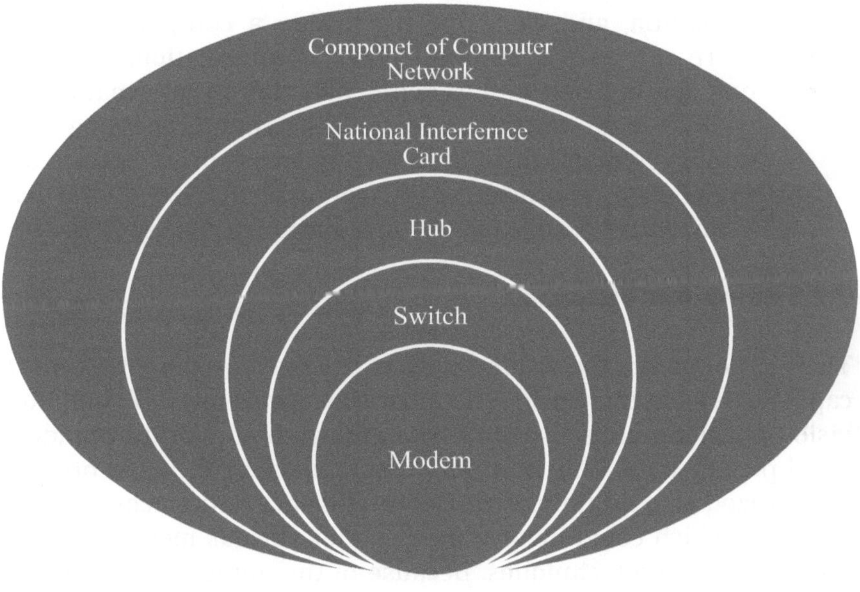

Figure 7.3 *Various Parts of Computer Network.*

Advantages of computer network using Artificial intelligence:

(i) **Operating capacity:** While network efficiency is essential to the smooth process of a computer network, obtaining fuzzy data can be challenging due to its volatility. Clients use only their brains to efficiently extract sophisticated and disorganized data information by using a computer network. In order to improve information processing effectiveness and shorten data recovery times, artificial intelligence could use fuzzy logic control method to obtain valuable information from huge amounts of data.

(ii) **Improve hierarchical management:** However, because of the network's high level of complexity, management tasks require a robust technical system. Even though conventional hierarchical management lacks good communication skills, it was unable to fully exploit its effectiveness inside the face of such a sophisticated computer network system.

(iii) **Enhance non-linear processing and learning capabilities:** Artificial intelligence does have a strong capacity for learning because it can replicate human thought. Inside a computer system, there are several different types of information, and each one has inside a computer system, there are several different types of information, and each one can be used to automate collection of information, as well as data transmission analysis and processing that mimics human function, to give business users better customer data. The positive environment in cyber world has been made possible by Ai's high learning capacity [13].

7.2.5 Computer Vision

The goal of computer vision study is to develop computers more capable viewing the world. The most conceptual level of computer vision problems aims to deduce something about world from identified picture data. This is an interdisciplinary field that is broadly categorized as a group of artificial intelligence and machine learning, both which could involve using either various methods or all-purpose learning techniques. Because of the numerous uses it has discovered in industries such as healthcare, sports and recreation, prototype design, and self-driving cars, computer vision has recent years gained momentum and importance. Object recognition tasks like image order, constraint, and recognition are used in most of these application areas. Convolution Neural Networks (CNNs) have recently made strides that have led to amazing performances in such greatest vision tasks and paradigms, proving their potency. Deep neural networks have subsequently become the fundamental units of deep learning computations in computer vision [14]. In the today world the AI has so many applications as shown in Figure 7.4.

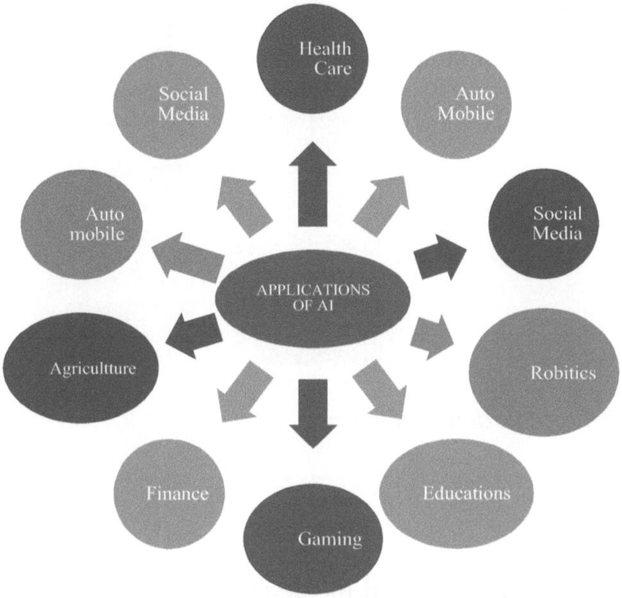

Figure 7.4 *Applications of AI.*

7.3 Emerging Non Volatile Memory Device

The two types of memory are volatile memory and non-volatile memory, depending upon whether memories can retain data after becoming switched off. Static random access memory (SRAM) and dynamic random access memory are both types of volatile memory (DRAM). SRAM and DRAM are frequently utilized and meet the growing demands for memories performance or bandwidth of AI technologies because of their quick access speeds. Even so, total amount of integrated storage device capacity is not large because the enormous cell area or the requirement to fabricate extensive capacitor [15]. The volatile memory must also hold information while receiving a steady supply of power, resulting in high backup power consumption. Emerging memory has non-volatile, like FeFET, PCRAM, MRAM, and ReRAM as shown in Figure 7.5.

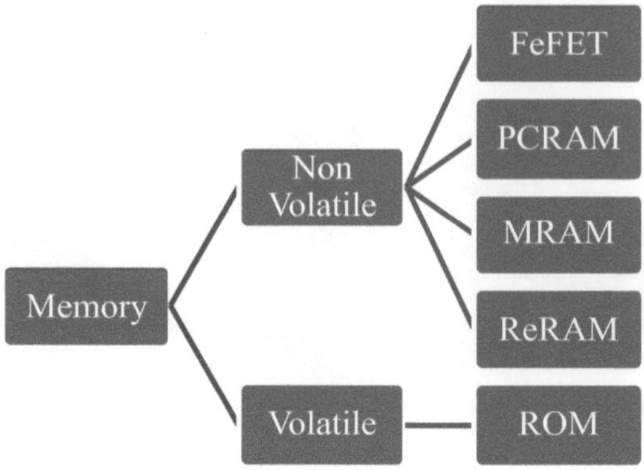

Figure 7.5 *Representation of Memory.*

7.3.1 FeFET

The FeFET was first proposed in the role of a non-volatile random-access memory. in the 1950s. The FeFET stands out by its high on/off current ratio, poor writing power, outstanding CMOS suitability, portable analogs/multi-bits integration. Data is stored as a lower and higher threshold voltage by altering the ferroelectric polarization. A 32MB FeFET-based memory has currently been produced by the Global- Foundries used a 22 nm procedure. Chinese Academy of Science presented a Fe- FET with a FinFET-based on/off current ratio of > 10^4 in the 10nm node. Zhejiang Institution's analogue 3-bit information may be encoded using FeFETs in central storage cells 40. Nevertheless, spite of the numerous benefits, a lack of endurance and high operating voltage of FeFET continue to constrain much further advancement [15]. Even with its suitability with CMOS, the ferroelectric characteristics of hafnium oxide (HfO_2) were discovered in 2011 and have attracted considerable attention. Ferroelectric HfO_2 has exhibited better remnant polarization, a vast coercive field, at a small dimension, and better performance features. Zirconium-doped HfO_2 (HZO) has surfaced as a promising material because of its lesser critical temperature and capable of adapting ferroelectric characteristics. The only material system that can be used to realize a new category of steep slope device known as a Negative Capacitance Field Effect Transistor (NCFET) is ferroelectric hafnium oxide, which is also used to realize memory. Because it can generate a sub threshold swing (SS) of 60 mV/decade for digital innovations, negative capacitance FETs have been researched both conceptually and empirically. Metal Ferroelectric Metal Insulator Semiconductor (MFMIS) configuration structure was the primary priority of earlier research on NCFET. The ferroelectric-insulator interface does have an unified potential, making this structure easier to design, but the leakage currents would make the ferroelectric FET's negative capacitance unpredictable. Due to this destabilization, MFMIS devices and circuit design are challenging to bias, which renders the MFMIS structure inappropriate for Rf circuits as shown in Figure 7.6. But at the other side, a metal-ferroelectric-insulator-semiconductor (MFIS) structure minimizes these problems and generates distinct device characteristics than MFMIS devices as a result of a non - stationary ferroelectric potential. The MFIS structure has therefore really been chosen for integrating cutting-edge techniques. There have been a few research have also looked into the RF efficiency of MFIS NCFETs. While simulation results besides revealed circuit-level figures of merit, test in showed a modest improved performance in FeFET [16]. The drain current and gate voltage characteristics of MFIS structure shown in Figure 7.7. The electric field representation of MFIS structure shown in Figure 7.8.

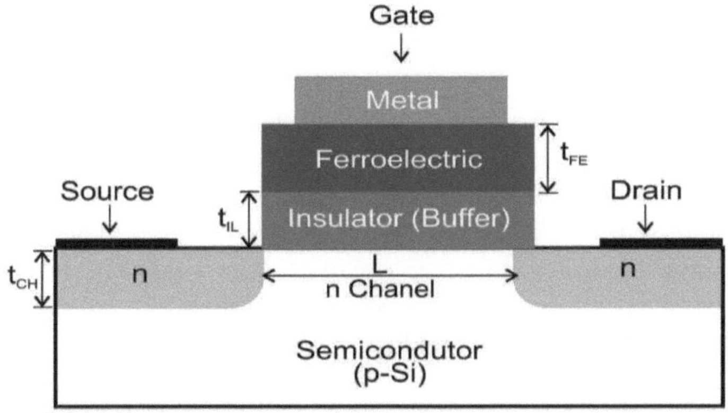

Figure 7.6 *Metal Ferroelectric Insulator Semiconductor Structure.*

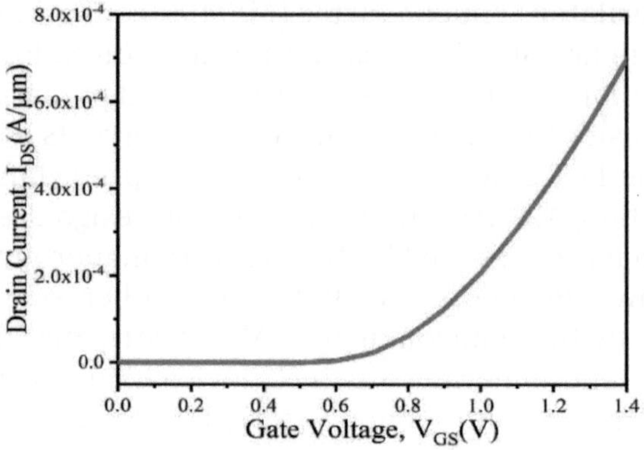

Figure 7.7 *Graph between the Drain Current (Id) and Gate Voltage (V).*

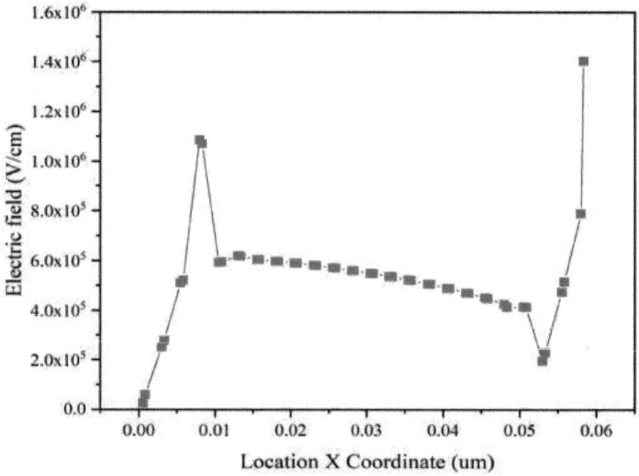

Figure 7.8 *Electric field graph of FeFET.*

The following Table 7.1 explain the various parameters of FeFET in the terms of I_{ON}, I_{OFF}, I_{ON}/I_{OFF}.

Table 7.1 *Parameters of FeFET*

Parameters	Units	FeFET (L=13nm)
I_{ON}	A/µm	$2.07x10^{-4}$
I_{OFF}	A/µm	$5.75x10^{-15}$
I_{ON}/I_{OFF}	-	$3.6x10^{10}$

7.3.2 PCRAM

PCRAM (Phase Change Random Memory), another novel category of memory, has a layered framework having a central layer that undergoes phase transition (GST is identical to the external memory substance). Such an substance has the ability to change from being isolated (relatively high condition) to being amorphous (high resistance state), and it uses the distinction in compatibility among these two states to record the data inside a system which can be broken down into two major stages, SET and RESET. A phase change material's resistivity varies through many orders of magnitude among its amorphous and crystalline states, making it extremely noise-tolerant enough yet to differentiate among both the "0" and "1" regions. The phase - changing materials currently employed by various businesses include sulphide (owned by Intel) or synthetic fabrics (GST) including germanium, antimony, or tellurium, including such Ge2Sb2Te5 (STMicroelectronics is presenting). Whenever the temperature of an amorphous material is increased to a level just above recrystallization that's below the melting point, and afterwards the temperature is gradually allowed to cool (one such procedure is the primary determinant of PCM speed), the amorphous material crystallizes (this step is called SET). The Figure 7.9 explains the internal structure of PCRAM. The material currently has a low resistivity due to its large conduction electrons densities as well as long-range energy development states. If a material is heated to a temperature just above its melting point while it is in the crystalline state, it quickly cools to its amorphous state after being cooled (this step is called RESET) [17].

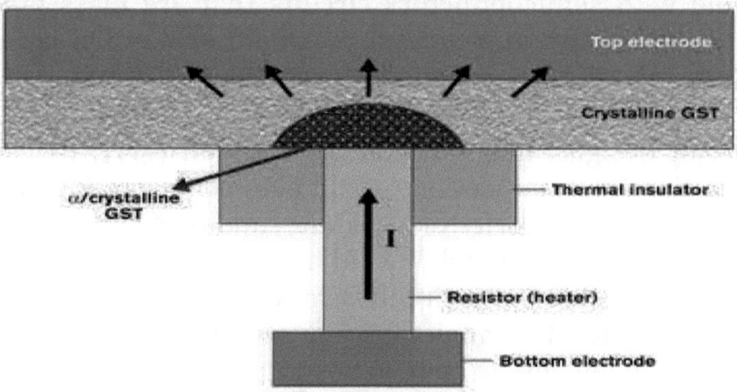

Figure 7.9 *Internal structure of PCRAM [9].*

Characteristics of PCRAM

(i) **Low latency, stable read as well as writes times:** Unlike NAND flash, PCM does not even have to erase old program codes prior to actually writing code changes, which increases accuracy and improves read and write times compared to NAND.

(ii) **Long life:** Because PCM read and writes operations are non-destructive, they have a much higher write resistance than flash memory, which allows PCM to more reliably eliminate mechanical hard disks than flash memory.

(iii) **Low power consumption:** Since PCM lacks mechanical orientation equipment and doesn't need a refresh current that can save program codes, it consumes less power than HDDs, NANDs, and DRAMs.

(iv) **High-density:-** storage is possible with some PCMs use to their non-transistor design.

(v) **Strong radiation resistance:-** Because PCM data storage is self - reliant of the charged particle properties of a material, it is able to satisfy the requirements of both space and defence.

7.3.3 Magnetic Random Access Memory (MRAM)

MRAM is a category of solid-state storage circuit design that stores information as magnetoresistive devices in stable magnetic regions, and reads information by detecting the magnetic states of the devices by assessing their resistance. To design chips that are highly compatible with semiconductors circuits that are mass manufactured, magneto resistive devices are actually integrated with complementary metal-oxide-semiconductor (CMOS) circuitry. While various magnetoresistive devices were used in the configuration of these circuit designs, magnetic tunnel junction (MTJ) devices are the foundation for commercially available MRAM product and services and the huge majority of MRAM technologies under development for future commercial MRAM innovations [10]. Regarding read operations, all of these circuits are resistive memories; however, the various MRAM technology types can be distinguished by the way in which the magnetic state is written. Instead of using electric charge to hold information, MRAM uses the polarization of magnetic fields. A tunnel gate layer, a fixed magnetic layer, and a free magnetic layer make up the memory cell. Whereas the magnetic field orientation of a fixed layer doesn't change, the polarization of the free magnetic layers. A memory cell exhibits poor resistance whenever the free layer's magnetic field direction is concurrent with that of the fixed layer; instead that, it exhibits high resistance. By identifying the memory cell's resistance, we can tell that whether stored information is a 0 or a 1 [18].

Characteristics of MRAM

(i) **The infinite number of reads and writes:** MRAM can be re - written indefinitely, just like DRAM, because the magnetism of ferromagnets really never reappears because it is not only unaffected by power outages.

(ii) **Non-volatile:** MRAM has non-volatility because ferromagnets' magnetic properties do not vanish in the absence of power.

(iii) **Fast write speeds and low power:** requirements allow for instant on/off and increased battery life in portable devices. The write time of MRAM is as little as 2.3 ns, and it uses extremely little power.

7.3.4 RRAM

As seen in figure a typical RRAM cell is made up of the multiple levels resemble a basic capacitor and is perfectly positioned with only a insulation layer on top implanted with both pair of connections known as the bottom electrode (BE) or top electrode (TE) as Metal-Insulator-Metal (MIM). RRAM is referred to as a two-terminal cell, and when voltage is applied, its characteristics change. The oxidation and reduction reaction is the basis for its operation, and it switches for both low resistance states (LRS) and high resistance states (HRS). There are two distinct binary regions in a digital memory storage case: logic "0" values correspond not to storing information, as well as logic "1" represents information stored. In RRAM, a conductive filament or bridge is created between the two conducting layers by applying an electrical field to the entire system. Thereby, a conductive bridge would be built as a result of redox reactions, and the two electrodes will also be connected directly. Consequently, the device would be at an LRS, which corresponds until rationality is reached "1" when a conductive filament (CF) is established. Throughout comparison, after the filament ruptured, the HRS is produced. According to Figure 7.10, HRS was equivalent to the logic state "0." The insulating, resistive switch layer in semiconductor-based RRAM is crucial to memory performance. So under influence of electric field, resistance switching is used by RRAM to store and erase data. RRAM typically uses two switching procedures to function. Under the impact of an electric stimulation, the SET process changes the device's resistance state from HRS to LRS (Vtop or Vset). As shown in Figure 7.10, the RESET procedure, on the other hand, returns the resistance to HRS under (Vtop or Vreset) [19].

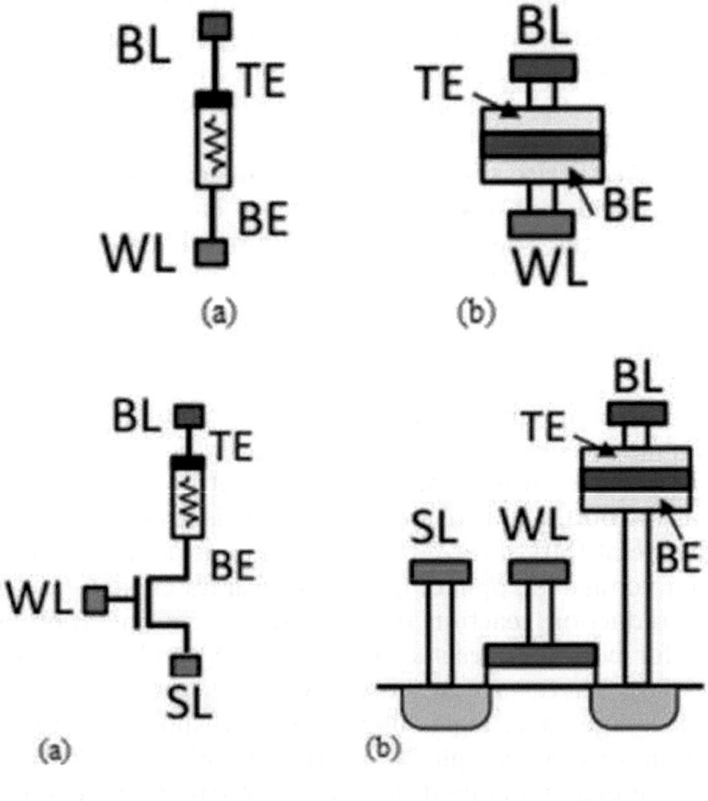

Figure 7.10 *(a) 1R RRAM cell; (b) cell cross-section view [19].*

The 1T1R structure is made up of one transistor and one memristor, as depicted in Figure 7.10. The transistor drain is attached to the bottom electrode of the memristor, BE, and the Bit Line (BL) of the RRAM is linked to the positive terminal, TE. In the RRAM memory, WL stands for the word line as well as SL for the source line. The transistor is utilized to obtain the chosen cell and keep it separate from unchosen cells. The transistor is also primarily used to constrain the write current (set the compliance). The selection of transistor specifications is crucial, though, as it affects the size of the cell, the performance of cells, and potential applications for the finished devices. The main problem with the 1T1R configuration would be that a single inability to effectively simultaneously higher and lower voltages via because NMOS devices are better at driving minimal voltages and PMOS devices are better at driving high voltages [20].

Characteristics of RRAM

(i) High speed: The trigger resistor's transition pulse width, which is typically less than the 100ns, determines the RRAM erase speed.

(ii) Robustness: RRAM read and write and NAND differ in that they use a reversible quasi mode that can significantly extend their useful lives.

(iii) Multi-layer memory functionality: A few RRAM materials have a range of resistance states, growing the storage density by enabling multi-layer data to be stored within a memory cell.

7.5 Conclusion

In this chapter the analysis of different emerging analysis using Artificial Intelligence is discussed. Since they have been developing for years, modern memory innovations are now at a crucial stage when efficiency is much more crucial in more tasks. In summary, the adoption of Technology, 5G, IoT, and Industry 4.0 has resulted in a rapid increase in the availability of data, and the modern computing demands are pushing memories innovation towards greater capacity, larger read/write times, quicker read/write rates, and less power consumption. The three popular memory technologies, DRAM, Flash, or SRAM, are projected to be replaced by new memory, which the semiconductors industry is currently funding.

References

1. Yudong Zhang, Saeed Balochian, Praveen Agarwal, Vishal Bhatnagar, and Orwa Jaber Housheya "Artificial Intelligence and Its Applications" Hindawi Publishing Corporation Mathematical Problems in Engineering Volume 2014, Article ID 840491, 10 pages http://dx.doi.org/10.1155/2014/840491
2. D. Poole, A. Mackworth, R. Goebel, Computational intelligence: a logical approach, Edition of the book, New York: Oxford University Press, 1998, p. 1.
3. Sudhir Allam "Artifical Intelligence and its Applications" International Engineering Journal For Research & Development 2020, Vol. 2 issue 6, E-ISSN NO:-2349-0721
4. Avneet Pannu "Artificial Intelligence and its Application in Different Areas" International Journal of Engineering and Innovative Technology (IJEIT) Volume 4, Issue 10, April 2015
5. Jean-Paul Haton "A brief introduction to artificial intelligence" 2006 IFAC
6. Hind Benbya, Stella Pachidi, Sirkka L. Jarvenpaa "Artificial Intelligence in Organizations: Implications for Information Systems Research" Journal of the Association for Information Systems (2021) 22(2), xxx-xxx doi: 10.17705/1jais.00491
7. Jatin Borana "Applications of Artificial Intelligence & Associated Technologies" Proceeding of International Conference on Emerging Technologies in Engineering, Biomedical, Management and Science [ETEBMS-2016], 5-6 March 2016
8. Iqbal H. Sarker "Machine Learning: Algorithms, Real World Applications and Research Directions" SN Computer Science (2021) 2:160 https://doi.org/10.1007/s42979-021-00592-x
9. Nur Farhana Hordri, Siti Sophiayati Yuhaniz, Siti Mariyam Shamsuddin "Deep Learning and Its Applications: A Review" Conference: Postgraduate Annual Research on Informatics Seminar 2016 At: University Teknologi Malaysia, Kuala Lumpur
10. Yingxu Wang "On Cognitive Computing" Int. J. of Software Science and Computational Intelligence, 1(3), 1-15, July-September 2009
11. Mr. Gopala Krishna Sriram "A Brief Overview of Cognitive Computing" International Research Journal of Modernization in Engineering Technology and Science, Vol. 04/Issue:03/March-2022

12. Anuj Kumar Bharti "A Research Paper on Basic of Computer Network" Journal of Emerging Technologies and Innovative Research (JETIR), Volume 6, Issue 3, 2019
13. Yifei Zhou "Research of artificial intelligence in computer network technology" 2nd International Conference on Applied Physics and Computing (ICAPC 2021), IOP Publishing, 2083 (2021) 042082 IOP. doi:10.1088/1742-6596/2083/4/042082.
14. Aryan Karn "ARTIFICIAL INTELLIGENCE IN COMPUTER VISION" International Journal of Engineering Applied Sciences and Technology, 2021 Vol. 6, Issue 1, ISSN No. 2455-2143, Pages 249-254
15. Zhou, K.-J. et al. "The trend of emerging non-volatile TCAM for parallel search and AI applications" Chip **1**, 100012 (2022). https://doi.org/10.1016/j.chip.2022.100012
16. Snehlata Yadav, Sonam Rewari, Rajeshwari "Pandey Junctionless Accumulation Mode Ferroelectric FET (JAM-FE-FET) for High Frequency Digital and Analog Applications" Journal off Silicon, Springer, 2021, Silicon https://doi.org/10.1007/s12633-021-01537-y
17. Emerging Storage Technologies: MRAM, RRAM, and PCRAM – Utmel, 2021.
18. Dmytro Apalkov, Bernard Dieny, J. M. Slaughter "Magnetoresistive Random Access Memory" Proceedings of the IEEE | Vol. 104, No. 10, October 2016
19. Usman Bature, Mohd Haris, I. Md Nawi, Furqan Zahoor "ZnO Based Resistive Random Access Memory Device: A Prospective Multifunctional Next-Generation Memory" IEEE Access 10.1109/ACCESS.2021.3098061
20. Basma Hajri, Hassen Aziza, Mohammad M. Mansour, and Ali Chehab RRAM Device Models: A Comparative Analysis with Experimental Validation" IEEE Access DOI 10.1109/ACCESS.2019.2954753
21. Anjana Bhardwaj, Pradeep Kumar, B. Raj, Sunny Anand, "Design and Performance Optimization of doping-less Vertical Nanowire TFET using Gate Stack Technique", Journal of Electronic Materials (JEMS) Springer, Vol. 41, Issuse 7, PP. 4005-4013, 2022.
22. Jeetendra Singh, BRaj, "Tunnel Current Model of Asymmetric MIM Structure Levying Various Image Forces to Analyze the

Characteristics of Filamentary Memristor", Applied Physics A, Springer, Vol. 125, No.3, pp. 203.1 to 203.11 Feb 2019.

23. Candy Goyal, Jagpal Singh Ubhi and B Raj, Low Leakage Zero Ground Noise Nanoscale Full Adder using Source Biasing Technique, "Journal of Nanoelectronics and Optoelectronics", American Scientific Publishers, Vol. 14, PP. 360–370, March 2019.
24. Gurmohan Singh, R. K. Sarin and B Raj, "A Novel Robust Exclusive-OR Function Implementation in QCA Nanotechnology with Energy Dissipation Analysis", Journal of Computational Electronics, Springer, Volume 15, Issue 2, PP. 455-465, June 2016.
25. Tanu Wadhera, Deepti Kakkar, Girish wadhwa, B Raj, " Recent Advances and Progress in Development of the Field Effect Transistor Biosensor: A Review" Journal of ELECTRONIC MATERIALS, Springer, Volume 48, Issue 12, pp 7635–7646, December 2019.
26. Girish Wadhwa, Priyanka Kamboj, Balwinder Raj" Design optimisation of junctionless TFET biosensor for high sensitivity" Advances in Natural Sciences: Nanoscience and Nanotechnology, Vol. 10, PP. 045001 (7), (2019).
27. Priya Bansal, BRaj, "MEMRISTOR MODELING AND ANALYSIS FOR LINEAR DOPANT DRIFT KINETICS," Journal of Nanoengineering and Nanomanufacturing, American Scientific Publishers, Vol. 6, pp. 1–7, 2016.
28. Amandeep Singh, Mamta Khosla, BRaj, "Circuit Compatible Model for Electrostatic Doped Schottky Barrier CNTFET,"Journal of Electronic Materials, Springer, Vol. 45, Issue 12, pp 4825-4835, 2016.
29. Ashima, D Vaithiyanathan, B Raj," Performance Analysis of Charge Plasma induced Graded Channel Si nanotube", Journal of Engineering Research (JER), EMSME Special Issue pp. 146-154 Aug 2021.
30. Abhishek Singh Tomar, Vijay Kumar Magraiya, and B Raj, Scaling of Access and Data Transistor for High Performance DRAM Cell Design, Quantum Matter, Vol. 2, PP. 412-416 Oct. 2013.
31. Neeraj Jain, BRaj, "Parasitic Capacitance and Resistance Model Development and Optimization of Raised Source/Drain SOI FinFET Structure for Analog Circuit Applications, Journal of

Nanoelectronics and Optoelectronins, ASP, USA, Vol. 13, PP. 531–539, Apl 2018.

32. Shradhya Singh, S. K. Vishvakarma and B Raj, "Analytical Modeling of Split-Gate Junction-Less Transistor for a Biosensor Application", Sensing and Bio-sensing, Elsevier, Vol. 18, Pages 31-36, April 2018.
33. Maisagalla Gopal, and Balwinder Raj, Low Power 8T SRAM Cell Design for High Stability Video Applications, ITSI Transaction on Electrical and Electronics Engineering, Vol. 1(5), pp. 91-97, 2013.
34. Balwinder Raj, Jatin Mitra, Deepak Kumar Bihani, Rangharajan V, A.K. Saxena and S. Dasgupta, "Analysis of Noise Margin, Power and Process Variation for 32 nm FinFET Based 6T SRAM Cell" Journal of Computer (JCP), Academy Publisher, FINLAND, vol. 5, issue 6, PP.1-8, 2010.
35. Divya Sharma, Rajesh Mehra and B Raj, "Comparative Analysis of Photovoltaic Technologies for High Efficiency Solar Cell Design", Superlattices and Microstructures, Elsevier, Volume 153, PP. 106861, May 2021.
36. Pawandeep Kaur, Avtar Singh Buttar, B Raj, "A comprehensive Analysis of Nanoscale Transistor based Biosensor: A Review, Indian Journal of Pure and Applied Physics, Vol. 59, pp. 304-318, April 2021.
37. Divya Yadav, Balwant Raj and B Raj, "Design and Simulation of Low Power Microcontroller for IoT Applications", Journal of Sensor Letters, ASP, Vol. 18, PP. 401–409, May 2020.
38. Shailendra Singh, B Raj,"A 2-D Analytical Surface Potential and Drain current Modeling of Double-Gate Vertical t-shaped Tunnel FET", Journal of Computational Electronics, Springer, Vol. 19, PP.1154–1163, Apl 2020.
39. Jeetendra Singh, B Raj, "An Accurate and Generic Window function for Non-linear Memristor Model" Journal of Computational Electronics, Springer, Volume 18 Issue 2, PP. 640-647, June 2019.
40. Manjit Kaur; Neena Gupta; Sanjeev Kumar; BRaj; Arun Kumar Singh, "Comparative RF and Crosstalk Analysis of Carbon Based Nano Interconnects" IET Circuits, Devices & Systems, Vol.15(6), PP. 493-503, Feb 2021.
41. Nehru Kandasamy, Firdous Ahmad, D. Ajitha, B Raj, and Nagarjuna Telagam "Quantum Dot Cellular Automata based Scan

Flip Flop and Boundary Scan Register, IETE Journal of Research, Volume 66, PP. 535-548, 2020.

42. Sharma, S. K.., Raj, B., Khosla, M., "Enhanced Photosensivity of Highly Spectrum Selective Cylindrical Gate In1-xGaxAs Nanowire MOSFET Photodetector, Modern Physics letter-B, Vol. 33, No. 12, PP. 1950144 (2019).
43. Jeetendra Singh, BRaj, "Design and Investigation of 7T2M NVSARM with Enhanced Stability and Temperature Impact on Store/Restore Energy", IEEE Transactions on Very Large Scale Integration Systems, Vol. 27, Issure 6, PP. 1322 - 1328 June 2019.
44. Anil Kumar Bhardwaj, Sumeet Gupta, BRaj, Amandeep Singh, "Impact of Double Gate Geometry on the Performance of Carbon Nanotube Field Effect Transistor Structures for Low Power Digital Design", Computational and Theoretical Nanoscience, ASP, Vol. 16, PP. 1813–1820, 2019.
45. Neeraj Jain, BRaj, "Thermal Stability Analysis and Performance Exploration of Asymmetrical Dual-k underlap Spacer (ADKUS) SOI FinFET for Security and Privacy Applications", Indian Journal of Pure & Applied Physics (IJPAP), Vol. 57, pp. 352-360, May 2019.
46. Amandeep Singh, Mamta Khosla, BRaj, "Design and Analysis of Dynamically Configurable Electrostatic Doped Carbon Nanotube Tunnel FET"," Microelectronics Journal, Elesvier, Volume 85, Pages 17-24, March 2019.
47. Neeraj Jain and Balwinder Raj, "Dual-k Spacer Region Variation at the Drain Side of Asymmetric SOI FinFET Structure: Performance Analysis towards the Analog/RF Design Applications", communicated in "Journal of Nanoelectronics and Optoelectronics", American Scientific Publishers, Vol. 14, PP. 349–359, March 2019.
48. Jeetendra Singh, Sanjeev Sharma, B Raj, Mamta Khosla Analysis of barrier layer thickness on performance of In1-xGaxAs based Gate Stack Cylindrical Gate Nanowire MOSFET, " JNO, ASP, Vol. 13, PP 1473-1477, Oct 2018.
49. Neeraj Jain, B Raj, "Analysis and Performance Exploration of High-k SOI FinFETs Over the Conventional Low-k SOI FinFET toward Analog/RF Design", Journal of Semiconductors (JoS), IOP Science, Vol.39, Issue 12, PP- 124002-1-7, Dec 2018.
50. Candy Goyal, Jagpal Singh Ubhi and BRaj, A reliable leakage reduction technique for approximate full adder with reduced

ground bounce noise, Journal of Mathematical Problems in Engineering, Hindawi, Volume 2018, Article ID 3501041, 16 pages, 15 Oct 2018.

51. Anuradha, Jeetendra Singh, B Raj and Mamta Khosla, "Design and Performance Analysis of Nano-scale Memristor-based Nonvolatile SRAM" Journal of Sensor Letter", American Scientific Publishers, Vol. 16, PP 798–805, Oct 2018.
52. Girish Wadhwa, BRaj, "Parametric Variation Analysis of Charge-Plasma-based Dielectric Modulated JLTFET for Biosensor Application" IEEE Sensor Journal, VOL. 18, NO. 15, AUGUST 1, PP. 6070 – 6077, 2018 (SCI)
53. Jeetendra Singh, BRaj, "Comparative Analysis of Memristor Models for Memories Design", JoS, IoP, Vol. 39, No. 7, PP. 074006-1-12, July 2018.
54. Divya Yadav, Shailesh Singh Chouhan, Santosh Kumar Vishvakarma and BRaj, "Application Specific Microcontroller Design for IoT based WSN", Sensor Letter, ASP, Vol. 16, PP. 374–385, May 2018.
55. Gurmohan Singh, R. K. Sarin and BRaj, "Fault-Tolerant Design and Analysis of Quantum-Dot Cellular Automata Based Circuits", IEEE/IET Circuits, Devices & Systems, Vol. 12, PP. 638 – 64, 2018.
56. Jeetendra Singh, BRaj, "Modeling of Mean Barrier Height Levying Various Image Forces of Metal Insulator Metal Structure to Enhance the Performance of Conductive Filament Based Memristor Model", IEEE Nanotechnology, Vol. 17, No. 2, PP. 268-267, March 2018.
57. Aakash Jain, Sanjeev Sharma, BRaj, "Analysis of Triple Metal Surrounding Gate (TM-SG) III-V Nanowire MOSFET for Photosensing Application", Opto-electronics Journal, Elsevier, Volume 26, Issue 2, Pages 141-148, May 2018.
58. Aakash Jain, Sanjeev Sharma, B Raj, "Design and Analysis of High Sensitivity Photosensor Using Cylindrical Surrounding Gate MOSFET For Low Power Sensor Applications", Engineering Science and Technology, an International Journal, Elsevier's, Volume 19, Issue 4, Pages 1864–1870, December 2016.
59. Amandeep Singh, Mamta Khosla, B Raj, "Analysis of Electrostatic Doped Schottky Barrier Carbon Nanotube FET for Low Power Applications," Journal of Materials Science: Materials in Electronics, Springer, Vol.28, PP. 1762–1768, 2017.

60. G. Saiphani Kumar, Amandeep Singh, BRaj, "Design and Analysis of Gate All Around CNTFET based SRAM cell Design", Journal of Computational Electronics, Springer, Volume 17, Issue 1, pp 138–145, March 2018.
61. Gurinder pal Singh, B. S. Sohi, Balwinder Raj, "Material Properties Analysis of Graphene Base Transistor (GBT) for VLSI Analog Circuits, Indian Journal of Pure & Applied Physics (IJPAP), Vol. 55, pp. 896-902, Dec-2017.
62. Amandeep Singh, Mamta Khosla, B Raj, "Comparative Analysis of Carbon Nanotube Field Effect Transistor and Nanowire Transistor for Low Power Circuit Design," Journal of Nanoelectronics and Optoelectronics, American Scientific Publishers, USA, Vol. 11, PP. 388-393, June 2016.
63. Sunil Kumar and B Raj, "Estimation of Stability and Performance metric for Inward Access Transistor based 6T SRAM Cell Design using n-type/p-type DMDG-GDOV TFET" IEEE VLSI Circuits and Systems Letter, Volume3, Issue2, PP 25-39, June2017
64. Shashikant Sharma, Anjan Kumar, Manisha Pattanaik and B Raj, "Forward Body Biased Multimode Multi-Threshold CMOS Technique for Ground Bounce Noise Reduction in Static CMOS Adders" International Journal of Information and Electronics Engineering, pp. 567-572, Vol. 3, issue 3, 2013.
65. Hamendra Singh, Pankaj Kumar, Balwinder Raj, "Performance Analysis of Majority Gate SET Based 1-bit Full Adder", *International Journal of Computer and Communication Engineering* (IJCCE), IACSIT Press Singapore, ISSN: 2010-3743, Volume 2, No. 4, 2013.
66. Anil Kumar Bhardwaj, Sumeet Gupta and B Raj, "Investigation of Parameters for Schottky Barrier (SB) Height for Schottky Barrier Based Carbon Nanotube Field Effect Transistor Device", Journal of Nanoelectronics and Optoelectronics, ASP, Vol. 15, PP. 783–791, July 2020.
67. Priya Bansal, BRaj, "MEMRISTOR: A VERSATILE NONLINEAR MODEL FOR DOPANT DRIFT AND BOUNDARY ISSUES," JCTN, American Scientific Publishers, Vol. 14 (5), PP. 2319-2325, May 2017.
68. Neeraj Jain, BRaj, "An Analog and Digital Design Perspective Comprehensive Approach on Fin-FET (Fin-Field Effect transistor) Technology - A Review", Reviews in Advanced Sciences and Engineering (RASE), ASP, Vol. 5, PP. 1-14, 2016.

69. Sanjeev Sharma, B Raj, Mamta Khosla, "Subthreshold Performance of In1-xGaxAs based Dual Metal with Gate Stack Cylindrical/Surrounding Gate Nanowire MOSFET for Low Power Analog Applications, Journal of Nanoelectronics and Optoelectronics, American Scientific Publishers, USA, Vol.12, PP. 171–176, 2017.
70. Balwinder Raj, A. K. Saxena and S. Dasgupta, "Analytical Modeling for the Estimation of Leakage Current and Subthreshold Swing Factor of Nanoscale Double Gate FinFET Device" Microelectronics International, UK, vol. 26, pp. 53-63, 2009.
71. Soniya, Shailendra Singh, Girish Wadhwa and Balwinder Raj, "An Analytical Modeling for Dual Source Vertical Tunnel Field Effect Transistor" submitted to 'International Journal of Recent Technology and Engineering (IJRTE)' at Volume-8 Issue-2, PP. 603-608, July 2019.
72. Shailendra Singh, B Raj, "Design and analysis of hetrojunction Vertical T-shaped Tunnel Field Effect Transistor", Journal of Electronics Material, Springer, Volume 48, Issue 10, pp 6253–6260, October 2019.
73. Candy Goyal, Jagpal Singh Ubhi and B Raj, "A Low Leakage CNTFET based Inexact Full Adder for Low Power Image Processing Applications", International Journal of Circuit Theory and Applications, Wiley, Volume47, Issue9, Pages 1446-1458, September 2019.
74. B Raj, A. K. Saxena and S. Dasgupta, "A Compact Drain Current and Threshold Voltage Quantum Mechanical Analytical Modeling for FinFETs" Journal of Nanoelectronics and Optoelectronics (JNO), USA, vol. 3, no. 2, pp. 163-170, 2008.
75. Girish Wadhwa, B Raj, "An Analytical Modeling of Charge Plasma based Tunnel Field Effect Transistor with Impacts of Gate underlap Region" Superlattices and Microstructures, Elsevier, Vol. 142, PP.106512, June 2020.
76. Shailendra Singh, B Raj,"Modeling and Simulation analysis of SiGe hetrojunction Double GateVertical t-shaped Tunnel FET", Superlattices and Microstructures, Elsevier Volume 142, PP. 106496, June 2020.
77. Amandeep Singh, Dinesh Kumar Saini, Dinesh Agarwal, Sajal Aggarwal, Mamta Khosla, BRaj, "Modeling and Simulation of Carbon Nanotube Field Effect Transistor and its Circuit Application ," Journal of Semiconductors (JoS), IOP Science, Vol.37, PP. 074001-6, July 2016.

78. Neeraj Jain, Balwinder Raj, "Device and Circuit Co-Design Perspective Comprehensive Approach on FinFET Technology - A Review, Journal of Electron Devices, Vol. 23, Num. 1, pp 1890-1901, 2016.
79. Sunil Kumar and BRaj, "Analysis of I_{ON} and Ambipolar Current for Dual-Material Gate-drain Overlapped DG-TFET," Journal of Nanoelectronics and Optoelectronics, American Scientific Publishers, USA, Vol.11, PP. 323-333, June 2016.
80. Naveed Anjum, Tarun Bali and B Raj, Design And Simulation Of Handwritten Multiscript Character Recognition, International Journal of Advanced Research in Computer and Communication Engineering, Vol.2 (7), PP. 2544-2549, July 2013.
81. Sanjeev Sharma, BRaj, Mamta Khosla, "A Gaussian Approach for Analytical Subthreshold Current Model of Cylindrical Nanowire FET with Quantum Mechanical Effects, Microelectronics Journal, Elsevier, Vol. 53, PP. 65-72, April 2016.
82. Karmjit Singh, BRaj, "Performance and analysis of temperature dependent Multi-walled Carbon Nanotubes as Global Interconnects at different technology nodes," Journal of Computational Electronics, Springer, Volume 14 Issue 2, pp. 469-476, June 2015.
83. Sunil Kumar and BRaj, "Compact channel potential analytical Modeling of DG-TFET based on Evanescent–mode Approach," Journal of Computational Electronics, Springer Vol. 14, No. 2, pp.820-827, July 2015.
84. Karmjit Singh, BRaj, "Temperature Dependent Modeling and Performance Evaluation of Multi-Walled CNT and Single-Walled CNT as Global Interconnects," Journal of Electronic Materials, Springer, Vol. 44, Issue 12, pp 4825-4835, Dec 2015.
85. V.K. Sharma, M. Pattanaik and B.Raj, "INDEP approach for leakage reduction in nanoscale CMOS circuits", International Journal of Electronics, Taylor & Francis, Vol.102, No.2, pages 200-215, 2014.
86. Karmjit Singh, BRaj, "Influence of Temperature on MWCNT bundle, SWCNT bundle and Copper interconnects for Nanoscaled Technology nodes," Journal of Materials Science: Materials in Electronics. Springer, Vol. 26, No. 8, pp. 6134-6142, 2015.
87. Naveed Anjum, Tarun Bali and B Raj, "Design and Simulation of Handwritten Gurumukhi and Devanagri Numerical Recog-

nition, International Journal of Computer Applications, Published by Foundation of Computer Science, New York, USA, Vol. 73(12), pp. 16-21, 2013.

88. Khandelwal, S., Gupta, V., BRaj, and Gupta, R D., Process Variability Aware Low Leakage Reliable Nano Scale DG-FinFET SRAM Cell Design Technique, Journal of Nanoelectronics and Optoelectronics, Volume 10, Number 6, pp. 810-817, December 2015.

89. V.K. Sharma, M. Pattanaik and B Raj, "ONOFIC Approach: Low Power High Speed Nanoscale VLSI Circuits Design", International Journal of Electronics, Taylor & Francis, Vol. 101, issue 1, pages 61-73, 2014.

90. Khandelwal, S., Balwinder Raj, and Gupta, R D., FinFET based 6T SRAM Cell Design: Analysis of Performance Metric, Process Variation and Temperature Effect. Journal of Computational and Theoretical Nanoscience, ASP, USA, Vol. 12, PP. 2500-2506, 2015.

91. Sumit Singh; Yadav, Shekhar; Rahul, Jagdeep; Srivastava, Anurag; BRaj, Impact of HfO_2 in Graded Channel Dual Insulator Double Gate MOSFET, Journal of Computational and Theoretical Nanoscience, American Scientific Publishers, Volume 12, Number 6, pp. 950-953, April 2015.

92. Vijay Kumar Sharma, Manisha Pattanaik, BRaj, PVT variations aware low leakage INDEP approach for nanoscale CMOS Circuits, Microelectronics Reliability, Elsevier, Vol. 54, pp 90-99, 2014.

93. B Raj, A. K. Saxena and S. Dasgupta, "Quantum Mechanical Analytical Modeling of Nanoscale DG FinFET: Evaluation of Potential, Threshold Voltage and Source/Drain Resistance", Elsevier's Journal of Material Science in Semiconductor Processing, Elsevier, Vol. 16, issue 4, pp. 1131- 1137, 2013.

94. Maisagalla Gopal, Siva Sankar D Prasad and Balwinder Raj, 8T SRAM Cell Design for Dynamic and Leakage Power Reduction, International Journal of Computer Applications, Published by Foundation of Computer Science, New York, USA, Vol. 71(9), pp. 43-48, June 2013.

95. Manisha Pattanaik, B Raj, Shashikant Sharma and Anjan Kumar, "Diode Based Trimode Multi-Threshold CMOS Technique for Ground Bounce Noise Reduction in Static CMOS Adders", Advanced Materials Research, Trans Tech Publications, Switzerland, vol. 548, pp 885-889, 2012.

96. Balwinder Raj, A. K. Saxena and S. Dasgupta, "Nanoscale FinFET Based SRAM Cell Design: Analysis of Performance metric, Process variation, Underlapped FinFET and Temperature effect" IEEE Circuits and System Magazine, vol. 11, issue 2, pp. 38- 50, 2011.
97. V.K. Sharma, M. Pattanaik & Balwinder Raj, "Leakage Current ONOFIC Approach for Deep Submicron VLSI Circuit Design", International Journal of Electrical, Computer, Electronics and Communication Engineering, World Academy of Sciences, Engineering and Technology, Vol. 7, issue 4, pages 239-244, 2013.
98. Tulika Chawla, Mamta Khosla, BRaj, Design and simulation of Triple metal Double-gate Germanium on Insulator Vertical Tunnel Field Effect Transistor, Microelectronics Journal, Elsevier; Vol. 114, PP. 105125, August 2021.
99. Parminder Kaur, Sandeep Singh Gill and B Raj, "Comparative Analysis of OFETs Materials and Devices for Sensor Applications", Journal of Silicon, Springer, Vol 14, PP. 4463–4471, 2022.
100. Sanjeev Kumar Sharma, Parveen Kumar, Balwant Raj and BRaj, "$In_{1-x}Ga_xAs$ Double Metal Gate-Stacking Cylindrical Nanowire MOSFET for highly sensitive Photo detector", *Journal of Silicon, Springer Vol.*14, PP. 3535–3541, 2022.
101. B Raj, A. K. Saxena and S. Dasgupta, "Analytical Modeling of Quasi Planar Nanoscale Double Gate FinFET with Source/Drain Resistance and field dependent carrier mobility: A quantum mechanical study" Journal of Computer (JCP), Academy Publisher, FINLAND, vol. 4, issue 9, PP.1-8, 2009.
102. Bhushan, S., Khandelwal, S. and Raj, B., (2012). Analyzing Different Mode FinFET based Memory Cell at Different Power Supply for Leakage Reduction, Seventh International Conference on Bio-Inspired Computing: Theories and Application, (BIC-TA 2012) Advances in Intelligent Systems and Computing, Volume 202, p89-100 (2013).
103. Jeetendra Singh, BRaj, "Temperature Dependent Analytical Modeling and Simulations of Nanoscale Memristor, Journal: Engineering Science and Technology, an International Journal, Elsevier's, Vol. 21, PP- 862-868, Oct 2018.
104. Shradhya Singh, Shashi Bala, Balwant Raj, B Raj, "Improved Sensitivity of Dielectric Modulated Junctionless Transistor for

Nanoscale Biosensor Design", Sensor Letter, ASP, Vol. 18, PP.328–333, Apl 2020.

105. Vivek Kumar, Santosh Kumar Vishvakarma, and B Raj, "Design and Performance Analysis of ASIC for IoT Applications" Sensor Letter ASP, Vol. 18, PP. 31–38, Jan 2020.
106. Akanksha Jaiswal, R. K. Sarin, B Raj and Shikha Sukhija, "A Novel Circular Slotted Microstrip-fed Patch Antenna with Three Triangle Shape Defected Ground Structure for Multi-band Applications", Advanced Electromagnetic (AEM), Vol. 7, Issue 3, PP. 56-63, Aug 2018.
107. Girish Wadhwa, B Raj, "Label Free Detection of Biomolecules using Charge-Plasma-Based Gate Underlap Dielectric Modulated Junctionless TFET" Journal of Electronic Materials (JEMS), Springer, Volume 47, Issue 8, pp 4683–4693, August 2018.
108. Gurmohan Singh, R. K. Sarin and B Raj, "Design and Performance Analysis of a New Efficient Coplanar Quantum-Dot Cellular Automata Adder", Indian Journal of Pure & Applied Physics (IJPAP), Vol. 55, pp. 97-103, February 2017.
109. Amandeep Singh, Mamta Khosla, Balwinder Raj, "Design and Analysis of Electrostatic Doped Schottky Barrier CNTFET Based Low Power SRAM," International Journal of Electronics and Communications, (AEÜ), Elsevier, Vol. 80, PP.67–72, 2017.
110. Parminder Kaur, Vikas Pandey and B Raj, "Comparative Study of Efficient Design, Control and Monitoring of Solar Power using IoT", Sensor Letter, ASP Vol. 18, PP. 419–426, May 2020.
111. Anil Kumar Bhardwaj, Sumeet Gupta and B Raj, Development & Analysis of compact model for Double Gate Schottky Barrier CNTFET, Journal of Nanoelectronics and Optoelectronics, ASP, Vol. 15, pp. 1199–1208, Aug 2020.
112. Girish Wadhwa, Priyanka Kamboj; Jeetendra Singh, B Raj, "Design and Investigation of Junctionless DGTFET for Biological Molecule Recognition" Transactions on Electrical and Electronic Materials, Springer, Vol. , PP. 8 Aug, 2020.
113. Tulika Chawla, Mamta Khosla, B Raj, "Optimization of Double-gate Dual material GeOI-Vertical TFET for VLSI Circuit Design, IEEE VLSI Circuits and Systems Letter, Vol. 6, issue-2, PP.13-25, Aug 2020.
114. Sachin Kumar Verma, Shailendra Singh, Girish Wadhwa and B Raj, "Detection of Biomolecules using Charge-Plasma based

Gate Underlap Dielectric Modulated Dopingless TFET" Transactions on Electrical and Electronic Materials (TEEM), Springer, Vol. 21, PP. 528-535, June 2020.

115. Neeraj Jain, BRaj, "Impact of underlap spacer region variation on Electrostatic and analog/RF performance of symmetrical high-k SOI FinFET at 20 nm channel length", Journal of Semiconductors (JoS), IOP Science, Vol.38, Issue 12, PP- 122002, Dec 2017.
116. Shailendra Singh, B Raj," Analytical Modeling and Simulation analysis of T-shaped III-V heterojunction Vertical T-FET", Superlattices and Microstructures, Elsevier, Vol. 147, PP. 106717, Nov 2020.
117. Gurmohan Singh, R. K. Sarin and B Raj, "Design and Analysis of Area Efficient QCA Based Reversible Logic Gates", Journal of Microprocessors and Microsystems, Elsevier, Vol. 52, PP.59-68, May 2017.
118. Amandeep Singh, Mamta Khosla, BRaj, "Compact Model for Ballistic Single Wall CNTFET under Quantum Capacitance Limit," Journal of Semiconductors (JoS), IOP Science, Vol.37, PP. 104001-8, Oct 2016.
119. Sonal Singh; Mamta Khosla, Girish Wadhwa; B Raj, "Design and analysis of double-gate junctionless vertical TFET for gas sensing applications", Applied Physics A, Springer, Vol.127, No.16, 2 Jan 2021.
120. Inderjit Singh, BRaj, Mamta Khosla, and Brajesh Kumar Kaushik, Potential MRAM Technologies for Low Power SoCs, SPIN World Scientific Publisher, SCIE; Vol. 10, No. 04, PP.2050027, Dec 2020.

8

Recent Trends in Perovskite Based Solar Cells

Vivek Bhojak[1,2], Neha Sharma[3] and Praveen K. Jain[1]
[1]Department of Electronics and Communication Engineering, Swami Keshvanand Institute of Technology, Management and Gramothan, Rajasthan (Jaipur), India
[2]Anand International College of Engineering, Rajasthan (Jaipur), India
[3]Department of Physics, Vishwavidyalaya Engineering College, Ambikapur, Surguja, India

Abstract: Perovskite SC (solar cell) & Coming-generation renewable sources of energy have transformed the photovoltaic industry & new research. After initial research perovskite solar cell have a remarkable Power Conversion Efficiency (PCE) of 24.2%, which is higher than the solar cell based on thin films of multicrystalline silicon, thin films of crystalline silicon, cu, In, Ga, Se and cadmium telluride, based on plastic substrates, all of which are suitable for transitioning into flexible solar cell. Organic plus inorganic -hybrid perovskite material revolutionized solar energy conversion systems. PSCs are popular because of their high efficiency, balanced charge transport, robust optical absorption, and longer lengths for carrier diffusion. However, perovskite solar cells' poisonous lead prevents commercialization. Tin based perovskite solar cell, which is lead free, might solve this problem; however, several technical challenges must be resolved before they can equal the power-conversion efficiency (PCEs) & stability of lead-based ones. This article explores the evolution of power-conversion efficiency (PCEs) of conception into current developments. Discussed were device architectures and working, potential and horizons of lead-free low toxicity perovskite solar cell and fabrication techniques.

8.1 Introduction

From the first industrial revolution, fossil fuels were vital. Despite the fact that earlier industrial revolution propelled human-civilization into an unparalleled golden age. They have also used a lot of energy and resources and made the conflict between humans and nature worse, which has cost us a lot in terms of the environment and ecology. The global economy and society are rapidly expanding their productive capacities., and fossil fuels cannot supply our energy needs. In the 21st century, mankind faces tremendous energy, resource, environmental, ecological, and climate change issues. It has become inevitable that old energy sources will be phased out in favor of cleaner forms of energy that can be generated from renewable resources as long as the use of energy around the globe continues to rise.

These challenges started the 4th industrial revolution & green industrial revolution. Because of pollution free, renewable energy & other benefits, solar energy, the first way of all the power ineluctably supports human and societal progress. The great availability of solar energy has led to its widespread use as the source of power. SC (Solar cells) are the most significant method to utilize solar energy, and new kinds of solar cells using lead-free perovskites as its light-absorbing materials are evolving quickly. In Optoelectronic devices, SC (Solar cells) are very appropriate.

Research of inexpensive, high efficiency-η solar cell has been a priority since the last few years because of their superior properties. Synthetic perovskites have been found to be a possible cheap base material for good efficiency of SC (Solar cells) about or higher than 25 %. It has high potential for the future as PCE grew in the range 3.8% - 25 %. The lead (Pb) free PSCs has created interest as of its high light absorption coefficient, cheaper and easy preparation.
In this chapter, the most recent details on SC (Solar cells), the history of them, and different Perovskite solar cells (PSCs) are discussed to provide a comparative analysis on the performance.

8.2 Solar Cell Technology

Solar cells, commonly known as photovoltaic cells, directly convert sunlight into energy. The photovoltaic effect, which converts photons to voltage, gives photovoltaics (PV) its name. In 1954, silicon SC (Solar cells) produced electric current in sunlight at Bell Laboratories. Solar cells quickly powered space spacecraft and calculators and watches. Today, solar cell energy is cost effective in many locations, and PV (photovoltaic) systems are being implemented at huge proportions to strengthen the electric grid.

Figure 8.1 demonstrates solar cell assembly and operation. SC (Solar cell) is constructed with n type and p type silicon. Sun light attacks the SC's (Solar cell) surface with numbers of photons. Photons transfer cell energy. Photons power p-type electrons. Electrons may evacuate the circuit via the barrier into the top, n-type layer with this energy, which results in flow of current.

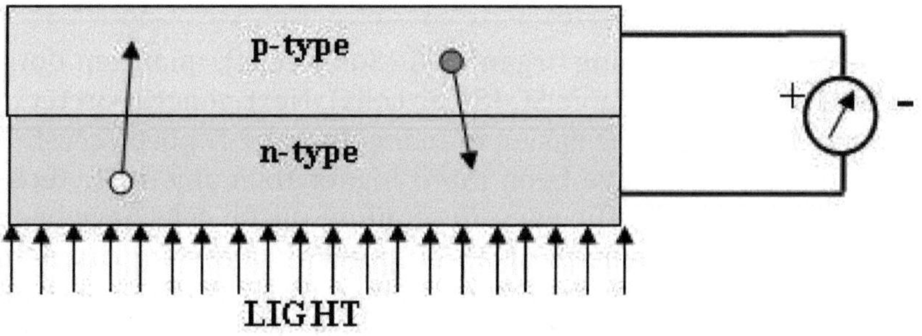

Figure 8.1 *Construction and working of solar cell.*

Following characteristics, necessary for a material to be considered suitable for use in solar cells:

- Cheap (cost < 1$/watt), simple (with long life > 20 years).
- Abundant material, integrated large-scale manufacturability.
- Direct band gap (~1.5 eV)
- The capacity for forming junctions
- High absorption coefficient > 10^5 cm^{-1}
- Good efficiency.
- Energy payback period < 2 years

Most silicon solar cells are affordable and efficient. These are incorporated into larger modules, which may then be fixed on the rooftops of any building or on racks that are attached to the ground, to construct large-scale utility-scale systems.

One more common type of solar cell also called photovoltaic (PV) solar cell which is made of a very thin layer of semiconductor material, such as cadmium telluride or Cu, In, Ga, Se etc. Thin film Solar cell are lightweight, flexible which makes them perfect for solar powered windows. Thin film SC (Solar cells) may also be manufactured with less energy and more easily scaled up than silicon solar cells.

Materials for these third generations are needed. Group III elements (including indium and gallium) and Group V elements (including arsenic and antimony) are used. It's more expensive to manufacture SC than it is to produce most other technologies. They are more effective at converting solar energy into usable forms. These SC are used in UAVs, Satellites, and other applications that need a high power-to-weight ratio (Solar cells).

Researchers are developing organic SC (Solar cells), quantum dots, and hybrid organic inorganic SC (Solar cells). Next-generation technology may be cheaper or easier to make. In most reports, crystalline silicon Solar cells have been rated higher than any other technology in terms of their efficiency. Crystalline silicon cells have been shown to achieve efficiencies of 25-30 percent in certain cases. In contrast, the efficiency of thin films ranges from 3 to 11 percent. Since 2009, metal halide PSCs have revolutionized PV [1–3]. The perovskite materials used in next-generation light harvesters have a high PCE because of their great properties [4-6]. The maximum PCE that has been validated by NREL is 25.6% [7–10]. This value is much closer to monocrystalline silicon solar cells than other photovoltaics as shown in Figure 8.2.

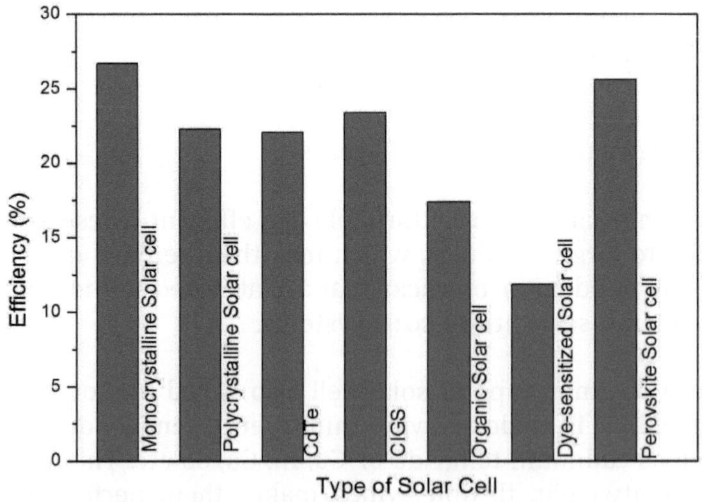

Figure 8.2 *Comparison of PCEs of different type of photovoltaics.*

8.3 Perovskite Solar Cell (PSCs)

Perovskite has the crystal-structure of ABX3, which consists of large cations A and B, as well as an anion X that binds to both of these cations. Common perovskite materials have layered structures that consist of a cubic lattice inside an octahedral lattice, and they offer exceptional thermal, optical, and electromagnetic properties. SC (Solar cell) utilizes organic-and-inorganic metal, halide-perovskite-materials. Octahedral assemblages of metal and halogen form robust three-dimensional networks.

Such materials have four characteristics. First, the materials having low exciton binding energy, good photoelectric characteristics, and good optical absorption coefficients (104 cm^{-1}) [13]. Second, perovskite absorbs solar energy well [14]. Third, materials have a high dielectric-constant (k) & transport and capture charge carriers (electron & hole) [15]. Finally, electrons and holes may be conveyed synchronously across 100 nm or even 1 μm [16–19].

This is why solar-cell materials feature high short-circuit current densities (J_S) and open-circuit voltages (V_O). Perovskite is able to generate electron and hole pairs in its main layer by absorbing of light from the sun. Because of perovskite material's exciton-binding-energy, Excitons have the ability to reform into new excitons or mix with free carriers to create a current. Because of their high carrier mobility and recombination probabilities, $CH_3NH_3PbI_3$-($MAPbI_3$) and other such materials have a longer carrier lifespan and a higher carrier diffusion distance in comparison to other materials. At least one hundred nanometers and one meter are required for carrier diffusion in MAPbI3 and Clx, respectively. [16, 17]. Due to carrier diffusion and longevity, Perovskite solar cells perform better. ETMs and HTMs collect these electrons and holes that are free. The metal electrode is responsible for collecting holes from the HTM layer. External circuit creates a photocurrent with the FTO and metal electrodes are connected.

The highest efficiency of solar-energy conversion occurs in perovskite solar cell (PSCs) [11]. Inexpensive materials may be used to create halide-perovskites via the use of highly prolific organic electronics deposition techniques [12]. Kojima, Kim and Lee tell about the solar cell potential for halide-perovskite [1,2]. Since this fundamental study, PSC-power conversion efficiency has increased more than-22% [13], making ABX3 halide-perovskites a top contender for the next solar cell generation. The remarkable optoelectronic qualities of semiconductor halide perovskites, which include their good properties as -adjustable bandgap, good absorption-coefficient, high mobility, extended carrier recombination lifetimes, and low exciton binding energy, are what make them successful [20–22]. Pb-based perovskite materials exhibit substantial spectral absorption and long-carrier-diffusion-lengths for electrons-holes. [23-24] The PSC device also eliminates the requirement for high heat and pressure during silicon-based solar cell manufacturing. Perovskite solar cells are simpler to make and less susceptible to internal faults. [25] To produce conventional silicon cells, expensive, time-consuming processes requiring cleanrooms, high temperatures (>1000 °C), and vacuum are required. [26]. The hybrid organic inorganic perovskite material can be produced using wet chemistry which can be used in the laboratory. There are many different ways that hybrid perovskites, methylammonium, and formamidinium lead trihalides may be synthesized. Spin-coating, slot-die-coating, blade-coating, spray-coating, inkjet-printing, screen-printing, electro-deposition, and vapor-deposition are some examples of the various methods that may be used [27-31]. These advantages would reduce manufacturing costs and simplify industrial operating requirements. Compared to Si based solar cell, PbI2 and MAI are less expensive also.

8.4 Perovskite Solar Cell Evolution

8.4.1 Architectural Considerations for Perovskite Solar Cell

The inclusion of free electrons and holes into the electron transport layer (ETL) and the hole transport layer (HTL) is made possible by these cells. Glass is used for the anodes and metal with FTO- Fluorine Doped Tin Oxide is used as cathodes. These cells are used to generate electricity.

Characteristics of these depend on whether the transport-medium—an electron-or-a hole meets the light first. Electron ETL is deposited first by these, then electron HTL The ETL of this solar cell-SC is exposed to light first. The following are the two architectures that may be found in standard n-i-p.

Mesoscopic arrangement N-i-P, where perovskite material is positioned between HTL and ETL. The anode is made up of a metallic cap and a transparent glass cathode.

Table 8.1 *History*

Year	Remark
1792	The German mineralogist Gustav Rose was the one who made the first discovery of this mineral, it was the Russian mineralogist Lev Perovski who was honored with its naming in 1839.
2009	In 2009, Kojima et al. presented the first Hybrid Organic Inorganic Perovskites (HOIP)-based photovoltaic system by combining liquid electrolyte with (CH3NH3PbBr3, MAPbBr3) and (CH3NH3PbI3, MAPbI3). The device's instability and low (< 3%) PCEs make it inappropriate. [1]
2012	In 2012, a solid-state hole transport substance was employed instead of a liquid electrolyte because it was more efficient. This modification improved MAPbI3 solar cell PCE by over 10%. [2]
2017	PCE progressed in short amount of time to 22.1 percent. Solar cells utilizing hybrid lead halide perovskites were found to have an impressively high-power conversion efficiency. [21,25]
2019	Using lead-based perovskite, it is feasible to achieve efficiencies in excess of 25%. The tandem solar cell, Oxford Photovoltaics (PV) , gives a record efficiency of 27.3%. This statistic was demonstrated by Oxford Photovoltaics. Germany's Fraunhofer ISE verified the accuracy of this record. [9]

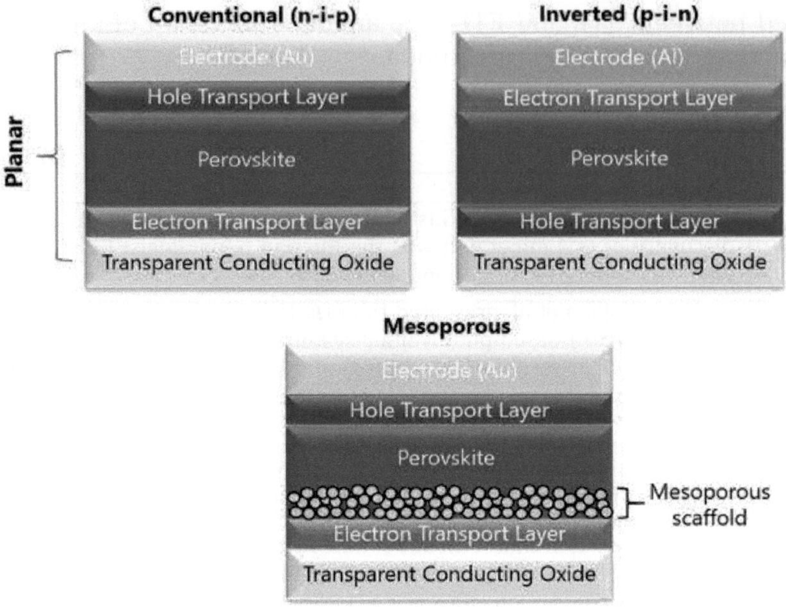

Figure 8.3 *Architectures.*

N-i-P planar arrangement, which simplifies production by removing the mesoporous metal oxide layer. In n-i-p-planar architectures, perovskite-light-harvesting layers are located in-between ETL-HTL.

8.4.2 Solar Cell Made Using Inorganic Perovskite

Recently, there has been a lot of interest in all-inorganic perovskites having Cs+ as the A-site cation (CsPbIxBr3x, where x ranges from 0 to 3). This is due to the fact that these perovskites are thermally stable [32-33]. Quantum dots, dimensional control, compositional optimization, and interfacial engineering have been used to enhance inorganic CsPbIxBr3−x PVSC performance [34-35] compositional optimization [36-37], dimensional control [38-39], and interfacial engineering [40-41]. Recent developments have resulted in the creation of a solvent-controlled growth technique for the production of high-quality black-phase CsPbI3 perovskite film. Further, phenyltrimethylammonium bromide surface termination of the perovskite layer yielded a 17.06% PCE CsPbI3 PVSC17. Interface engineering and perovskite passivation improvement boosted CsPbI3 PVSC PCE to 18.4%6. These critical studies demonstrate the potential of CsPbIxBr3−x PVSCs despite their lower PCEs than hybrid equivalents.

Most high-performance inorganic PVSCs use the classic n-i-p design with doped hole transport layer. Due to dopants' instability, electronics incorporating such doped HTLs have degraded severely [42-45]. In this, the development of inorganic p-i-n inverted PV-SC is extremely promising because it may exempt the use of unstable doped HTL and is very comfort with the fabrication of tandem solar cell, given that the majority of perovskite tandem solar cell is made using the p-i-n design [46-47]. It can improve operational stability and help build these cells to boost PCE.

Han et al. [48] stabilized cubic phase CsPbI3 by meticulously managing the distortion in the lattice using a succession of organic terminal groups in order to attain a stable PCE of 13.2% for the consequent inverted PVSC. This was done in order to achieve the goal. Mai et al. created an inverted CsPbI2Br PVSC with a 13.3% PCE using a ZnO/C60 bilayer ETL [49]. These cases showed that bilayer ETL improves carrier extraction and reduces device leakage. The efficiency of the inverted CsPbI2Br PV-SC was raised to 13.74% after being doped with InCl3, which was incorporated into the perovskite. All of these inverted inorganic PVSCs still perform badly in comparison to their conventional counterparts. This is mostly because of the significant open-circuit voltage (V0c) loss [50].

Imperfections on the surface of the perovskite that are positively or negatively charged and flaws on the grain boundaries can be the source of significant V0c losses in PVSCs [51-52]. In order to passivate perovskite surface & grain boundaries, Lewis base cations [53–54], polymer [55–56] and quaternary ammonium cations [57–58] have all been utilized as methods.

8.4.3 Organic Perovskite Solar Cell

OPV is a speedily expanding PV technology that offers high cell efficiency (18.2% certified), a long performance lifetime (>10 years unencapsulated), and the ability to produce rolls of solar panels using solution processing. OPV may be attractive to the building-integrated PV sector since it can produce effective transparent devices and utilise any colour absorber.

The range of organic materials that may be created and manufactured for absorber, acceptor, and interfaces is one of OPV's strongest points.

Waste disposal is challenging due to the inability of perovskites containing metal to breakdown. Because wearable technology is so popular, certain metal cations cannot be used because they are harmful. Perovskites made without organic metals may solve the issue [59]. Organic perovskites weren't known until 2002. There were no new organic perovskites discovered after 15 years. Metal-free A(NH4)X3 perovskites were introduced by Xiong et al. [60]. Similar to conventional perovskites in terms of their crystal structure, metal-free perovskites have lattice ions organized differently. Six anion (X) form an octahedron with a highly charged cation in the center in traditional perovskites. In certain other perovskite, anions like Li3OBr occupy the A and B sites while six cations form the octahedron. These perovskite include antiperovskite. The metal-free perovskites disclosed by Xiong et al. differ from conventional and antiperovskites in having an A-site cation with more charges than a B-site cation. The inorganic "inverse perovskite" Cui Q, Liu Y [61] is similar to metal-free perovskite.

Xiong et al. found that MDABCO-NH4I3 was ferroelectric at 293 K. Similar to BaTiO3's 26 C/cm2, MDABCO-NH4 I3 has a 22 C/cm^2 spontaneous polarisation. These results suggest that organic perovskites could take the role of inorganic perovskites in ferroelectric devices. Organic perovskites are used in X-ray imaging. Massive DABCO-NH4Br3 crystals were used by Zhao et al. to produce X-ray detectors [62].

Unlike inorganic ones, organic perovskites do not fracture when heated to a high temperature. Organic perovskites are suitable for use in implanted biosensors and bioelectronics due to their diversity in chemical composition.

8.4.4 Hybrid Perovskite Solar Cell

Research on organic materials is often restricted to low-performance applications. Charge carriers in molecular materials don't move around very effectively, which is the root of this issue. Strong chemical bonds between metal electrodes and organic molecules stop charge carriers from getting into organic compounds [63–66]. Organic electronic compounds' characteristics may be altered by resonant interactions in hybrid structures having inorganic and organic components [64, 65]. The inorganic semiconductor's high conductivity and the organic component's strong light-matter interactions might be combined in a hybrid structure. They were classified as organic-inorganic hybrid materials because of the high exciton binding energy (about several hundred meV) that they had. This was made possible by the dielectric confinement that they possessed (approximately several hundred meV). In these layered organic-inorganic perovskites, layers are inserted between organic ammonium insulator layers. [67–71].

To create the first dye-sensitized solar cell, the light absorber was employed in 2009. (DSSC). The new material was exceedingly crystalline and had special optical and electrical qualities, although its initial efficiency was just 3.8%. Energy conversion efficiency for these cells is more than 22%. This is as a result of the manufacturing process being optimized, new core materials being employed, and enhanced photovoltaic device structure. Perovskite material 2 RAMX3 is a light-harvesting material. RA is a monovalent organic cation such as (HN = CHNH3+, FA+) or (CH3NH3+, MA+), M is a bivalent metallic ion such as Sn2+ or Pb2+, and X is often a negatively charged halide in this 3D organic-inorganic hybrid molecule. For single junction solar cell or tandem solar cell, either cations or anions (which may be coupled with RAMX3 in crystal units) are required. In a good band gap, one would be able to see the majority of visible light spectrum, which spans from 400 to 800 nanometers, and perhaps even the infrared. 5 The composition of the solar cell has an effect on the amount of time required for diffusion, the lifetimes of charge carriers, and how mobile they are [72 - 75].

Lead halide-derived organic-inorganic perovskites may be utilised to create novel kinds of light sources [78–79] and nonlinear optical devices. Perovskites made of organometallic halides have shown potential for use in high-efficiency nanostructured electronics [80]. Organometallic trihalide perovskite absorbers have made it feasible to create high-efficiency solar cells, which has increased interest in photovoltaics research recently [81–84].

Two-dimensional layers of inorganic material and one-dimensional layers of ammonium make up the organic-inorganic hybrid crystal. These layers are made up of a two-dimensional sheet of [MX6-4] octahedra that have halide ions joining their corners. To create an octahedral cluster, the M2+ ion was encircled by six X- halogen ions from both sides. The inorganic layer is composed of several atomic layers all working together to form the overall structure. The -NH3+ terminals of cations align with the layers of [MX6 - 4 anion] by hydrogen bonding with the equatorial and axial halide ions. Alkylammonium ions cancel each other out [MX6-4], creating a multi-layered structure [85].

8.4.5 Heterojunction Perovskite Solar Cell

In heterojunction solar cells, a crystalline silicon cell is positioned between two "thin-film" amorphous silicon layers. Compared to silicon solar panels, this improves energy conversion and efficiency.
Recently, it is more interested in inverted planar heterojunction structures for perovskite solar cells (PSCs), which consist of a polycrystalline perovskite film sandwiched between a hole-extraction layer and an electron-extraction layer [86-89]. Perovskite solar cells are used to generate electricity.. These PSCs are simple to fabricate, can function with flexible substrates, can be engineered for different energy bands, and can be used to create multijunction cells. As an added bonus, their PCEs are above 20%.

8.4.6 Flexible Perovskite Solar Cells (FPSCs)

FPSCs may be made in a variety of forms and used in photovoltaic-integrated buildings, portable electronics, and wearable technology. In recent years, there has been a discernible uptick in FPSCs with a maximum documented PCE of 20%-21%. Low processing temperatures and instability caused by frequent mechanical bending are merely two of the technological hurdles confronted by flexible printed circuit boards (FPSCs).

The substrates of flexible solar cells impact both structural stability and photovoltaic efficiency (Jung et al., 2019). The three different kinds of FPSC substrates are flexible ultrathin glass, metallic (copper or stainless-steel foil, Ti) and polymers (PEN or PET). Most flexible substrate demands may be satisfied by flexible glass substrates (thickness in microns). It is desired to have a high temperature tolerance of more than 600 °C, good conductivity, exceptional chemical stability. Comparatively speaking to other contenders, fragility, weight, and cost may be problems. Despite having excellent charge conductivity, thermal stability and corrosion resistance, the metal substrate's intrinsic opacity reduces power conversion efficiency. The substrate is made of plastic, which is flexible, translucent, and reasonably priced. The primary limitations of plastic substrates are their low processing tolerance temperature (below 150 °C) and weak oxygen and moisture gas barrier characteristics [56]. An acceptable temperature should be maintained during all activities. Flexible electronics hold a lot of potential for FPSCs on plastic substrates [94-99].

Additionally, electrodes must have low-sheet resistance, transparency, durability, and mechanical flexibility (Li et al., 2018). As flexible PSC electrodes, ITO, graphene, polymer, and Ag-NWs have all been employed (summarized in Table 1). NiOx and ZnO were used by Najafi et al. as HTL and ETL on the flexible substrate PEN/ITO. 16.6% PCE was present in three single-step cation-based perovskite fllms. Using a triple cation perovskite, devised a novel low-temperature method to develop high-performance flexible planar PSCs for implementation on substrate of ITO/PEN, which increases FPSC PCE by up to 17%. Using a flexible ITO/PEN substrate, constructed effective, long-lasting FPSCsc[100-106].

A wide range of applications are available because of F-PSCs' excellent efficiency and bending resistance. These bendable solar cells are suitable for building integrated photovoltaics since they can be put on curved surfaces [107-113]. They can also attain a high power density of 2 W/g,67 because of their low weight and excellent efficiency, which qualifies them for usage in high-altitude unmanned aerial vehicles. To make this F-PSC commercially viable, a number of challenges must be overcome: (1) developing a uniform large area coating technique; (2) ensuring long term stability through encapsulation technology; and (3) utilizing newly emerging, distinctive flexible cells for specialized applications [114-120].

8.5 Prospect for lead-free Photovoltaic Devices

Due to their high 22% PCE, lead halide perovskites (APbX3) with A being an organic or inorganic positively charged ion and X being a halide (e.g. I–, Br–, or Cl–) are promising photovoltaic materials. These semiconducting materials have high absorption coefficients, low cost, easy synthesis, high dielectric constant, powerful photoluminescence, small exciton binding energies, and intrinsic ferroelectric polarization [121-126]. Various "hybrid" architectures can integrate inorganic and organic cations (A), central elements (other than Pb), and anions (X) for total optical tunability across the visible spectrum. Spin-coating, vapour deposition, or solution-based precipitation produce lead halide perovskites [127-131].

Lead is a hazardous heavy metal with serious health risks. The EPA's hazardous waste standard for lead is 5 mg/L, whereas the drinking water MCL is 0.015 mg/L. Pb-containing perovskite solar cells breakdown quickly in the environment. Wang and colleagues found that Pb-based PSCs can withstand 45% humidity for 2000 hours. (3) PSCs produce more defective goods without PCE than successful things. Damaged PSCs were often thrown away without treatment and ended up in a landfill or other waste disposal. In accordance with research studies, It is possible for lead ions that have been filtered out of PSCs to travel through the soil, be absorbed by plants, and then make their way into the food chain. Even worse, it was shown that leached discarded PSCs had ten times the amount of mobile lead as other Pb compounds [132-137].

Synthesizing lead-free perovskite photovoltaics is a potential area of study for the discovery, processing, and development of materials and devices. Two classes of materials, double perovskites and tin-halide perovskites, are alternatives to lead-halide perovskites. The high effective mass and indirect band gap of perovskites containing double halides may limit their PV efficiency. Deep antisite-related traps in their defect chemistry may further limit their PV potential. Pb substitution with Sn, Ge, and Bi has been studied theoretically [57–62].

Sn-based perovskites possess smaller optical bandgaps (1.2 eV - 1.4 eV) and higher carrier motilities than Pb-based materials [68-69], low toxicity makes them suitable for perovskite absorbers for high-performance lead-free PSCs [68–70].

Ge-based perovskites with MAGeI3, CsGeI3, and FAGeI3 can also withstand 150 °C. These findings led Mathews et al. to propose Ge as a lead-free perovskite material [64]. In 2015, Kanatzidis et al. synthesised AGeI3 perovskite molecules and characterised their structural, electrical, and optical characteristics [61]. Room-temperature AGeI3 perovskites are rhombohedral [60]. Kanatzidis et al. [61] synthesised Ge-based perovskites with modified A-site cations using the pyramidal [GeI3] – building block. The experimental results showed that CsGeI3, MAGeI3, FAGeI3, and CH3C(NH2)2GeI3 are three dimensional perovskite structures with direct bandgaps of 1.6 eV, 1.9 eV, 2.2 eV, and 2.5 eV, respectively. On the other hand the C(NH2)3GeI3, (CH3)2C(H)NH3GeI3, and (CH3)3NHGeI3 are one dimensional chain structures with indirect bandgaps of 2.7, 2.5, and 2.8 eV, respectively [138-143].

Divalent transition metal cations (Cu2+, Fe2+, Zn2+) may be synthesized to adjust photovoltaic characteristics in Pb-free perovskites [63, 159]. Cu2+, a lead-free perovskite, is stable, abundant, cheap, and absorbs near-infrared light. Solar cells utilise many Cu-based perovskites [63]. Bi-based perovskites are lead-free, air-stable, and highly tunable. Bi and Sb also have greater environmental stability in the +3 valence state than Sn2+. Bi- and Sb-based PSCs remain very inefficient. Optimizing bandgaps, film morphology, manufacturing methods, and device designs improves device performance. To improve PSC efficiency, Cu-based transitional metal perovskites must lower their wide optical bandgap and improve charge transfer [144-149].

Theoretical simulations have improved our knowledge of lead-free PSC efficiency, stability, and carrier transport mechanics. Theoretical scientific calculations and actual solar cell fabrication techniques should soon turn lead-free PSC research around. Lead-free perovskite materials and PSCs will progress here. To meet commercial development needs, lead-free perovskites, especially 2D stacked ones, may be continuously improved. Despite the slower advancement of lead-free perovskites compared to their lead-based counterparts, we should stay confident that their efficiency may exceed 15% with more study.

8.6 Perovskite Solar Cell Fabrication Methods

Blade coating, Inkjet printing, spray coating, screen printing, Slot-die coating, Electro deposition, and Vapour-phase deposition are all scalable Perovskite Solar Cell deposition processes.

Coating: The deposition technique of blade-coating distributes a precursor solution over surfaces to produce wet, thin coatings. The thickness of the film may vary depending on the precursor ink concentration, the distance between the blade and the substrate, and the blade speed. Continuous manufacturing may be accomplished by the use of roll-to-roll arrangements, in which the blade remains stationary and flexible substrates are transported by a rolling roller [90].

Slot-die coating: This method transfers ink to the substrate using a reservoir with a slit. The ink flow can be managed more successfully with slot-die coating, but because it takes more ink to fill the reservoir and supply pipe, it is not ideal for the development of new ink chemistries. As a consequence of this, blade coating has gotten a great deal more attention than slot-die coating, which results in the production of Perovskite Solar Cells with a considerably higher PCE. Slot-die coating trumps blade coating in terms of yield and repeatability after the ink has been entirely produced. Consequently, it may be employed in future roll-to-roll manufacturing [90].

Spraying: Small liquid droplets are sprayed onto surfaces using a spray coating nozzle. Spray coating droplets are produced via pneumatic spraying, ultrasonic spraying, and electrospraying. Layers of dense oxide are deposited by PSCs by the use of pneumatic and ultrasonic spraying. Utilizing ultrasonic spraying, perovskite is deposited. Due to the random nature of spraying droplet position and size, numerous droplets must overlap in a limited area to provide total coverage. New droplets may dissolve previously deposited components, complicating processing. Keeping the substrate heated while spraying expedites solvent evaporation and inhibits redissolution. adjusting the solvent ratio and substrate temperature regulates solvent removal and material redissolution [90]

Inkjet: Nozzles govern the size and direction of droplets in inkjet printing in order to disperse the ink precursor. Digital printers provide ultrafine lateral resolution with miniaturized nozzles and a tight substrate-nozzle distance. Inkjet printing for high-volume, large-area perovskite solar module production depends on device structure and speed. [90]

Screen Printing: In this, ink is held and transferred to the substrate using a patterned mesh screen. With a lateral resolution of 75–100 m, this method may produce effective patterns. The viscous ink is contained inside the open pores of the mesh screen while it is being distributed by a squeegee, and the photosensitive polymer emulsion that is exposed blocks the unwanted region. The design is made by applying ink to the substrate in the appropriate places. The thickness of a film may be determined by looking at both the size of the mesh and the thickness of the emulsion layer [90].

Electrodeposition. For large-area thin film deposition, electrodeposition is a second scalable solution deposition method. There are no known instances of the electrochemical deposition of halide perovskites in a direct manner, and the electrochemical deposition of perovskite thin films often occurs in a pair of stages. The application of PbO2 or PbO via electrodeposition onto a conducting substrate. PbI2 is produced when lead oxide combines with organic halides to yield PbI2, then converted into a perovskite. Electrodeposition is advantageous due to its non-toxic solvent [90].

Vapour-phase deposition: Physical vapour deposition (PVD) is used to deposit CIGS and CdTe, whilst chemical vapour deposition (CVD) is utilised to dope silicon. CVD has been developed for large-scale solar cell manufacturing. Using PVD and CVD, thin perovskite films have been deposited [90].

8.7 Constraints to scaling Perovskite solar cell technology

Deposited materials and manufacturing processes are the primary factors preventing the commercialisation of perovskite solar cells. This is because deposited materials prevent large-scale placement and manufacturing processes have a negative impact on device layers, module connectivity, material cost, and stability. The fundamental disadvantage of these cell is that module performance and stability may be affected by temperature, light, and humidity. Reduce the energy consumption and cost, material synthesis, and film processing for Perovskite Solar Cells and PV modules to reduce their cost and effect on the environment [91].

PSC efficiency decreases as device size increases. Each functional layer's film lacks coverage, homogeneity, and flatness as a result of solution processing [92]. Scaling up results in stability problems. Perovskite Solar Cells are susceptible to instability. Stability relies on temperature, light, and environmental (moisture, oxygen) exposure. The device may degenerate due to the transport layer, active layer, or electrode. The stability of the device is impacted by the material's inherent qualities as well as the film quality of each functional layer [92]. They provide module stability and weather resistance. Researchers in the field of perovskite photovoltaics must agree on methods for assessing stability and provide standardized testing methodology [3]

Commercialisation of Perovskite solar cells is influenced by cost. Due to the fact that the Perovskite Solar Cell's key components may be manufactured utilizing inexpensive coating techniques and abundant materials. Device amplification must solve the tradeoff between cheap cost and excellent performance of these [92].

8.8 Conclusion

The current energy crisis is the most significant issue facing the world. Solar cells are one potential answer to the energy dilemma. Between the years of 2012 and 2016, hybrid organic–inorganic perovskites-based solar cells were taken into consideration as a photovoltaic (PV) technology. This chapter provides perovskite materials and device engineering knowledge to develop photovoltaic activities in organo-metal halide perovskites. This chapter discusses perovskite materials, solar cell assembly, and electron transport layer developments (PSCs). This study extends perovskite device design and manufacturing procedures to provide flexible and high-performance PSC devices. To stabilize perovskite solar cells, researchers must improve the absorber layer's inherent stability, carefully design the device structure, and find moisture-resistant encapsulant materials.

References

[1] Kojima, A.; Teshima, K.; Shirai, Y.; Miyasaka, T. Organometal Halide Perovskites as Visible-Light Sensitizers for Photovoltaic Cells. J. Am. Chem. Soc. 2009, 131, 6050–6051.

[2] Kim, H.-S.; Lee, C.-R.; Im, J.-H.; Lee, K.-B.; Moehl, T.; Marchioro, A.; Moon, S.-J.; Humphry-Baker, R.; Yum, J.-H.; Moser, J.E.; et al. Lead Iodide Perovskite Sensitized All-Solid-State Submicron Thin Film Mesoscopic Solar Cell with Efficiency Exceeding 9%. Sci. Rep. 2012, 2, 591.

[3] Lee, M.M.; Teuscher, J.; Miyasaka, T.; Murakami, T.N.; Snaith, H.J. Efficient Hybrid Solar Cells Based on Meso-Superstructured Organometal Halide Perovskites. Science 2012, 338, 643–647.

[4] Xing, G.; Mathews, N.; Sun, S.; Lim, S.S.; Lam, Y.M.; Gratzel, M.; Mhaisalkar, S.; Sum, T.C. Long-Range Balanced Electron- and Hole-Transport Lengths in Organic-Inorganic CH3NH3PbI3 . Science 2013, 342, 344–347.

[5] Lee, J.-W.; Seol, D.-J.; Cho, A.-N.; Park, N.-G. High-Efficiency Perovskite Solar Cells Based on the Black Polymorph of HC(NH2) 2PbI3 . Adv. Mater. 2014, 26, 4991–4998.

[6] Yin, W.-J.; Shi, T.; Yan, Y. Unusual defect physics in CH3NH3PbI3 perovskite solar cell absorber. Appl. Phys. Lett. 2014, 104, 063903.

[7] Jeong, M.; Choi, I.W.; Go, E.M.; Cho, Y.; Kim, M.; Lee, B.; Jeong, S.; Jo, Y.; Choi, H.W.; Lee, J.; et al. Stable perovskite solar cells with efficiency exceeding 24.8% and 0.3-V voltage loss. Science 2020, 369, 1615–1620.

[8] Jeong, J.; Kim, M.; Seo, J.; Lu, H.; Ahlawat, P.; Mishra, A.; Yang, Y.; Hope, M.A.; Eickemeyer, F.T.; Kim, M.; et al. Pseudo-halide anion engineering for α-FAPbI3 perovskite solar cells. Nature 2021, 592, 381–385.

[9] Min, H.; Kim, M.; Lee, S.-U.; Kim, H.; Kim, G.; Choi, K.; Lee, J.H.; Seok, S., II. Efficient, stable solar cells by using inherent bandgap of α-phase formamidinium lead iodide. Science 2019, 366, 749–753.

[10] Kim, M.; Kim, G.-H.; Lee, T.K.; Choi, I.W.; Choi, H.W.; Jo, Y.; Yoon, Y.J.; Kim, J.W.; Lee, J.; Huh, D.; et al. Methylammonium Chloride Induces Intermediate Phase Stabilization for Efficient Perovskite Solar Cells. Joule 2019, 3, 2179–2192.

[11] Saliba, M., Correa-Baena, J.-P., Graetzel, M., Hagfeldt, A., and Abate, A. (2017). Perovskite solar cells from the atomic to the film level.

[12] Tzounis, L., Stergiopoulos, T., Zachariadis, A., Gravalidis, C., Laskarakis, A., and Logothetidis, S. (2017). Perovskite solar cells from small scale spin coating process towards roll-to-roll printing: optical and morphological studies. Mater. Today 4, 5082–5089.

[13] S. Sun, T. Salim, N. Mathews et al., "The origin of high efficiency in low-temperature solution-processable bilayer organometal halide hybrid solar cells," Energy Environ. Sci., vol. 7, no. 1, pp. 399–407, 2014.

[14] C. C. Stoumpos, C. D. Malliakas, and M. G. Kanatzidis, "Semiconducting tin and lead iodide perovskites with organic cations: phase transitions, high mobilities, and near-infrared photoluminescent properties," Inorganic Chemistry, vol. 52, no. 15, pp. 9019–9038, 2013.

[15] T. Baikie, Y. Fang, J. M. Kadro et al., "Synthesis and crystal chemistry of the hybrid perovskite (CH3NH3)PbI3 for solid-state sensitised solar cell applications," Journal of Materials Chemistry A, vol. 1, no. 18, pp. 5628–5641, 2013.

[16] M. A. Green, A. Ho-Baillie, and H. J. Snaith, "The emergence of perovskite solar cells," Nature Photonics, vol. 8, no. 7, pp. 506–514, 2014.

[17] S. P. Singh and P. Nagarjuna, "Organometal halide perovskites as useful materials in sensitized solar cells," Dalton Transactions, vol. 43, no. 14, pp. 5247–5251, 2014.

[18] M. Luan, J. Song, X. Wei, F. Chen, and J. Liu, "Controllable growth of bulk cubic-phase CH3NH3PbI3 single crystal with exciting room-temperature stability," CrystEngComm, vol. 18, no. 28, pp. 5257–5261, 2016.

[19] I. Chung, B. Lee, J. He, R. P. H. Chang, and M. G. Kanatzidis, "All-solid-state dye-sensitized solar cells with high efficiency," Nature, vol. 485, no. 7399, pp. 486–489, 2012.

[20] Yao ZB, Wang WL, Shen HP, et al. CH3 NH3 PbI3 grain growth and interfacial properties in mesostructured perovskite solar cells fabricated by two-step deposition. Sci Technol Adv Mater. 2017;18:253–262.

[21] Han JH, Yin XW, Nan H, et al. Enhancing the performance of perovskite solar cells by hybridizing SnS quantum dots with CH3 NH3 PbI3. Small. 2017;13:1700953.

[22] Yin X, Yao Z, Luo Q, et al. High efficiency inverted planar perovskite solar cells with solution-processed NiOx hole contact. ACS appl mater interfaces. 2017;9(3):2439–2448.

[23] M. M. Lee, J. Teuscher, T. Miyasaka, T. N. Murakami, H. J. Snaith, Science 2012, 338, 643.

[24] Q. Dong, Y. Fang, Y. Shao, P. Mulligan, J. Qiu, L. Cao, J. Huang, Science 2015, 347, 967.

[25] *Jun, Kang (10 January 2017).* "High Defect Tolerance in Lead Halide Perovskite CsPbBr3"*. The Journal of Physical Chemistry Letters. 8 (2): 489–493.*

[26] Perovskite the Future of Solar Cells?. engineering.com. December 6, 2013

[27] Saidaminov, Makhsud I.; Abdelhady, Ahmed L.; Murali, Banavoth; Alarousu, Erkki; Burlakov, Victor M.; Peng, Wei; Dursun, Ibrahim; Wang, Lingfei; He, Yao; MacUlan, Giacomo; Goriely, Alain; Wu, Tom; Mohammed, Omar F.; Bakr, Osman M. (2015). "High-quality bulk hybrid perovskite single crystals within minutes by inverse temperature crystallization". Nature Communications. 6: 7586.

[28] Howard, I. A.; Abzieher, T.; Hossain, I. M.; Eggers, H.; Schackmar, F.; Ternes, S.; Richards, B. S.; Lemmer, U.; Paetzold, U. W. (2019). "Coated and Printed Perovskites for Photovoltaic Applications". Advanced Materials. 31 (26): e1806702.

[29] Snaith, Henry J. (2013). "Perovskites: The Emergence of a New Era for Low-Cost, High-Efficiency Solar Cells". The Journal of Physical Chemistry Letters. 4 (21): 3623–3630.

[30] Jung, Yen-Sook; Hwang, Kyeongil; Heo, Youn-Jung; Kim, Jueng-Eun; Vak, Doojin; Kim, Dong-Yu (2018). "Progress in Scalable Coating and Roll-to-Roll Compatible Printing Processes of Perovskite Solar Cells toward Realization of Commercialization". Advanced Optical Materials. 6 (9): 1701182.

[31] Li, Zhen; Klein, Talysa R.; Kim, Dong Hoe; Yang, Mengjin; Berry, Joseph J.; Hest, Maikel F. A. M. van; Zhu, Kai (2018). "Scalable fabrication of perovskite solar cells". Nature Reviews Materials. 3 (4): 18017. Bibcode:2018NatRM...318017L.

[32] Wang, Y. et al. Thermodynamically stabilized β-cspbi3-based perovskite solar cells with efficiencies >18. Science 365, 591–595 (2019).

[33] Wang, H. et al. Cesium lead mixed-halide perovskites for low-

energy loss solar cells with efficiency beyond 17%. Chem. Mater. 31, 6231–6238 (2019).
[34] Swarnkar, A. et al. Quantum dot–induced phase stabilization of α-cspbi3 perovskite for high-efficiency photovoltaics. Science 354, 92–95 (2016).
[35] Sanehira, E. M. et al. Enhanced mobility cspbi3 quantum dot arrays for record-efficiency, high-voltage photovoltaic cells. Sci. Adv. 3, eaao4204 (2017).
[36] Wang, Q. et al. Stabilizing the α-phase of cspbi3 perovskite by sulfobetaine zwitterions in one-step spin-coating films. Joule 1, 371–382 (2017).
[37] Li, B. et al. Surface passivation engineering strategy to fully-inorganic cubic cspbi 3 perovskites for high-performance solar cells. Nat. Commun. 9, 1076 (2018).
[38] Zhang, T. et al. Bication lead iodide 2d perovskite component to stabilize inorganic α-cspbi3 perovskite phase for high-efficiency solar cells. Sci. Adv. 3, e1700841 (2017).
[39] Jiang, Y. et al. Reduced-dimensional α-cspbx3 perovskites for efficient and stable photovoltaics. Joule 2, 1356–1368 (2018).
[40] Zeng, Q. et al. Polymer-passivated inorganic cesium lead mixed-halide perovskites for stable and efficient solar cells with high open-circuit voltage over 1.3 v. Adv. Mater. 30, 1705393 (2018).
[41] Tian, J. et al. Dual interfacial design for efficient cspbi2br perovskite solar cells with improved photostability. Adv. Mater. 31, 1901152 (2019).
[42] Wang, P. et al. Solvent-controlled growth of inorganic perovskite films in dry environment for efficient and stable solar cells. Nat. Commun. 9, 2225 (2018).
[43] Wang, Y., Zhang, T., Kan, M. & Zhao, Y. Bifunctional stabilization of allinorganic α-cspbi3 perovskite for 17% efficiency photovoltaics. J. Am. Chem. Soc. 140, 12345–12348 (2018).
[44] Li, Y. et al. Li dopant induces moisture sensitive phase degradation of an allinorganic cspbi 2 br perovskite. Chem. Commun. 54, 9809–9812 (2018).
[45] Liao, H. C. et al. Dopant free hole transporting polymers for high efficiency, environmentally stable perovskite solar cells. Adv. Energy Mater. 6, 1600502 (2016).
[46] Bastos, J. P. et al. Light-induced degradation of perovskite solar cells: the influence of 4-tert-butyl pyridine and gold. Adv. Energy Mater. 8, 1800554 (2018).

[47] Magomedov, A. et al. Pyridination of hole transporting material in perovskite solar cells questions the long-term stability. J. Mater. Chem. C 6, 8874–8878 (2018).
[48] Beal, R. E. et al. Cesium lead halide perovskites with improved stability for tandem solar cells. J. Phys. Chem. Lett. 7, 746–751 (2016).
[49] Zhang, S. et al. Solvent engineering for efficient inverted perovskite solar cells based on inorganic cspbi2br light absorber. Mater. Today Energy 8, 125–133 (2018).
[50] Wu, T. et al. Efficient and stable cspbi3 solar cells via regulating lattice distortion with surface organic terminal groups. Adv. Mater. 31, 1900605 (2019).
[51] Liu, C. et al. All-inorganic cspbi2br perovskite solar cells with high efficiency exceeding 13%. J. Am. Chem. Soc. 140, 3825–3828 (2018).
[52] Liu, C. et al. Structurally reconstructed cspbi2br perovskite for highly stable and square-centimeter all-inorganic perovskite solar cells. Adv. Energy Mater. 9, 1803572 (2019).
[53] Jiang, Q. et al. Surface passivation of perovskite film for efficient solar cells. Nat. Photon. 13, 460–466 (2019).
[54] Ran, C., Xu, J., Gao, W., Huang, C. & Dou, S. Defects in metal triiodide perovskite materials towards high-performance solar cells: origin, impact, characterization, and engineering. Chem. Soc. Rev. 47, 4581–4610 (2018).
[55] Niu, T. et al. Stable high-performance perovskite solar cells via grain boundary passivation. Adv. Mater. 1706576 (2018).
[56] Guo, Y., Sato, W., Shoyama, K. & Nakamura, E. Sulfamic acid-catalyzed lead perovskite formation for solar cell fabrication on glass or plastic substrates. J. Am. Chem. Soc. 138, 5410–5416 (2016).
[57] Qin, P.-L. et al. Stable and efficient organo-metal halide hybrid perovskite solar cells via π-conjugated lewis base polymer induced trap passivation and charge extraction. Adv. Mater. 30, 1706126 (2018).
[58] Zheng, X. et al. Defect passivation in hybrid perovskite solar cells using quaternary ammonium halide anions and cations. Nat. Energy 2, 17102 (2017).
[59] Song X, Hodes G, Zhao K, et al. Metal-free organic halide perovskite: A new class for next optoelectronic generation devices. Advanced Energy Materials, 2021, 11
[60] Ye H Y, Tang Y Y, Li P F, et al. Metal-free three-dimensional

perovskite ferroelectrics. Science, 2018, 361, 151
[61] Akkerman Q A, Manna L. What defines a halide perovskite. ACS Energy Lett, 2020, 5, 604,2003331
[62] Song X, Cui Q, Liu Y, et al. Metal-free halide perovskite single crystals with very long charge lifetimes for efficient X-ray imaging. Adv Mater, 2020, 32, 2003353
[63] Pope M, Kalmann H, Magnante P. Electroluminescence in Organic Crystals, The Journal of Chemical Physics 1963; 38 2042–2043.
[64] Agranovich V, Atanasov R, Bassani F. Hybrid interface excitons in organic-inorganic quantum wells. Solid State Communications 1994; 92 295–301.
[65] Agranovich V, La Rocca G C, Bassani F. Efficient electronic energy transfer from a semiconductor quantum well to an organic material. JETP Letters 1997; 66 748–751.
[66] Agranovich V, Benisty H, Weisbuch C. Organic and inorganic quantum wells in a microcavity: Frenkel-Wannier-Mott excitons hybridization and energy transforma- tion. Solid State Communications 1997; 102 631–636.
[67] Dammak T, Elleuch S, Bougzhala H, Mlayah A, Chtourou R, Abid Y. Synthesis, vi- brational and optical properties of a new three-layered organic inorganic perovskite (C4H9NH3)4Pb3 I4Br6 . Journal of Luminescence 2009; 129 893–897.
[68] Mitzi D, Field C, Harrlson W, Guloy A. Conducting tin halides with a layered organ- ic-based perovskite structure. Nature 1994; 369 467–469.
[69] Hong X, Ishihara T, Nurmikko A V. Photoconductivity and electroluminescence in lead iodide based natural quantum well structures. Solid State Communications 1992; 84 657-661.
[70] Cheng Z Y, Wang Z, Xing R B, Han Y C, Lin J. Patterning and photoluminescent properties of perovskite-type organic/inorganic hybrid luminescent films by soft lith- ography. Chemical Physics Letters 2003; 376 481–486.
[71] Tabuchi Y, Asai K, Rikukawab M, Sanuib K Ishigure K. Preparation and characteri- zation of natural lower dimensional layered perovskite-type compounds. Journal of Physics and Chemistry of Solids 2000; 61 837–845.
[72] National Renewable Energy Laboratory, http://www.nrel.gov/pv/assets/images/efficiency_chart.jpg, accessed December 2016.

[73] C. D. Bailie, M. G. Christoforo, J. P. Mailoa, A. R. Bowring, E. L. Unger, W. H. Nguyen, J. Burschka, N. Pellet, J. Z. Lee, M. Gratzel, R. Noufi, T. Buonassisi, A. Salleo and M. D. McGehee, Energy Environ. Sci., 2015, 8, 956–963

[74] G. Xing, N. Mathews, S. S. Lim, N. Yantara, X. Liu, D. Sabba, M. Grätzel, S. Mhaisalkar and T. C. Sum, Nat. Mater., 2014, 13, 476–480

[75] J. H. Noh, S. H. Im, J. H. Heo, T. N. Mandal and S. I. Seok, Nano Lett., 2013, 13, 1764–1769 (at ref first paragraph on page 13 & 14)

[76] Ema K, Ishi J, Kunugita H, Ban T, Kondo T. All-optical serial-to-parallel conversion of Tbits/s signals using a four-wave-mixing process. Optical and Quantum Electron- ics 2001; 33 1077–1087.

[77] Brehier A, Parashkov R, Lauret J S, Deleporte E. Strong exciton–photon coupling in a microcavity containing layered perovskites semiconductors. Applied Physics Letters 2006; 89 171110-1.

[78] Era M, Morimoto S, Tsutsui T, Saito S. Organic–inorganic heterostructure electrolu- minescent device using a layered perovskite semiconductor (C6H5C2H4NH3)2PbI4. Applied Physics Letters 1994; 65 676–678.

[79] Shibuya K, Koshimizu M, Takeoka Y, Asai, K. Scintillation properties of (C6H13NH3)2PbI4: exciton luminescence of an organic/inorganic multiple quantum well structure compound induced by 2.0 MeV protons. Nuclear Instruments and Methods B 2002; 194 207–212.

[80] Liu M, Johnston M B, Snaith H J. Efficient planar heterojunction perovskite solar cells by vapour deposition. Nature 2013; 501 395–398.

[81] Snaith H J. Perovskites: The Emergence of a New Era for Low-Cost, High- Efficiency Solar Cells. The Journal of Physical Chemistry Letters 2013; 4 3623–3630.

[82] Edri E, Kirmayer S, Cahen D Hodes G. High Open-Circuit Voltage Solar Cells Based on Organic–Inorganic Lead Bromide Perovskite The Journal of Physical Chemistry Letters 2013; 4 897–902.

[83] Heo J H, Im S H, Noh J H, Mandal T N, Lim C S, Chang J A, Lee Y H, Kim H J, Sarkar A, Nazeeruddin M K, Gratzel M, Seok S. Efficient inorganic–organic hybrid hetero- junction solar cells containing perovskite compound and polymeric hole conduc-

tors. Nature photonics 2013; 7 486–491.

[84] Burschka J, Pellet N, Moon S J, Humphry-Baker R, Gao P, Nazeeruddin M K, Gratzel M. Sequential deposition as a route to high-performance perovskite-sensitized solar cells; Nature 2013. 499, 316– 319.

[85] Tabuchi Y, Asai K, Rikukawa M, Sanui K Ishigure K. Preparation and characteriza- tion of natural lower dimensional layered perovskite-type compounds; Journal of Physics and Chemistry of Solids 2000. 61 837–845.

[86] 1. K. A. Bush et al., 23.6%-efficient monolithic perovskite/silicon tandem solar cells with improved stability. Nat. Energy 2, 17009 (2017).

[87] 2. J. Zhao et al., Is Cu a stable electrode material in hybrid perovskite solar cells for a 30-year 5 lifetime? Energy Environ. Sci. 9, 3650–3656 (2016).

[88] 3. X. Zheng et al., Defect passivation in hybrid perovskite solar cells using quaternary ammonium halide anions and cations. Nat. Energy 2, 17102 (2017).

[89] 4. W. Chen et al., Efficient and stable large-area perovskite solar cells with inorganic charge extraction layers. Science 350, 944–948 (2015).

[90] Zhen Li, Talysa R. Klein, Dong Hoe Kim, Mengjin Yang, Joseph J. Berry, Maikel F. A. M. van Hest and Kai Zhu, "Scalable fabrication of perovskite solar cells", NATURE REVIEWS | MATERIALS,2018

[91] Ahmed Esmail Shalan, "Challenges and Approaches Towards Upscaling Assembly of Hybrid Perovskite Solar Cells", Materials Advances, May,2020

[92] Fangfang Wang, Yezhou Cao, Cheng Chen, Qing Chen, Xiao Wu, Xinguo Li, Tianshi Qin, and Wei Huang, "Materials toward the Upscaling of Perovskite Solar Cells: Progress, Challenges, and Strategies", Adv. Funct. Mater. 2018,

[93] Yaoguang Rong, Yue Hu, Anyi Mei, Hairen Tan, Makhsud I. Saidaminov, Sang Il Seok, Michael D. McGehee, Edward H. Sargent, Hongwei Han, "Challenges for commercializing perovskite solar cells", SCIENCE, 2018

[94] Manjit Kaur; Neena Gupta; Sanjeev Kumar; BRaj; Arun Kumar Singh, "Comparative RF and Crosstalk Analysis of Carbon Based Nano Interconnects" IET Circuits, Devices & Systems, Vol.15(6), PP. 493-503, Feb 2021.

[95] Nehru Kandasamy, Firdous Ahmad, D. Ajitha, B Raj, and Nagarjuna Telagam "Quantum Dot Cellular Automata based Scan Flip Flop and Boundary Scan Register, IETE Journal of Research, Volume 66, PP. 535-548, 2020.

[96] Sharma, S. K.., Raj, B., Khosla, M., "Enhanced Photosensivity of Highly Spectrum Selective Cylindrical Gate In1-xGaxAs Nanowire MOSFET Photodetector, Modern Physics letter-B, Vol. 33, No. 12, PP. 1950144 (2019).

[97] Jeetendra Singh, BRaj, "Design and Investigation of 7T2M NVSARM with Enhanced Stability and Temperature Impact on Store/Restore Energy", IEEE Transactions on Very Large Scale Integration Systems, Vol. 27, Issure 6, PP. 1322 - 1328 June 2019.

[98] Anil Kumar Bhardwaj, Sumeet Gupta, BRaj, Amandeep Singh, "Impact of Double Gate Geometry on the Performance of Carbon Nanotube Field Effect Transistor Structures for Low Power Digital Design", Computational and Theoretical Nanoscience, ASP, Vol. 16, PP. 1813–1820, 2019.

[99] Neeraj Jain, BRaj, "Thermal Stability Analysis and Performance Exploration of Asymmetrical Dual-k underlap Spacer (ADKUS) SOI FinFET for Security and Privacy Applications", Indian Journal of Pure & Applied Physics (IJPAP), Vol. 57, pp. 352-360, May 2019.

[100] Amandeep Singh, Mamta Khosla, BRaj, "Design and Analysis of Dynamically Configurable Electrostatic Doped Carbon Nanotube Tunnel FET"," Microelectronics Journal, Elesvier, Volume 85, Pages 17-24, March 2019.

[101] Neeraj Jain and Balwinder Raj, "Dual-k Spacer Region Variation at the Drain Side of Asymmetric SOI FinFET Structure: Performance Analysis towards the Analog/RF Design Applications", communicated in "Journal of Nanoelectronics and Optoelectronics", American Scientific Publishers, Vol. 14, PP. 349–359, March 2019.

[102] Jeetendra Singh, Sanjeev Sharma, B Raj, Mamta Khosla Analysis of barrier layer thickness on performance of In1-xGaxAs based Gate Stack Cylindrical Gate Nanowire MOSFET," JNO, ASP, Vol. 13, PP 1473-1477, Oct 2018.

[103] Neeraj Jain, B Raj, "Analysis and Performance Exploration of High-k SOI FinFETs Over the Conventional Low-k SOI FinFET toward Analog/RF Design", Journal of Semiconductors (JoS), IOP Science, Vol.39, Issue 12, PP- 124002-1-7, Dec 2018.

[104] Candy Goyal, Jagpal Singh Ubhi and BRaj, A reliable leakage reduction technique for approximate full adder with reduced ground bounce noise, Journal of Mathematical Problems in Engineering, Hindawi, Volume 2018, Article ID 3501041, 16 pages, 15 Oct 2018.

[105] Anuradha, Jeetendra Singh, B Raj and Mamta Khosla, "Design and Performance Analysis of Nano-scale Memristor-based Nonvolatile SRAM" Journal of Sensor Letter", American Scientific Publishers, Vol. 16, PP 798–805, Oct 2018.

[106] Girish Wadhwa, BRaj, "Parametric Variation Analysis of Charge-Plasma-based Dielectric Modulated JLTFET for Biosensor Application" IEEE Sensor Journal, VOL. 18, NO. 15, AUGUST 1, PP. 6070 – 6077, 2018 (SCI)

[107] Jeetendra Singh, BRaj, "Comparative Analysis of Memristor Models for Memories Design", JoS, IoP, Vol. 39, No. 7, PP. 074006-1-12, July 2018.

[108] Divya Yadav, Shailesh Singh Chouhan, Santosh Kumar Vishvakarma and BRaj, "Application Specific Microcontroller Design for IoT based WSN", Sensor Letter, ASP, Vol. 16, PP. 374–385, May 2018.

[109] Gurmohan Singh, R. K. Sarin and BRaj, "Fault-Tolerant Design and Analysis of Quantum-Dot Cellular Automata Based Circuits", IEEE/IET Circuits, Devices & Systems, Vol. 12, PP. 638 – 64, 2018.

[110] Jeetendra Singh, BRaj, "Modeling of Mean Barrier Height Levying Various Image Forces of Metal Insulator Metal Structure to Enhance the Performance of Conductive Filament Based Memristor Model", IEEE Nanotechnology, Vol. 17, No. 2, PP. 268-267, March 2018.

[111] Aakash Jain, Sanjeev Sharma, BRaj, "Analysis of Triple Metal Surrounding Gate (TM-SG) III-V Nanowire MOSFET for Photosensing Application", Opto-electronics Journal, Elsevier, Volume 26, Issue 2, Pages 141-148, May 2018.

[112] Aakash Jain, Sanjeev Sharma, B Raj, "Design and Analysis of High Sensitivity Photosensor Using Cylindrical Surrounding Gate MOSFET For Low Power Sensor Applications", Engineering Science and Technology, an International Journal, Elsevier, Volume 19, Issue 4, Pages 1864–1870, December 2016.

[113] Amandeep Singh, Mamta Khosla, B Raj, "Analysis of Electrostatic Doped Schottky Barrier Carbon Nanotube FET for Low

Power Applications," Journal of Materials Science: Materials in Electronics, Springer, Vol.28, PP. 1762–1768, 2017.

[114] G. Saiphani Kumar, Amandeep Singh, BRaj, "Design and Analysis of Gate All Around CNTFET based SRAM cell Design", Journal of Computational Electronics, Springer, Volume 17, Issue 1, pp 138–145, March 2018.

[115] Gurinder pal Singh, B. S. Sohi, Balwinder Raj, "Material Properties Analysis of Graphene Base Transistor (GBT) for VLSI Analog Circuits, Indian Journal of Pure & Applied Physics (IJPAP), Vol. 55, pp. 896-902, Dec-2017.

[116] Amandeep Singh, Mamta Khosla, B Raj, "Comparative Analysis of Carbon Nanotube Field Effect Transistor and Nanowire Transistor for Low Power Circuit Design," Journal of Nanoelectronics and Optoelectronics, American Scientific Publishers, USA, Vol. 11, PP. 388-393, June 2016.

[117] Sunil Kumar and B Raj, "Estimation of Stability and Performance metric for Inward Access Transistor based 6T SRAM Cell Design using n-type/p-type DMDG-GDOV TFET" IEEE VLSI Circuits and Systems Letter, Volume3, Issue2, PP 25-39, June2017

[118] Shashikant Sharma, Anjan Kumar, Manisha Pattanaik and B Raj, "Forward Body Biased Multimode Multi-Threshold CMOS Technique for Ground Bounce Noise Reduction in Static CMOS Adders" International Journal of Information and Electronics Engineering, pp. 567-572, Vol. 3, issue 3, 2013.

[119] Hamendra Singh, Pankaj Kumar, Balwinder Raj, "Performance Analysis of Majority Gate SET Based 1-bit Full Adder", *International Journal of Computer and Communication Engineering* (IJCCE), IACSIT Press Singapore, ISSN: 2010-3743, Volume 2, No. 4, 2013.

[120] Anil Kumar Bhardwaj, Sumeet Gupta and B Raj, "Investigation of Parameters for Schottky Barrier (SB) Height for Schottky Barrier Based Carbon Nanotube Field Effect Transistor Device", Journal of Nanoelectronics and Optoelectronics, ASP, Vol. 15, PP. 783–791, July 2020.

[121] Priya Bansal, BRaj, "MEMRISTOR: A VERSATILE NONLINEAR MODEL FOR DOPANT DRIFT AND BOUNDARY ISSUES," JCTN, American Scientific Publishers, Vol. 14 (5), PP. 2319-2325, May 2017.

[122] Neeraj Jain, BRaj, "An Analog and Digital Design Perspective Comprehensive Approach on Fin-FET (Fin-Field Effect tran-

sistor) Technology - A Review", Reviews in Advanced Sciences and Engineering (RASE), ASP, Vol. 5, PP. 1-14, 2016.
[123] Sanjeev Sharma, B Raj, Mamta Khosla, "Subthreshold Performance of In1-xGaxAs based Dual Metal with Gate Stack Cylindrical/Surrounding Gate Nanowire MOSFET for Low Power Analog Applications, Journal of Nanoelectronics and Optoelectronics, American Scientific Publishers, USA, Vol.12, PP. 171–176, 2017.
[124] Balwinder Raj, A. K. Saxena and S. Dasgupta, "Analytical Modeling for the Estimation of Leakage Current and Subthreshold Swing Factor of Nanoscale Double Gate FinFET Device" Microelectronics International, UK, vol. 26, pp. 53-63, 2009.
[125] Soniya, Shailendra Singh, Girish Wadhwa and Balwinder Raj, "An Analytical Modeling for Dual Source Vertical Tunnel Field Effect Transistor" submitted to 'International Journal of Recent Technology and Engineering (IJRTE)' at Volume-8 Issue-2, PP. 603-608, July 2019.
[126] Shailendra Singh, B Raj, "Design and analysis of hetrojunction Vertical T-shaped Tunnel Field Effect Transistor", Journal of Electronics Material, Springer, Volume 48, Issue 10, pp 6253–6260, October 2019.
[127] Candy Goyal, Jagpal Singh Ubhi and B Raj, "A Low Leakage CNTFET based Inexact Full Adder for Low Power Image Processing Applications", International Journal of Circuit Theory and Applications, Wiley, Volume47, Issue9, Pages 1446-1458, September 2019.
[128] B Raj, A. K. Saxena and S. Dasgupta, "A Compact Drain Current and Threshold Voltage Quantum Mechanical Analytical Modeling for FinFETs" Journal of Nanoelectronics and Optoelectronics (JNO), USA, vol. 3, no. 2, pp. 163-170, 2008.
[129] Girish Wadhwa, B Raj, "An Analytical Modeling of Charge Plasma based Tunnel Field Effect Transistor with Impacts of Gate underlap Region" Superlattices and Microstructures, Elsevier, Vol. 142, PP.106512, June 2020.
[130] Shailendra Singh, B Raj,"Modeling and Simulation analysis of SiGe hetrojunction Double GateVertical t-shaped Tunnel FET", Superlattices and Microstructures, Elsevier Volume 142, PP. 106496, June 2020.
[131] Amandeep Singh, Dinesh Kumar Saini, Dinesh Agarwal, Sajal Aggarwal, Mamta Khosla, BRaj, "Modeling and Simulation of Carbon Nanotube Field Effect Transistor and its Circuit Appli-

cation," Journal of Semiconductors (JoS), IOP Science, Vol.37, PP. 074001-6, July 2016.

[132] Neeraj Jain, Balwinder Raj, "Device and Circuit Co-Design Perspective Comprehensive Approach on FinFET Technology - A Review, Journal of Electron Devices, Vol. 23, Num. 1, pp 1890-1901, 2016.

[133] Sunil Kumar and BRaj, "Analysis of I_{ON} and Ambipolar Current for Dual-Material Gate-drain Overlapped DG-TFET," Journal of Nanoelectronics and Optoelectronics, American Scientific Publishers, USA, Vol.11, PP. 323-333, June 2016.

[134] Naveed Anjum, Tarun Bali and B Raj, Design and Simulation Of Handwritten Multiscript Character Recognition, International Journal of Advanced Research in Computer and Communication Engineering, Vol.2 (7), PP. 2544-2549, July 2013.

[135] Sanjeev Sharma, BRaj, Mamta Khosla, "A Gaussian Approach for Analytical Subthreshold Current Model of Cylindrical Nanowire FET with Quantum Mechanical Effects, Microelectronics Journal, Elsevier, Vol. 53, PP. 65-72, April 2016.

[136] Karmjit Singh, BRaj, "Performance and analysis of temperature dependent Multi-walled Carbon Nanotubes as Global Interconnects at different technology nodes," Journal of Computational Electronics, Springer, Volume 14 Issue 2, pp. 469-476, June 2015.

[137] Sunil Kumar and BRaj, "Compact channel potential analytical Modeling of DG-TFET based on Evanescent–mode Approach," Journal of Computational Electronics, Springer Vol. 14, No. 2, pp.820-827, July 2015.

[138] Karmjit Singh, BRaj, "Temperature Dependent Modeling and Performance Evaluation of Multi-Walled CNT and Single-Walled CNT as Global Interconnects," Journal of Electronic Materials, Springer, Vol. 44, Issue 12, pp 4825-4835, Dec 2015.

[139] V.K. Sharma, M. Pattanaik and B. Raj, "INDEP approach for leakage reduction in nanoscale CMOS circuits", International Journal of Electronics, Taylor & Francis, Vol.102, No.2, pages 200-215, 2014.

[140] Karmjit Singh, BRaj, "Influence of Temperature on MWCNT bundle, SWCNT bundle and Copper interconnects for Nanoscaled Technology nodes," Journal of Materials Science: Materials in Electronics. Springer, Vol. 26, No. 8, pp. 6134-6142, 2015.

[141] Naveed Anjum, Tarun Bali and B Raj, "Design and Simulation of Handwritten Gurumukhi and Devanagri Numerical Recognition, International Journal of Computer Applications, Published by Foundation of Computer Science, New York, USA, Vol. 73(12), pp. 16-21, 2013.

[142] Khandelwal, S., Gupta, V., BRaj, and Gupta, R D., Process Variability Aware Low Leakage Reliable Nano Scale DG-FinFET SRAM Cell Design Technique, Journal of Nanoelectronics and Optoelectronics, Volume 10, Number 6, pp. 810-817, December 2015.

[143] V.K. Sharma, M. Pattanaik and B Raj, "ONOFIC Approach: Low Power High Speed Nanoscale VLSI Circuits Design", International Journal of Electronics, Taylor & Francis, Vol. 101, issue 1, pages 61-73, 2014.

[144] Khandelwal, S., Balwinder Raj, and Gupta, R D., FinFET based 6T SRAM Cell Design: Analysis of Performance Metric, Process Variation and Temperature Effect. Journal of Computational and Theoretical Nanoscience, ASP, USA, Vol. 12, PP. 2500-2506, 2015.

[145] Sumit Singh; Yadav, Shekhar; Rahul, Jagdeep; Srivastava, Anurag; BRaj, Impact of HfO_2 in Graded Channel Dual Insulator Double Gate MOSFET, Journal of Computational and Theoretical Nanoscience, American Scientific Publishers, Volume 12, Number 6, pp. 950-953, April 2015.

[146] Vijay Kumar Sharma, Manisha Pattanaik, BRaj, PVT variations aware low leakage INDEP approach for nanoscale CMOS Circuits, Microelectronics Reliability, Elsevier, Vol. 54, pp 90-99, 2014.

[147] B Raj, A. K. Saxena and S. Dasgupta, "Quantum Mechanical Analytical Modeling of Nanoscale DG FinFET: Evaluation of Potential, Threshold Voltage and Source/Drain Resistance", Journal of Material Science in Semiconductor Processing, Elsevier, Vol. 16, issue 4, pp. 1131- 1137, 2013.

[148] Maisagalla Gopal, Siva Sankar D Prasad and Balwinder Raj, 8T SRAM Cell Design for Dynamic and Leakage Power Reduction, International Journal of Computer Applications, Published by Foundation of Computer Science, New York, USA, Vol. 71(9), pp. 43-48, June 2013.

[149] Manisha Pattanaik, B Raj, Shashikant Sharma and Anjan Kumar, "Diode Based Trimode Multi-Threshold CMOS Technique for Ground Bounce Noise Reduction in Static CMOS Adders",

Advanced Materials Research, Trans Tech Publications, Switzerland, vol. 548, pp 885-889, 2012.

9

CNTFET-based SRAM cell for High Performance Computers

Meenakshi Devi[1,], Akhil Gorkha[2] and Simran[3]*
[1]School of Engineering, Computing and Mathematics, Oxford Brookes University, Oxford, United Kingdom
[2]Beant College of Engineering and Technology, Gurdaspur, Punjab
[3]Sant Longowal Institute of Engineering and Technology Sangrur, Punjab
**dmeenakshi2596@gmail.com*

Abstract: This study developed an SRAM cell based on carbon nanotube field effect transistor (CNTFET) technology and evaluated its performance using different technologies. A comparison has been made between the proposed SRAM cell and established CMOS-based SRAM cell technologies. HSPICE simulation software is used for SRAM cell design and performance analysis. According to the simulation results, the proposed CNTFET-based SRAM cell outperforms the conventional CMOS-based SRAM cell technology regarding speed, noise immunity, and power consumption. The study also discusses several techniques that can be used to successfully implement the proposed SRAM cell. Finally, analysis has been performed on CNTFET-based SRAM cells.

9.1 Introduction

SRAM is a flash memory widely used in everything from desktops to cell phones and various electronic devices such as microprocessors, digital signal processors, and graphics processing units [1]. SRAM cells store information in non-volatile storage cells that store data as long as the device is powered on and is distinguished by its reasonably fast speed, low power consumption, and inexpensive cost. SRAM has been produced for decades using metal-oxide-semiconductor transistors and other conventional transistors. But, as the need for increased performance, highly efficient, and cost-effective memory devices develops, researchers have begun to investigate alternatives such as carbon nanotube field-effect transistors (CNTFETs) [2]. This paper aims to overview CNTFET-based SRAM cells and their potential applications. Carbon Nanotube Field Effect Transistor (CNTFET) is one of these technologies that have the potential to improve the performance of SRAM cells.

Basic Physics of CNTFET

The electrons in the carbon atom design are in the shell state in the $1s^2 2s^2 2p^2$ configuration. In graphene, the sp^2 inter-propagation occurs by maintaining a covalent bond between two distant shells. Carbon molecules are concentrated in isolated sheets of graphene with a hexagonal cross-section. The distance between the carbon particles in the hexagonal lattice (d) is approximately 1.44 Å, and between C-C atoms (ie, σ bonds) is 120°. The lattice constant (a) value is $\sqrt{3}d$ = 2.49 Å. The 2p electrons from each carbon molecule in the cross-section form escape from the π orbitals between two adjacent sheets. The layers between the carbon molecules that divide the different graphene sheets are about 3.35 Å.

Carbon nanotubes are tubes made of carbon atoms folded into a hexagonal shape but empty inside. The diameter of the CNT tube is 1-5 nm, 100,000 times thinner than a human hair, and the length can reach hundreds of microns. In this way, they created a closed, stable system along the CNT axis that facilitates the ballistic transport of electrons and holes. CNTs exist in two types, single-walled CNTs (SWCNTs) or multi-walled CNTs (MWCNTs), as shown in Figure 9.1 (a,b), and can be grown in different diameters.

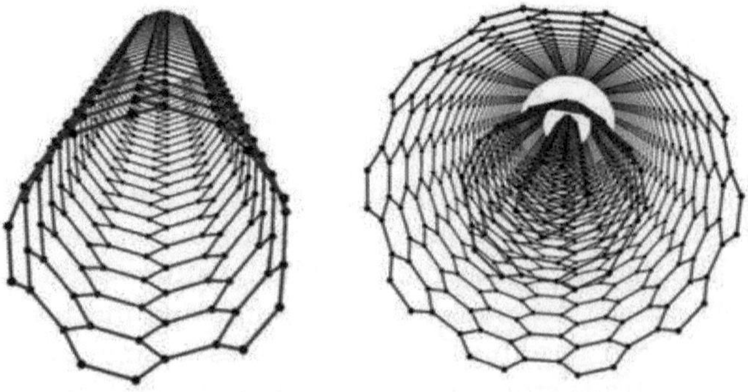

(a) (b)

Figure 9.1 *(a) Single Walled CNT (b) Multi-Walled CNT.*

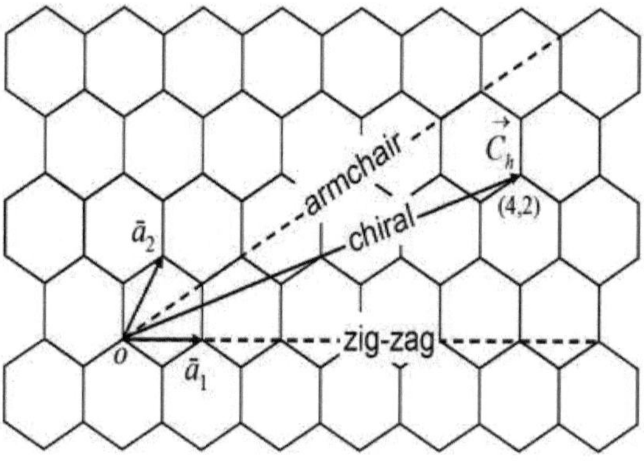

Figure 9.2 *Sheet made of Graphene [4].*

Single-walled carbon nanotubes (SWCNTs) are formed as graphite sheets rolled up and connected into cylinders along chiral vectors [ā1, ā2], as shown in Figure 9.2, called lattice unit vectors and subscripts (m, n) is a positive integer characterising the chirality of the tube. CNT circumference, i.e. The length of Ch is determined by

$$Ch=a\sqrt{(m^2+mn+n^2)} \quad (1)$$

$$\text{Chiral vector } (Ch) = m.\bar{a}1 + n.\bar{a}2 \quad (2)$$

As shown in Figure 9.2, single-walled CNTs are classified into three categories according to their chirality indices (m, n).
(1) Armchair (m = n)
(2) Zigzag (m = 0 or n = 0)
(3) Chirality (any other value of m, n).

The formula for calculating the CNT diameter is given in [5-10] and the following is determined using the chiral index

$$d = Ch/\pi. \quad (3)$$

$$d=(a/\pi)*\sqrt{(m^2+mn+n^2)} \quad (4)$$

SWCNTs have exceptional electrical and mechanical properties. In terms of electrical conductivity, SWCNTs are either metallic or semiconducting because their band gap is zero for 3 times |m−n| exhibit metallic behaviour or exhibit semiconductor behaviour with a limited band gap [11-16].

Carbon Nanotube field-effect transistors (CNTFET)

The main phenomenon of carbon nanotube field-effect transistors (CNTFETs) is related to standard silicon devices. But in this three-terminal device, the channel material must be replaced by semiconducting carbon nanotubes, the bridge elements that form the connection between the source and drain. The device is now electrostatically enabled or disabled using the latch voltage. Figure 9.3 shows the basic structure of CNTFET [17-22].

Figure 9.3 *CNTFET basic structure.*

The basic architecture of a CNTFET is shown in Figure 9.3, and CNTFETs are still unsustainable for very large-scale integration (VLSI) design and industrial commercial applications. First, we need to create tools that facilitate circuit design and performance benchmarking [23-28]. This way, CNTFET can enter the path of progress in this technology era.

In terms of how the device works, CNFETs are two different types of Schottky Barrier Controlled FETs (SB-CNTFETs) or Ballistic CNTFETs. Conduction of the SB-CNTFET is performed at the end contacts of the SB by tunnelling the bulk carriers. The ballistic conduction of CNTFETs is caused by the suppression of p-type electron or n-type hole transport, and the gate voltage affects the heavily doped source/drain [29-33]. There is a non-tunnelling voltage barrier in the channel region. Figures 9.4 (a, b) show the energy band diagram.

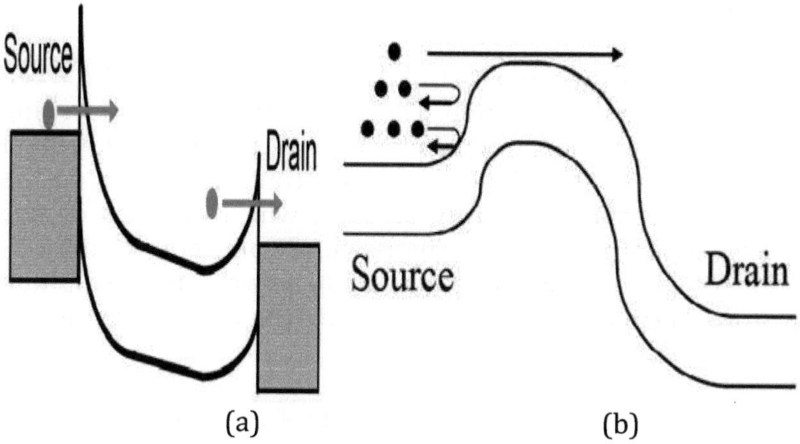

Figure 9.4 *Energy band diagram of (a) SB-CNTFET and (b) Ballistic CNT-FET[7].*

Conventional Static Random Access Memory

SRAM cells are typically used in cross-coupled CMOS IC circuitry in embedded memories to store some data. The accuracy, low power consumption, and high speed of SRAM memory make it a new-generation memory [34-40]. Because they are quick, reliable, and simple to produce using conventional logic manufacturing techniques, they are always found on the same die as microcontrollers and microprocessors. SRAM has mainly three functions: Hold, Read and Write [41-46].

READ function

An SRAM cell's data is accessed and output to the data bus when a read operation is conducted on the cell. The following steps comprise the read operation. The memory controller is given the row and column addresses of the SRAM cell. After that, the memory controller turns on the word line connected to the row address, which depends on the SRAM cell's access transistors. Then, by turning on the bit lines connected to the column address, the state of the flip-flop circuit is detected. After being amplified, the sensed state is output to the data bus [47-53].

WRITE function

The data that is to be written is stored in the cell when an SRAM cell is subjected to a write operation. The steps involved in the writing operation are: The memory controller is given the row and column addresses of the SRAM cell. The memory controller turns on the word line connected to the row address, which depends on the SRAM cell's access transistors. The bit lines connected to the column address are then applied to the data to be written. The flip-flop circuit stores the data in the new state after the write driver amplifies it and writes it there [54-60].

In summary, the read operation of an SRAM cell entails detecting the state of the flip-flop circuit and outputting it to the data bus. In contrast, the write operation entails storing new data in the flip-flop circuit [61-64].

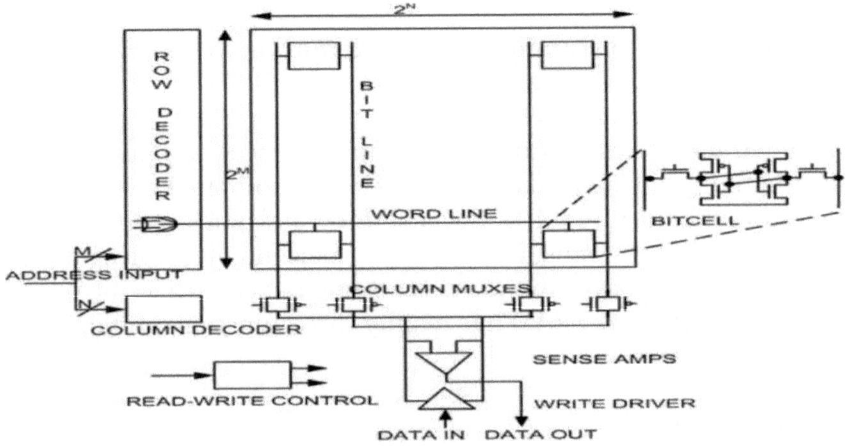

Figure 9.5 *A general SRAM array structure.*

SRAM cell operations:

A flip-flop circuit, which stores a single bit of data, is made up of six transistors in SRAM cells. The binary values 0 and 1 are represented by the flip-flop circuit's two stable states. These states are changed during read-and-write operations on SRAM cells [65-71].

An SRAM cell can be in one of three states: standby, where the circuit is inactive, reading, when data is requested, and writing when the contents are being updated. An SRAM cell's readability and write stability must be tested in both read and write modes [72-77]. The three states' respective operating systems are as follows:

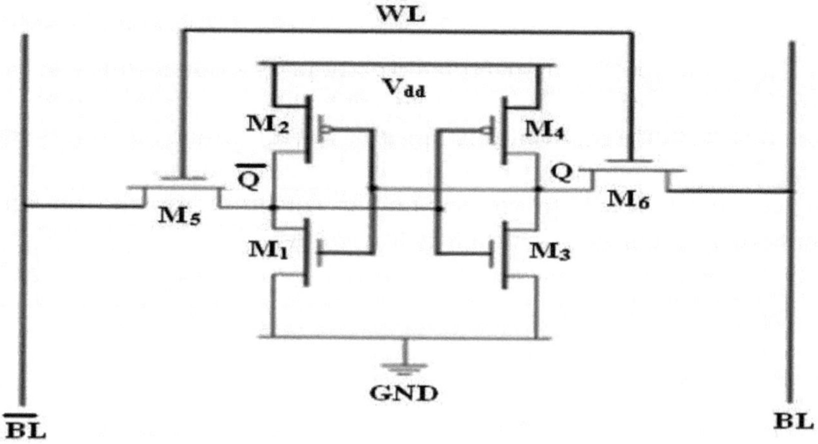

Figure 9.6 Conventional SRAM cell.

Hold Operation

As shown in Figure 9.6, Word-Line is not in contact with Vdd, but no hold operation has been taken into account in the SRAM memory cell. As a result, the driver transistors (M_5, M_6) are no longer connected to the circuit and the internal circuitry holds the data until a power supply is available [78-84]. There is no column pre-charged circuit charging either of the bit lines (BL and BLB). For example, If the transistor named M_1 is switched off and the M_4 transistor is switched on, the '0' value will have previously been stored in the M_3 transistors as a result. The result is at node Q zero [85-88].

Read Operation

Theoretically, a bit line, a controller transistor, and an affirmed word line are all that are required to retrieve data from SRAM. The parasitic capacitances on the bit line are, however, excessively high and enormous for the circuitry. It is a little difficult and more complicated in practice to expedite the read operation [89-93]. The WL is turned on after both input lines (BL, BLB) have been pre-charged. Because the word line links the access transistors to the internal circuitry of the cell, as seen in Figure 9.7, the voltage on the one-bit line slightly drops when the word line is activated. As a result, the sense amplifier can tell whether a "1" or a "0" was stored at the Q node [14] or [16] by identifying which bit line has a higher potential. The read operation speed is directly proportional to the sense amplifier's sensitivity. The read current will easily find a path to discharge because the NMOS is more powerful, which facilitates the voltage drop along the bit line. This is a goal that many researchers are working toward; to reduce power consumption, the bit line should be pre-charged at a slightly low voltage.

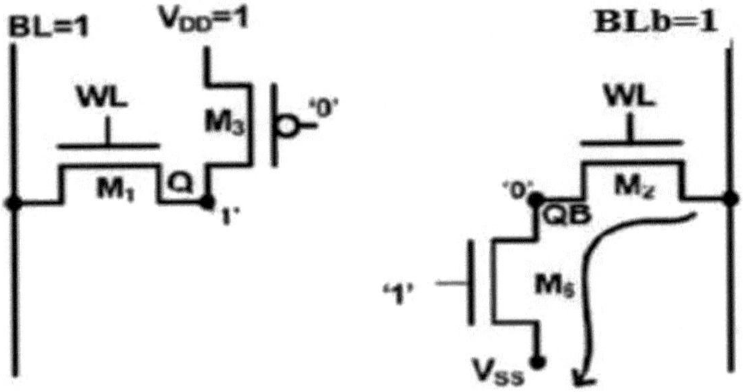

Figure 9.7 *Read '0' operation.*

Write operation

Writing a "1" or a "0" into an SRAM bit cell is shown in Figure 9.8. In the first instance, the write cycle only starts after writing any value to the bit-line. If we need to program "0" in the memory cell, we would also need to program "0" on the BL line i. e. Bit Line Bar=1 and Bit Line=1. The stored value is changed in the same way as when an SR latch continuously applies a reset pulse [94-98]. The external supply Vdd makes the word line functional. The reason why it functions in this way is likely because the controller transistors mounted in the bit lines are kept at a higher strength than the transistors housed inside the inverted cell. This makes it simple for us to replace a new number in a cell [99-104]. In reality, the NMOS transistors (M_5, M_6) that are connected to the bit lines are more powerful than any PMOS or N-MOS transistors that are incorporated into the memory cell's internal circuitry. Even though they are both the same size, this will be even simpler if we use inferior P-MOS rather than NMOS. As a result, the cross-coupling inverter design causes the voltage supply of 1-pair of transistors (M_2, M_1) in the cell to change, which then causes the voltage supply of the other pair of transistors (M_3, M_4) to change as well. The new value can now be entered into it even more easily as a result. Cross-coupled inverters amplify writing as a result [105-108].

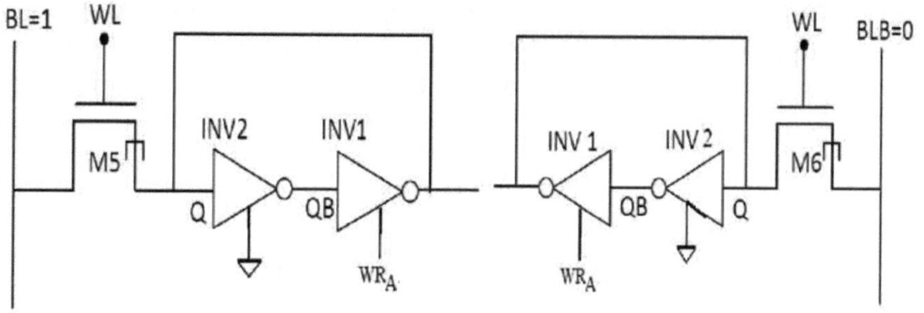

Figure 9.8 *(a) Writing value "1" (b) Writing value '0' [18].*

CNTFET-Based SRAM Cell:

Due to their distinctive electrical characteristics, such as high mobility, ballistic transport, and high current density, CNTFETs are viewed as a good replacement for conventional MOSFETs. The proposed SRAM cell in this study is based on a 6T configuration with CNTFET technology [109-113]. Two access transistors and two cross-coupled inverters make up a 6T SRAM cell. The read and write operations of the SRAM cell are managed by the CNTFET-based access transistors. The proposed CNTFET-based SRAM cell can be implemented using a variety of technologies, including CNTFETs with various channel lengths and diameters as well as various fabrication methods. The fabrication of CNTFETs can be optimised, and other technologies like spintronics and memristors can be integrated further to boost the performance of the CNTFET-based SRAM cell [114-120].

The 6T SRAM Cell is designed using the CNTFET model file method

The channel's physical length (Lch) and the gate's length of the transistors are the same in the simulation setup for the 6-T SRAM CNTFET cell. e. When the dielectric material is 4 nm thick and the doping length of the nanotubes is 32 nm, the Fermi level is maintained at 0 point 6 eV. Set the CNTFET tube's chirality to (19, 0), pitch to 20 nm, and the middle mean path between the absence of ideal elastic scattering and the intrinsic CNT (Lgeff) to 200 nm. As a result, the work function of the CNT operates at 4.5 eV while the work function of the source and drain operates at 4.6 eV. With a power supply of 0.9V, [121-125], transistor width shouldn't fall below 48nm. The cell design with six-transistor SRAM for simulation is shown in Figure 9.9.

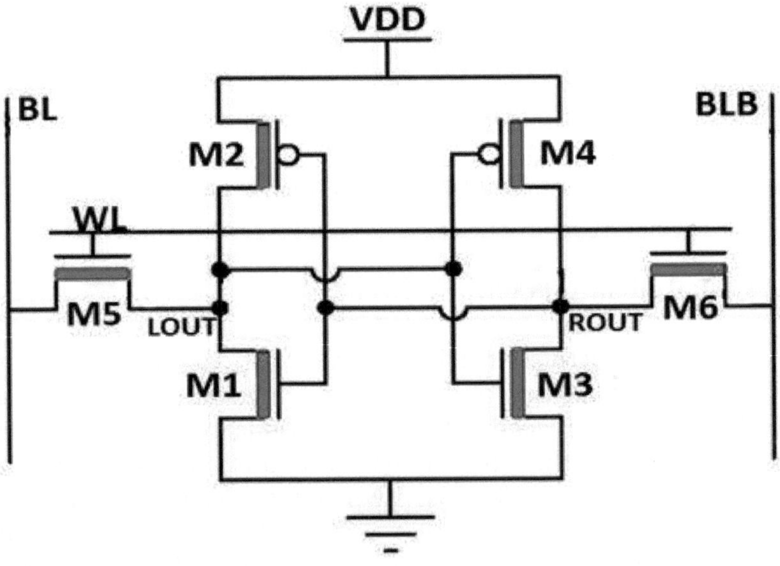

Figure 9.9 *6-T SRAM based on CNTFET cell design.*

The Stanford model file was used to help us design this CNTFET 6T SRAM memory cell. By adjusting the diameter of the Nanotube in this model, we can track the voltage of the threshold. By altering the tube diameter in CNTFETs, different designs can be produced, which is also feasible at variousvoltages. For the best-optimised performance in this method, we can use standard CNTs with fixed chirality vectors. However, all P-CNTFETs and N-CNTFETs use the same chirality of CNTs and chirality vectors, respectively.

Evaluation of performance:

HSPICE simulation tools are used to evaluate the proposed SRAM cell's performance [3]. The simulation results demonstrate as power consumption, speed, and noise immunity, the proposed CNTFET-based SRAM cell has a significant advantage over the established CMOS-based SRAM cell technology. The lower leakage current in CNTFETs enables the CNTFET-based SRAM cell to have a lower standby power consumption. Due to the higher mobility of CNTFETs, the CNTFET-based SRAM cell also has a quicker write time and read access time. The CNTFET-based SRAM cell also has improved noise immunity thanks to lower parasitic capacitance and inductance.

Static Noise margin (SNM)

In this section, simulation results and graphical representation to determine SNM by using the butterfly method for different scaled technologies at 0.9 voltage. When the cell was designed at 180 nm, the operations performed by SRAM were highly stable but as soon as technology nodes started to get lower, the stability of the SRAM cell started becoming unstable gradually. However, when the cell was designed at 32nm, the stability had a steep decline. It can also be noticed with the help of a comparison graph in Figure 9.10. Due to this stability decline a new SRAM cell has been designed with new material i.e., carbon nanotube so that stability can be enhanced.

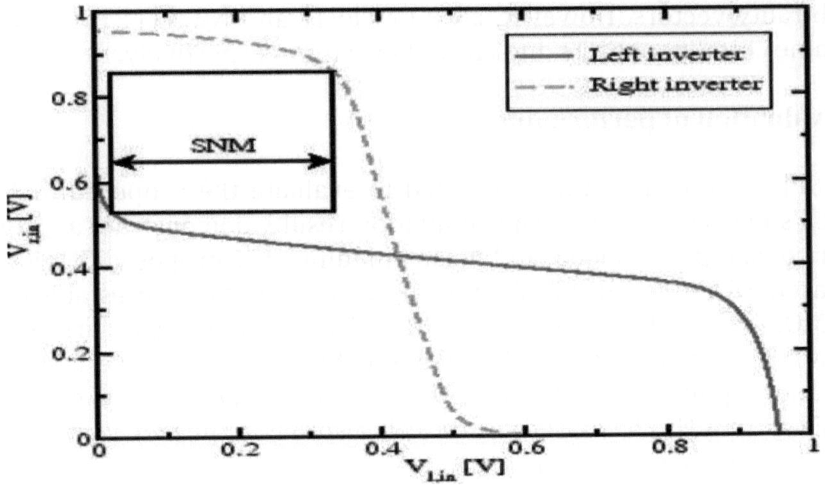

Figure 9.10 *Static noise margin analysis in the butterfly method.*

Delay

Write delay is the measurement of the amount of time period required to program data to an SRAM cell. The time it takes to write logic "0" from bit lines to an SRAM cell when that cell is holding data with logic "1" is referred to as the "logic "0" write delay. The write delay is greater than the read delay, as shown in Figure 9.11. Integrating the current pulse over time allows for the calculation of the critical charge. We thought about using the storage node marked "1" for the current injection because the charge needed for the 1-0 transition is less than that for the 0-1 transition. The time needed to change the node voltage by 10% to 90% is used to calculate the write performance. We took into account the average power used during a write operation when measuring power consumption. The capacitor that connects the nodes in the design of an SRAM cell adds a significant delay to the write operation. The other SRAM cells' additional components increase the Q node levels of this cell at the expense of a longer write time, and the performance characteristics of this cell improve uniformly with an incremental.

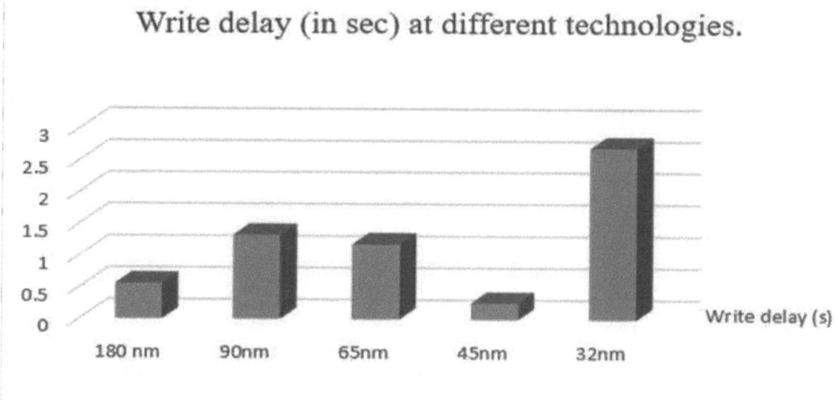

Figure 9.11 *Write logic '0' delay for different scaled technologies at 0.9 voltage.*

Power Consumption

When an SRAM cell is performing a holding operation, some power supply current is flowing to hold data in the cell. Static power consumption is the term used to describe this type of power usage. The threshold voltage also decreased as a result of the further reduction in process technology, which increased circuit speed with a lower supply voltage. The threshold voltage increases the leakage power because it is exponentially inversely proportional to the subthreshold leakage current. All leakage sources of devices below 65 nm are increasingly being dominated by power consumption. The variation of static power consumption with respect to various scaled technologies at 0.9 voltage can also be seen with the aid of the comparison graph in Figure 9.12.

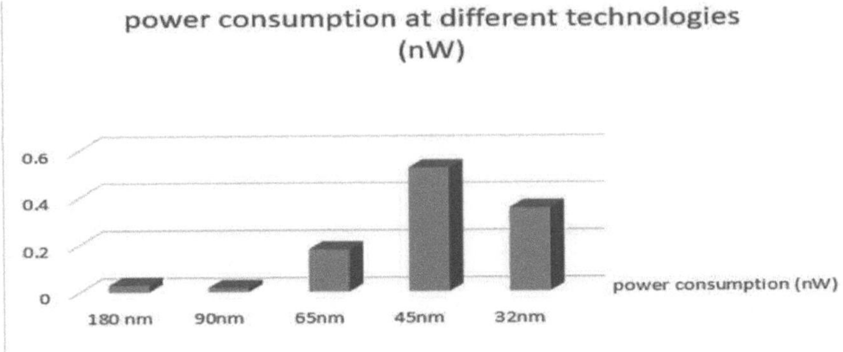

Figure 9.12 *Static power consumption for different scaled technologies at 0.9 voltage.*

The SRAM cell was designed with the Stanford CNTFET model which was released in 2008 [28]. This SRAM cell is designed by maintaining the same parameters as previous designs. The development of logic applications using carbon nanotubes (CNT) has been found to have a highly significant risk ratio, rendering CNTFET superior to CMOS. The highlighted point about is CNTFET also has a significantly smaller.

OFF current which greatly reduces the power consumption in the standby state. Figure 9.13, Figure 9.14 and Figure 9.15 show the simulation results of the static noise margin, read Static noise margin and write noise margin respectively.

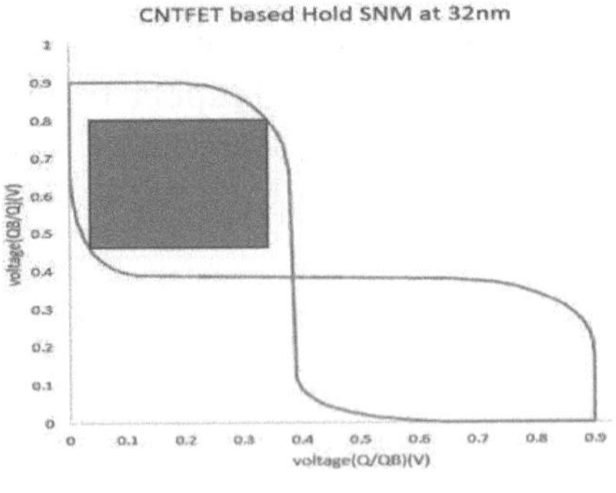

Figure 9.13 *Static Noise Margin*

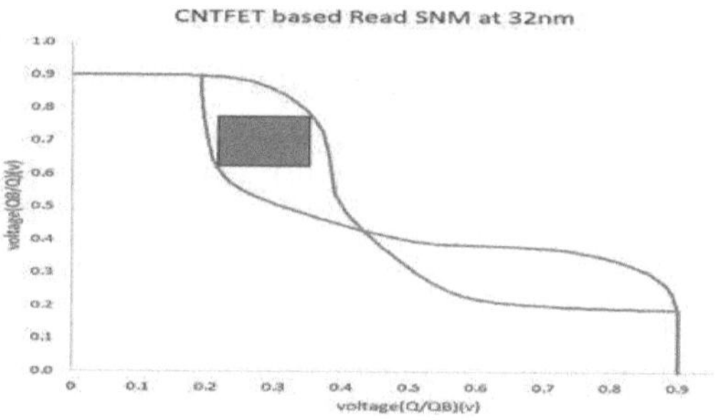

***Figure 9.14** Read Static Noise Margin*

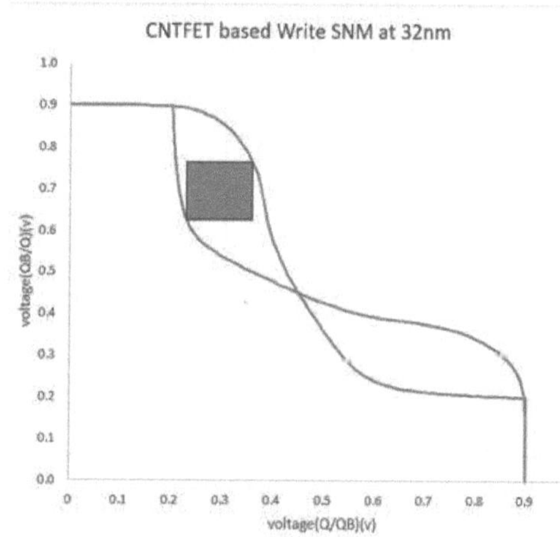

***Figure 9.15** Write Static Noise Program*

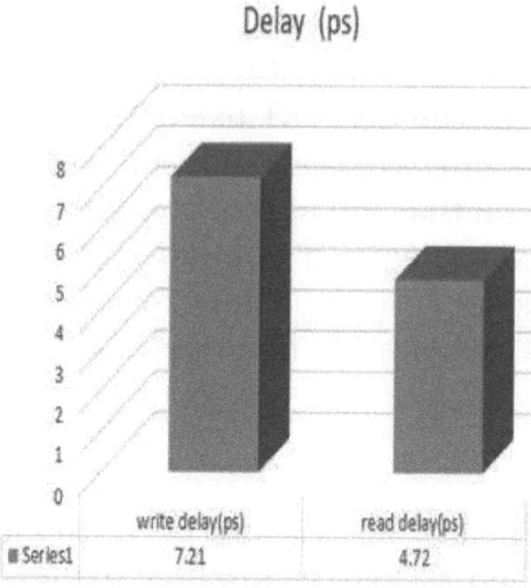

***Figure 9.16** Read and Write delay*

This research specifically concentrates on the stability, delay and power consumption of SRAM cells considering the technology variations. Table 9.1 shows the comparison of the characteristics of CMOS-FET (different channel lengths) and its CNTFET counterpart (32 nm channel length).

Table 9.1 *Simulation results of designed 6T SRAM cell parameters at different scaled technologies with 0.9V power supply.*

Technologies	SNM	RSNM	WSNM	Read Delay	Write Delay	Power Consumption
180nm	350 mV	0.12 V	1.5 V	0.44 ns	0.56 ns	0.0284 nW
90nm	300 mV	0.07 V	1.2 V	0.11 ns	1.32 ns	0.0172 nW
65nm	275 mV	0.065 V	1 V	0.25 ns	1.18 ns	0.0178 nW
45nm	245 mV	0.05 V	0.8 V	0.034 ns	0.26 ns	0.524 nW
32nm	127 mV	0.017 V	0.14 V	1.22 ns	2.7 ns	0.3522 nW
Stanford CNTFET Model at 32nm	0.32 V	0.17 V	0.14 V	4.72 ps	7.21 ps	0.01127 nW

The table's primary goal is to illustrate how characteristics change when technology nodes are switched from 180 nm to 32 nm. Sweeping the gate voltage between 0V and 0.9V (which denotes the "OFF" and "ON" phases) applies the gate voltage. This would make it easier to comprehend the transistor's OFF state and represent its leakage characteristics. The CNTFET uses three tubes in its channel at a wavelength of 32 nm, whereas we used a MOSFET with W=48 nm and L=32 nm. Given that the access transistors in the SRAM cell have the same measurements, a comparison between these two transistor sizes is crucial. It is important to remember that the 'ON' current in CNTFET is higher at 32 nanometers than the CMOS.

Conclusion

In this study, we present a design and performance assessment of an SRAM cell based on CNTFET technology using various technologies. The simulation results demonstrate that in terms of power consumption, speed, and noise immunity, the proposed CNTFET-based SRAM cell has a significant advantage over the established CMOS-based SRAM cell technology. The proposed CNTFET-based SRAM cell can be used in various electronic devices, including microprocessors, digital signal processors, and graphics processing units, to enhance performance and lower power consumption. To enhance the performance of the suggested CNTFET-based SRAM cell even more, spintronics and memristors can be added.

Future Work

In earlier memory design, we can observe exponential growth in machine learning applications like neural networks. Therefore, many new opportunities will present themselves if we reconsider the idea of memory design using machine learning algorithms. This report shows significant advantages to having memory and computation implemented simultaneously because some applications do not require memory and computations to be placed separately. The effect of process variation on device parameters, which affects the stability of the SRAM, can also be studied while designing the SRAM using the CNTFET. So, when designing cells for the SRAM, one must be process-variation tolerant. The device leakage below 65 nm is dominated by all other leakage sources. It can also be seen with the aid of the comparison graph of static power consumption with respect to various scaled technologies at 0.9 voltage.

References

1. Mohita, T. Newar, T. Roy, J. Chowdhury, and J. K. Das, 'Design and stability analysis of CNTFET based SRAM Cell', *2016 IEEE Students' Conf. Electr. Electron. Comput. Sci. SCEECS 2016*, pp. 1–5, 2016.
2. C. Science, 'Intelligent SRAM (ISRAM) for Improved Embedded System Performance', June 2003.
3. 'HSPICEQuickStart', 2011.
4. S. Dresselhaus, 'Physics of Carbon Nanotubes', 1995.
5. M. Elangovan and K. Gunavathi, 'Stability Analysis Of 6t CntfetSram Cell for Single and Multiple CNTs', *2018 4th Int. Conf. Devices, Circuits Syst.*, vol. 2, pp. 63–67, 2018.

6. S. Lin, S. Member, Y. Kim, S. Member, and F. Lombardi, 'Design of a CNTFET-Based SRAM Cell by Dual-Chirality Selection', vol. 9, no. 1, pp. 30–37, 2010.
7. R. P. S, B. K. Madhavi, and K. L. Kishore, 'A New Low Power 9T SRAM Cell based on CNTFET at 32nm Technology Node', vol. 2, no. 6, pp. 2761–2764, 2011.
8. A. Lin, G. Wan, and J. Deng, 'Carbon Nanotube Field Effect Transistors (CNFET) HSPICE Model', pp. 1–13, 2009.
9. H. Hamsa and T. Natarajan, 'A Study of Semiconductor Memory Technology by Comparing Volatile and Non-Volatile Memories', no. July 2018.
10. S. K. Yadav, 'Comparative Analysis of High Speed SRAM Cell for 90nm CMOS Technology', pp. 316–321, 2018.
11. P. Raikwal, V. Neema, and A. Verma, 'Design and analysis of low power single ended 8T SRAM ARRAY (4×4) at 180nm technology', *Int. Conf. Signal Process. Commun. Power Embed. Syst. SCOPES 2016 - Proc.*, pp. 312–316, 2017.
12. M. Jagasivamani and D. S. Ha, 'Development of a low-power SRAM compiler', *ISCAS 2001 - 2001 IEEE Int. Symp. Circuits Syst. Conf. Proc.*, vol. 4, pp. 498–501, 2001.
13. A. Pathak, D. Sachan, H. Peta, and M. Goswami, 'A Modified SRAM Based Low Power Memory Design', *Proc. IEEE Int. Conf. VLSI Des.*, vol. 2016-March, pp. 122–127, 2016.
14. A. Agal and B. Krishan, '6T SRAM Cell: Design and Analysis', *Int. J. Eng. Res. Appl.*, vol. 4, no. 3, pp. 574–577, 2014.
15. N. S. Bhat, 'Design and Modelling of Different Sram'S Based on Cntfet 32 Nm Technology', vol. 3, no. 1, pp. 69–83, 2012.
16. 'To appear in IEEE Transactions on VLSI Systems 1', no. iii, p. file:///C:/Users/cdac/Desktop/srsam papers/bsim330.
17. Z. Zhang, M. A. Turi, and J. G. Delgado-frias, 'SRAM Leakage in CMOS,nFinFET and CNTFET Technologies', pp. 267–270.
18. S. Joshi and U. Alabawi, 'CNTFET Based SRAM', vol. 2017, 2017.
19. I. O. P. C. Series and M. Science, 'Design and performance analysis of 6T SRAM cell on different CMOS technologies with stability characterization', 2019.
20. W. Liu *et al.*, 'BSIM3v3. 3 MOSFET Model Manual', 2005.
21. Anjana Bhardwaj, Pradeep Kumar, B. Raj, Sunny Anand, "Design and Performance Optimization of doping-less Vertical Nanowire TFET using Gate Stack Technique", Journal of Elec-

tronic Materials (JEMS) Springer, Vol. 41, Issuse 7, PP. 4005-4013, 2022.
22. Jeetendra Singh, BRaj, "Tunnel Current Model of Asymmetric MIM Structure Levying Various Image Forces to Analyze the Characteristics of Filamentary Memristor", Applied Physics A, Springer, Vol. 125, No.3, pp. 203.1 to 203.11 Feb 2019.
23. Candy Goyal, Jagpal Singh Ubhi and B Raj, Low Leakage Zero Ground Noise Nanoscale Full Adder using Source Biasing Technique, "Journal of Nanoelectronics and Optoelectronics", American Scientific Publishers, Vol. 14, PP. 360–370, March 2019.
24. Gurmohan Singh, R. K. Sarin and B Raj, "A Novel Robust Exclusive-OR Function Implementation in QCA Nanotechnology with Energy Dissipation Analysis", Journal of Computational Electronics, Springer, Volume 15, Issue 2, PP. 455-465, June 2016.
25. Tanu Wadhera, Deepti Kakkar, Girish wadhwa, B Raj, " Recent Advances and Progress in Development of the Field Effect Transistor Biosensor: A Review" Journal of ELECTRONIC MATERIALS, Springer, Volume 48, Issue 12, pp 7635–7646, December 2019.
26. Girish Wadhwa, Priyanka Kamboj, Balwinder Raj" Design optimisation of junctionless TFET biosensor for high sensitivity" Advances in Natural Sciences: Nanoscience and Nanotechnology, Vol. 10, PP. 045001 (7), (2019).
27. Priya Bansal, BRaj, "MEMRISTOR MODELING AND ANALYSIS FOR LINEAR DOPANT DRIFT KINETICS," Journal of Nanoengineering and Nanomanufacturing, American Scientific Publishers, Vol. 6, pp. 1–7, 2016.
28. Amandeep Singh, Mamta Khosla, BRaj, "Circuit Compatible Model for Electrostatic Doped Schottky Barrier CNTFET,"Journal of Electronic Materials, Springer, Vol. 45, Issue 12, pp 4825-4835, 2016.
29. Ashima, D Vaithiyanathan, B Raj," Performance Analysis of Charge Plasma induced Graded Channel Si nanotube", Journal of Engineering Research (JER), EMSME Special Issue pp. 146-154 Aug 2021.
30. Abhishek Singh Tomar, Vijay Kumar Magraiya, and B Raj, Scaling of Access and Data Transistor for High Performance DRAM Cell Design, Quantum Matter, Vol. 2, PP. 412-416 Oct. 2013.

31. Neeraj Jain, BRaj, "Parasitic Capacitance and Resistance Model Development and Optimization of Raised Source/Drain SOI FinFET Structure for Analog Circuit Applications, Journal of Nanoelectronics and Optoelectronins, ASP, USA, Vol. 13, PP. 531–539, Apl 2018.
32. Shradhya Singh, S. K. Vishvakarma and B Raj, "Analytical Modeling of Split-Gate Junction-Less Transistor for a Biosensor Application", Sensing and Bio-sensing, Elsevier, Vol. 18, Pages 31-36, April 2018.
33. Maisagalla Gopal, and Balwinder Raj, Low Power 8T SRAM Cell Design for High Stability Video Applications, ITSI Transaction on Electrical and Electronics Engineering, Vol. 1(5), pp. 91-97, 2013.
34. Balwinder Raj, Jatin Mitra, Deepak Kumar Bihani, Rangharajan V, A.K. Saxena and S. Dasgupta, "Analysis of Noise Margin, Power and Process Variation for 32 nm FinFET Based 6T SRAM Cell" Journal of Computer (JCP), Academy Publisher, FINLAND, vol. 5, issue 6, PP.1-8, 2010.
35. Divya Sharma, Rajesh Mehra and B Raj, "Comparative Analysis of Photovoltaic Technologies for High Efficiency Solar Cell Design", Superlattices and Microstructures, Elsevier, Volume 153, PP. 106861, May 2021.
36. Pawandeep Kaur, Avtar Singh Buttar, B Raj, "A comprehensive Analysis of Nanoscale Transistor based Biosensor: A Review, Indian Journal of Pure and Applied Physics, Vol. 59, pp. 304-318, April 2021.
37. Divya Yadav, Balwant Raj and B Raj, "Design and Simulation of Low Power Microcontroller for IoT Applications", Journal of Sensor Letters, ASP, Vol. 18, PP. 401–409, May 2020.
38. Shailendra Singh, B Raj,"A 2-D Analytical Surface Potential and Drain current Modeling of Double-Gate Vertical t-shaped Tunnel FET", Journal of Computational Electronics, Springer, Vol. 19, PP.1154–1163, Apl 2020.
39. Jeetendra Singh, B Raj, "An Accurate and Generic Window function for Non-linear Memristor Model" Journal of Computational Electronics, Springer, Volume 18 Issue 2, PP. 640-647, June 2019.
40. Manjit Kaur; Neena Gupta; Sanjeev Kumar; BRaj; Arun Kumar Singh, "Comparative RF and Crosstalk Analysis of Carbon Based Nano Interconnects" IET Circuits, Devices & Systems, Vol.15(6), PP. 493-503, Feb 2021.

41. Nehru Kandasamy, Firdous Ahmad, D. Ajitha, B Raj, and Nagarjuna Telagam "Quantum Dot Cellular Automata based Scan Flip Flop and Boundary Scan Register, IETE Journal of Research, Volume 66, PP. 535-548, 2020.
42. Sharma, S. K.., Raj, B., Khosla, M., "Enhanced Photosensivity of Highly Spectrum Selective Cylindrical Gate In1-xGaxAs Nanowire MOSFET Photodetector, Modern Physics letter-B, Vol. 33, No. 12, PP. 1950144 (2019).
43. Jeetendra Singh, BRaj, "Design and Investigation of 7T2M NVSARM with Enhanced Stability and Temperature Impact on Store/Restore Energy", IEEE Transactions on Very Large Scale Integration Systems, Vol. 27, Issure 6, PP. 1322 - 1328 June 2019.
44. Anil Kumar Bhardwaj, Sumeet Gupta, BRaj, Amandeep Singh, "Impact of Double Gate Geometry on the Performance of Carbon Nanotube Field Effect Transistor Structures for Low Power Digital Design", Computational and Theoretical Nanoscience, ASP, Vol. 16, PP. 1813–1820, 2019.
45. Neeraj Jain, BRaj, "Thermal Stability Analysis and Performance Exploration of Asymmetrical Dual-k underlap Spacer (ADKUS) SOI FinFET for Security and Privacy Applications", Indian Journal of Pure & Applied Physics (IJPAP), Vol. 57, pp. 352-360, May 2019.
46. Amandeep Singh, Mamta Khosla, BRaj, "Design and Analysis of Dynamically Configurable Electrostatic Doped Carbon Nanotube Tunnel FET"," Microelectronics Journal, Elesvier, Volume 85, Pages 17-24, March 2019.
47. Neeraj Jain and Balwinder Raj, "Dual-k Spacer Region Variation at the Drain Side of Asymmetric SOI FinFET Structure: Performance Analysis towards the Analog/RF Design Applications", communicated in "Journal of Nanoelectronics and Optoelectronics", American Scientific Publishers, Vol. 14, PP. 349–359, March 2019.
48. Jeetendra Singh, Sanjeev Sharma, B Raj, Mamta Khosla Analysis of barrier layer thickness on performance of In1-xGaxAs based Gate Stack Cylindrical Gate Nanowire MOSFET," JNO, ASP, Vol. 13, PP 1473-1477, Oct 2018.
49. Neeraj Jain, B Raj, "Analysis and Performance Exploration of High-k SOI FinFETs Over the Conventional Low-k SOI FinFET toward Analog/RF Design", Journal of Semiconductors (JoS), IOP Science, Vol.39, Issue 12, PP- 124002-1-7, Dec 2018.

50. Candy Goyal, Jagpal Singh Ubhi and BRaj, A reliable leakage reduction technique for approximate full adder with reduced ground bounce noise, Journal of Mathematical Problems in Engineering, Hindawi, Volume 2018, Article ID 3501041, 16 pages, 15 Oct 2018.
51. Anuradha, Jeetendra Singh, B Raj and Mamta Khosla, "Design and Performance Analysis of Nano-scale Memristor-based Nonvolatile SRAM" Journal of Sensor Letter", American Scientific Publishers, Vol. 16, PP 798–805, Oct 2018.
52. Girish Wadhwa, BRaj, "Parametric Variation Analysis of Charge-Plasma-based Dielectric Modulated JLTFET for Biosensor Application" IEEE Sensor Journal, VOL. 18, NO. 15, AUGUST 1, PP. 6070 – 6077, 2018 (SCI)
53. Jeetendra Singh, BRaj, "Comparative Analysis of Memristor Models for Memories Design", JoS, IoP, Vol. 39, No. 7, PP. 074006-1-12, July 2018.
54. Divya Yadav, Shailesh Singh Chouhan, Santosh Kumar Vishvakarma and BRaj, "Application Specific Microcontroller Design for IoT based WSN", Sensor Letter, ASP, Vol. 16, PP. 374–385, May 2018.
55. Gurmohan Singh, R. K. Sarin and BRaj, "Fault-Tolerant Design and Analysis of Quantum-Dot Cellular Automata Based Circuits", IEEE/IET Circuits, Devices & Systems, Vol. 12, PP. 638 - 64, 2018.
56. Jeetendra Singh, BRaj, "Modeling of Mean Barrier Height Levying Various Image Forces of Metal Insulator Metal Structure to Enhance the Performance of Conductive Filament Based Memristor Model", IEEE Nanotechnology, Vol. 17, No. 2, PP. 268-267, March 2018.
57. Aakash Jain, Sanjeev Sharma, BRaj, "Analysis of Triple Metal Surrounding Gate (TM-SG) III-V Nanowire MOSFET for Photosensing Application", Opto-electronics Journal, Elsevier, Volume 26, Issue 2, Pages 141-148, May 2018.
58. Aakash Jain, Sanjeev Sharma, B Raj, "Design and Analysis of High Sensitivity Photosensor Using Cylindrical Surrounding Gate MOSFET For Low Power Sensor Applications", Engineering Science and Technology, an International Journal, Elsevier's, Volume 19, Issue 4, Pages 1864–1870, December 2016.
59. Amandeep Singh, Mamta Khosla, B Raj, "Analysis of Electrostatic Doped Schottky Barrier Carbon Nanotube FET for Low

Power Applications," Journal of Materials Science: Materials in Electronics, Springer, Vol.28, PP. 1762–1768, 2017.

60. G. Saiphani Kumar, Amandeep Singh, BRaj, "Design and Analysis of Gate All Around CNTFET based SRAM cell Design", Journal of Computational Electronics, Springer, Volume 17, Issue 1, pp 138–145, March 2018.
61. Gurinder pal Singh, B. S. Sohi, Balwinder Raj, "Material Properties Analysis of Graphene Base Transistor (GBT) for VLSI Analog Circuits, Indian Journal of Pure & Applied Physics (IJPAP), Vol. 55, pp. 896-902, Dec-2017.
62. Amandeep Singh, Mamta Khosla, B Raj, "Comparative Analysis of Carbon Nanotube Field Effect Transistor and Nanowire Transistor for Low Power Circuit Design," Journal of Nanoelectronics and Optoelectronics, American Scientific Publishers, USA, Vol. 11, PP. 388-393, June 2016.
63. Sunil Kumar and B Raj, "Estimation of Stability and Performance metric for Inward Access Transistor based 6T SRAM Cell Design using n-type/p-type DMDG-GDOV TFET" IEEE VLSI Circuits and Systems Letter, Volume3, Issue2, PP 25-39, June2017
64. Shashikant Sharma, Anjan Kumar, Manisha Pattanaik and B Raj, "Forward Body Biased Multimode Multi-Threshold CMOS Technique for Ground Bounce Noise Reduction in Static CMOS Adders" International Journal of Information and Electronics Engineering, pp. 567-572, Vol. 3, issue 3, 2013.
65. Hamendra Singh, Pankaj Kumar, Balwinder Raj, "Performance Analysis of Majority Gate SET Based 1-bit Full Adder", *International Journal of Computer and Communication Engineering* (IJCCE), IACSIT Press Singapore, ISSN: 2010-3743, Volume 2, No. 4, 2013.
66. Anil Kumar Bhardwaj, Sumeet Gupta and B Raj, "Investigation of Parameters for Schottky Barrier (SB) Height for Schottky Barrier Based Carbon Nanotube Field Effect Transistor Device", Journal of Nanoelectronics and Optoelectronics, ASP, Vol. 15, PP. 783–791, July 2020.
67. Priya Bansal, BRaj, "MEMRISTOR: A VERSATILE NONLINEAR MODEL FOR DOPANT DRIFT AND BOUNDARY ISSUES," JCTN, American Scientific Publishers, Vol. 14 (5), PP. 2319-2325, May 2017.
68. Neeraj Jain, BRaj, "An Analog and Digital Design Perspective Comprehensive Approach on Fin-FET (Fin-Field Effect tran-

sistor) Technology - A Review", Reviews in Advanced Sciences and Engineering (RASE), ASP, Vol. 5, PP. 1-14, 2016.
69. Sanjeev Sharma, B Raj, Mamta Khosla, "Subthreshold Performance of In1-xGaxAs based Dual Metal with Gate Stack Cylindrical/Surrounding Gate Nanowire MOSFET for Low Power Analog Applications, Journal of Nanoelectronics and Optoelectronics, American Scientific Publishers, USA, Vol.12, PP. 171–176, 2017.
70. Balwinder Raj, A. K. Saxena and S. Dasgupta, "Analytical Modeling for the Estimation of Leakage Current and Subthreshold Swing Factor of Nanoscale Double Gate FinFET Device" Microelectronics International, UK, vol. 26, pp. 53-63, 2009.
71. Soniya, Shailendra Singh, Girish Wadhwa and Balwinder Raj, "An Analytical Modeling for Dual Source Vertical Tunnel Field Effect Transistor" submitted to 'International Journal of Recent Technology and Engineering (IJRTE)' at Volume-8 Issue-2, PP. 603-608, July 2019.
72. Shailendra Singh, B Raj, "Design and analysis of hetrojunction Vertical T-shaped Tunnel Field Effect Transistor", Journal of Electronics Material, Springer, Volume 48, Issue 10, pp 6253–6260, October 2019.
73. Candy Goyal, Jagpal Singh Ubhi and B Raj, "A Low Leakage CNTFET based Inexact Full Adder for Low Power Image Processing Applications", International Journal of Circuit Theory and Applications, Wiley, Volume47, Issue9, Pages 1446-1458, September 2019.
74. B Raj, A. K. Saxena and S. Dasgupta, "A Compact Drain Current and Threshold Voltage Quantum Mechanical Analytical Modeling for FinFETs" Journal of Nanoelectronics and Optoelectronics (JNO), USA, vol. 3, no. 2, pp. 163-170, 2008.
75. Girish Wadhwa, B Raj, "An Analytical Modeling of Charge Plasma based Tunnel Field Effect Transistor with Impacts of Gate underlap Region" Superlattices and Microstructures, Elsevier, Vol. 142, PP.106512, June 2020.
76. Shailendra Singh, B Raj,"Modeling and Simulation analysis of SiGe hetrojunction Double GateVertical t-shaped Tunnel FET", Superlattices and Microstructures, Elsevier Volume 142, PP. 106496, June 2020.
77. Amandeep Singh, Dinesh Kumar Saini, Dinesh Agarwal, Sajal Aggarwal, Mamta Khosla, BRaj, "Modeling and Simulation of Carbon Nanotube Field Effect Transistor and its Circuit Appli-

cation," Journal of Semiconductors (JoS), IOP Science, Vol.37, PP. 074001-6, July 2016.

78. Neeraj Jain, Balwinder Raj, "Device and Circuit Co-Design Perspective Comprehensive Approach on FinFET Technology - A Review, Journal of Electron Devices, Vol. 23, Num. 1, pp 1890-1901, 2016.

79. Sunil Kumar and BRaj, "Analysis of I_{ON} and Ambipolar Current for Dual-Material Gate-drain Overlapped DG-TFET," Journal of Nanoelectronics and Optoelectronics, American Scientific Publishers, USA, Vol.11, PP. 323-333, June 2016.

80. Naveed Anjum, Tarun Bali and B Raj, Design and Simulation Of Handwritten Multiscript Character Recognition, International Journal of Advanced Research in Computer and Communication Engineering, Vol.2 (7), PP. 2544-2549, July 2013.

81. Sanjeev Sharma, BRaj, Mamta Khosla, "A Gaussian Approach for Analytical Subthreshold Current Model of Cylindrical Nanowire FET with Quantum Mechanical Effects, Microelectronics Journal, Elsevier, Vol. 53, PP. 65-72, April 2016.

82. Karmjit Singh, BRaj, "Performance and analysis of temperature dependent Multi-walled Carbon Nanotubes as Global Interconnects at different technology nodes," Journal of Computational Electronics, Springer, Volume 14 Issue 2, pp. 469-476, June 2015.

83. Sunil Kumar and BRaj, "Compact channel potential analytical Modeling of DG-TFET based on Evanescent–mode Approach," Journal of Computational Electronics, Springer Vol. 14, No. 2, pp.820-827, July 2015.

84. Karmjit Singh, BRaj, "Temperature Dependent Modeling and Performance Evaluation of Multi-Walled CNT and Single-Walled CNT as Global Interconnects," Journal of Electronic Materials, Springer, Vol. 44, Issue 12, pp 4825-4835, Dec 2015.

85. V.K. Sharma, M. Pattanaik and B.Raj, "INDEP approach for leakage reduction in nanoscale CMOS circuits", International Journal of Electronics, Taylor & Francis, Vol.102, No.2, pages 200-215, 2014.

86. Karmjit Singh, BRaj, "Influence of Temperature on MWCNT bundle, SWCNT bundle and Copper interconnects for Nanoscaled Technology nodes," Journal of Materials Science: Materials in Electronics. Springer, Vol. 26, No. 8, pp. 6134-6142, 2015.

87. Naveed Anjum, Tarun Bali and B Raj, "Design and Simulation of Handwritten Gurumukhi and Devanagri Numerical Recognition, International Journal of Computer Applications, Published by Foundation of Computer Science, New York, USA, Vol. 73(12), pp. 16-21, 2013.

88. Khandelwal, S., Gupta, V., BRaj, and Gupta, R D., Process Variability Aware Low Leakage Reliable Nano Scale DG-FinFET SRAM Cell Design Technique, Journal of Nanoelectronics and Optoelectronics, Volume 10, Number 6, pp. 810-817, December 2015.

89. V.K. Sharma, M. Pattanaik and B Raj, "ONOFIC Approach: Low Power High Speed Nanoscale VLSI Circuits Design", International Journal of Electronics, Taylor & Francis, Vol. 101, issue 1, pages 61-73, 2014.

90. Khandelwal, S., Balwinder Raj, and Gupta, R D., FinFET based 6T SRAM Cell Design: Analysis of Performance Metric, Process Variation and Temperature Effect. Journal of Computational and Theoretical Nanoscience, ASP, USA, Vol. 12, PP. 2500-2506, 2015.

91. Sumit Singh; Yadav, Shekhar; Rahul, Jagdeep; Srivastava, Anurag; BRaj, Impact of HfO_2 in Graded Channel Dual Insulator Double Gate MOSFET, Journal of Computational and Theoretical Nanoscience, American Scientific Publishers, Volume 12, Number 6, pp. 950-953, April 2015.

92. Vijay Kumar Sharma, Manisha Pattanaik, BRaj, PVT variations aware low leakage INDEP approach for nanoscale CMOS Circuits, Microelectronics Reliability, Elsevier, Vol. 54, pp 90-99, 2014.

93. B Raj, A. K. Saxena and S. Dasgupta, "Quantum Mechanical Analytical Modeling of Nanoscale DG FinFET: Evaluation of Potential, Threshold Voltage and Source/Drain Resistance", Elsevier's Journal of Material Science in Semiconductor Processing, Elsevier, Vol. 16, issue 4, pp. 1131- 1137, 2013.

94. Maisagalla Gopal, Siva Sankar D Prasad and Balwinder Raj, 8T SRAM Cell Design for Dynamic and Leakage Power Reduction, International Journal of Computer Applications, Published by Foundation of Computer Science, New York, USA, Vol. 71(9), pp. 43-48, June 2013.

95. Manisha Pattanaik, B Raj, Shashikant Sharma and Anjan Kumar, "Diode Based Trimode Multi-Threshold CMOS Technique for Ground Bounce Noise Reduction in Static CMOS Adders",

Advanced Materials Research, Trans Tech Publications, Switzerland, vol. 548, pp 885-889, 2012.

96. Balwinder Raj, A. K. Saxena and S. Dasgupta, "Nanoscale FinFET Based SRAM Cell Design: Analysis of Performance metric, Process variation, Underlapped FinFET and Temperature effect" IEEE Circuits and System Magazine, vol. 11, issue 2, pp. 38- 50, 2011.
97. V.K. Sharma, M. Pattanaik & Balwinder Raj, "Leakage Current ONOFIC Approach for Deep Submicron VLSI Circuit Design", International Journal of Electrical, Computer, Electronics and Communication Engineering, World Academy of Sciences, Engineering and Technology, Vol. 7, issue 4, pages 239-244, 2013.
98. Tulika Chawla, Mamta Khosla, BRaj, Design and simulation of Triple metal Double-gate Germanium on Insulator Vertical Tunnel Field Effect Transistor, Microelectronics Journal, Elsevier; Vol. 114, PP. 105125, August 2021.
99. Parminder Kaur, Sandeep Singh Gill and B Raj, "Comparative Analysis of OFETs Materials and Devices for Sensor Applications", Journal of Silicon, Springer, Vol 14, PP. 4463–4471, 2022.
100. Sanjeev Kumar Sharma, Parveen Kumar, Balwant Raj and BRaj, "$In_{1-x}Ga_xAs$ Double Metal Gate-Stacking Cylindrical Nanowire MOSFET for highly sensitive Photo detector", *Journal of Silicon, Springer Vol.*14, PP. 3535–3541, 2022.
101. B Raj, A. K. Saxena and S. Dasgupta, "Analytical Modeling of Quasi Planar Nanoscale Double Gate FinFET with Source/Drain Resistance and field dependent carrier mobility: A quantum mechanical study" Journal of Computer (JCP), Academy Publisher, FINLAND, vol. 4, issue 9, PP.1-8, 2009.
102. Bhushan, S., Khandelwal, S. and Raj, B., (2012). Analyzing Different Mode FinFET based Memory Cell at Different Power Supply for Leakage Reduction, Seventh International Conference on Bio-Inspired Computing: Theories and Application, (BIC-TA 2012) Advances in Intelligent Systems and Computing, Volume 202, p89-100 (2013).
103. Jeetendra Singh, BRaj, "Temperature Dependent Analytical Modeling and Simulations of Nanoscale Memristor, Journal: Engineering Science and Technology, an International Journal, Elsevier's, Vol. 21, PP- 862-868, Oct 2018.

104. Shradhya Singh, Shashi Bala, Balwant Raj, B Raj, "Improved Sensitivity of Dielectric Modulated Junctionless Transistor for Nanoscale Biosensor Design", Sensor Letter, ASP, Vol.18, PP.328–333, Apl 2020.
105. Vivek Kumar, Santosh Kumar Vishvakarma, and B Raj, "Design and Performance Analysis of ASIC for IoT Applications" Sensor Letter ASP, Vol. 18, PP. 31–38, Jan 2020.
106. Akanksha Jaiswal, R. K. Sarin, B Raj and Shikha Sukhija, "A Novel Circular Slotted Microstrip-fed Patch Antenna with Three Triangle Shape Defected Ground Structure for Multiband Applications", Advanced Electromagnetic (AEM), Vol. 7, Issue 3, PP. 56-63, Aug 2018.
107. Girish Wadhwa, B Raj, "Label Free Detection of Biomolecules using Charge-Plasma-Based Gate Underlap Dielectric Modulated Junctionless TFET" Journal of Electronic Materials (JEMS), Springer, Volume 47, Issue 8, pp 4683–4693, August 2018.
108. Gurmohan Singh, R. K. Sarin and B Raj, "Design and Performance Analysis of a New Efficient Coplanar Quantum-Dot Cellular Automata Adder", Indian Journal of Pure & Applied Physics (IJPAP), Vol. 55, pp. 97-103, February 2017.
109. Amandeep Singh, Mamta Khosla, Balwinder Raj, "Design and Analysis of Electrostatic Doped Schottky Barrier CNTFET Based Low Power SRAM," International Journal of Electronics and Communications, (AEÜ), Elsevier, Vol. 80, PP.67–72, 2017.
110. Parminder Kaur, Vikas Pandey and B Raj, "Comparative Study of Efficient Design, Control and Monitoring of Solar Power using IoT", Sensor Letter, ASP Vol. 18, PP. 419–426, May 2020.
111. Anil Kumar Bhardwaj, Sumeet Gupta and B Raj, Development & Analysis of compact model for Double Gate Schottky Barrier CNTFET, Journal of Nanoelectronics and Optoelectronics, ASP, Vol. 15, pp. 1199–1208, Aug 2020.
112. Girish Wadhwa, Priyanka Kamboj; Jeetendra Singh, B Raj, "Design and Investigation of Junctionless DGTFET for Biological Molecule Recognition" Transactions on Electrical and Electronic Materials, Springer, Aug 2020.
113. Tulika Chawla, Mamta Khosla, B Raj, "Optimization of Double-gate Dual material GeOI-Vertical TFET for VLSI Circuit Design, IEEE VLSI Circuits and Systems Letter, Vol. 6, issue-2, PP.13-25, Aug 2020.

114. Sachin Kumar Verma, Shailendra Singh, Girish Wadhwa and B Raj, "Detection of Biomolecules using Charge-Plasma based Gate Underlap Dielectric Modulated Dopingless TFET" Transactions on Electrical and Electronic Materials (TEEM), Springer, Vol. 21, PP. 528-535, June 2020.
115. Neeraj Jain, BRaj, "Impact of underlap spacer region variation on Electrostatic and analog/RF performance of symmetrical high-k SOI FinFET at 20 nm channel length", Journal of Semiconductors (JoS), IOP Science, Vol.38, Issue 12, PP- 122002, Dec 2017.
116. Shailendra Singh, B Raj," Analytical Modeling and Simulation analysis of T-shaped III-V heterojunction Vertical T-FET", Superlattices and Microstructures, Elsevier, Vol. 147, PP. 106717, Nov 2020.
117. Gurmohan Singh, R. K. Sarin and B Raj, "Design and Analysis of Area Efficient QCA Based Reversible Logic Gates", Journal of Microprocessors and Microsystems, Elsevier, Vol. 52, PP.59-68, May 2017.
118. Amandeep Singh, Mamta Khosla, BRaj, "Compact Model for Ballistic Single Wall CNTFET under Quantum Capacitance Limit," Journal of Semiconductors (JoS), IOP Science, Vol.37, PP. 104001-8, Oct 2016.
119. Sonal Singh; Mamta Khosla, Girish Wadhwa; B Raj, "Design and analysis of double-gate junctionless vertical TFET for gas sensing applications", Applied Physics A, Springer, Vol.127, No.16, 2 Jan 2021.
120. Inderjit Singh, BRaj, Mamta Khosla, and Brajesh Kumar Kaushik, Potential MRAM Technologies for Low Power SoCs, SPIN World Scientific Publisher, SCIE; Vol. 10, No. 04, PP.2050027, Dec 2020.
121. Shailendra Singh, B Raj,"Parametric variation analysis on hetero-junction Vertical t-shape TFET for Supressing ambipolar conduction", Indian Journal of Pure and Applied Physics, Vol. 58,pp. 478-485 June 2020.
122. Soniya, Shailendra Singh, Girish Wadhwa and Balwinder Raj, Design and Analysis of Dual Source Vertical Tunnel Field Effect Transistor for high performance, Transactions on Electrical and Electronics Materials, Springer, Vol. 21, Pages 74–82, Oct 2019.
123. Manjit Kaur; Neena Gupta; Sanjeev Kumar; BRaj; Arun Kumar Singh, "RF Performance Analysis of Intercalated Graphene

Nanoribbon Based Global Level Interconnects" Journal of Computational Electronics, Springer, Vol. 19, PP.1002–1013, June 2020

124. Girish Wadhwa, BRaj, "Design and Performance Analysis of Junctionless TFET Biosensor for high sensitivity" IEEE Nanotechnology, Vol.18, PP. 567 – 574, 2019.
125. Jeetendra Singh, BRaj, "Enhanced Nonlinear Memristor Model Encapsulating Stochastic Dopant Drift, JNO, ASP 14, 958–963 (2019)

Printed by BoD™in Norderstedt, Germany

9 781922 617583